Stress and Strain in Epitaxy:
theoretical concepts, measurements and applications

Stress and Strain in Epitaxy: theoretical concepts, measurements and applications

Keynote lectures of the third Porquerolles school on special topics in Surface Science
Ile de Porquerolles, France, October 1-7, 2000

Edited by

Margrit Hanbücken

Marseille, France

Jean-Paul Deville

Strasbourg, France

2001

ELSEVIER

Amsterdam – London – New York – Oxford – Paris – Shannon – Tokyo

ELSEVIER SCIENCE B.V.
Sara Burgerhartstraat 25
P.O. Box 211, 1000 AE Amsterdam, The Netherlands

© 2001 Elsevier Science B.V. All rights reserved.

This work is protected under copyright by Elsevier Science, and the following terms and conditions apply to its use:

Photocopying
Single photocopies of single chapters may be made for personal use as allowed by national copyright laws. Permission of the Publisher and payment of a fee is required for all other photocopying, including multiple or systematic copying, copying for advertising or promotional purposes, resale, and all forms of document delivery. Special rates are available for educational institutions that wish to make photocopies for non-profit educational classroom use.

Permissions may be sought directly from Elsevier Science Global Rights Department, PO Box 800, Oxford OX5 1DX, UK; phone: (+44) 1865 843830, fax: (+44) 1865 853333, e-mail: permissions@elsevier.co.uk. You may also contact Global Rights directly through Elsevier's home page (http://www.elsevier.nl), by selecting 'Obtaining Permissions'.

In the USA, users may clear permissions and make payments through the Copyright Clearance Center, Inc., 222 Rosewood Drive, Danvers, MA 01923, USA; phone: (+1) (978) 7508400, fax: (+1) (978) 7504744, and in the UK through the Copyright Licensing Agency Rapid Clearance Service (CLARCS), 90 Tottenham Court Road, London W1P 0LP, UK; phone: (+44) 207 631 5555; fax: (+44) 207 631 5500. Other countries may have a local reprographic rights agency for payments.

Derivative Works
Tables of contents may be reproduced for internal circulation, but permission of Elsevier Science is required for external resale or distribution of such material.
Permission of the Publisher is required for all other derivative works, including compilations and translations.

Electronic Storage or Usage
Permission of the Publisher is required to store or use electronically any material contained in this work, including any chapter or part of a chapter.

Except as outlined above, no part of this work may be reproduced, stored in a retrieval system or transmitted in any form or by any means, electronic, mechanical, photocopying, recording or otherwise, without prior written permission of the Publisher.
Address permissions requests to: Elsevier Global Rights Department, at the mail, fax and e-mail addresses noted above.

Notice
No responsibility is assumed by the Publisher for any injury and/or damage to persons or property as a matter of products liability, negligence or otherwise, or from any use or operation of any methods, products, instructions or ideas contained in the material herein. Because of rapid advances in the medical sciences, in particular, independent verification of diagnoses and drug dosages should be made.

First edition 2001

Library of Congress Cataloging in Publication Data
A catalog record from the Library of Congress has been applied for.

ISBN: 0 444 50865 1

♾ The paper used in this publication meets the requirements of ANSI/NISO Z39.48-1992 (Permanence of Paper).
Printed in The Netherlands.

FOREWORD

This book contains keynote lectures which have been delivered at the 3rd Porquerolles' School on Surface Science, SIR2000 (Surfaces-Interfaces-Relaxation). This school was held between October 1st and October 7th 2000 on the Porquerolles island in the Mediterranean sea, near the french Riviera. It was organized jointly by the *Groupe Français de Croissance Cristalline* (GFCC, the French Crystal Growth Association) and the *Groupe de Recherche « Relaxation dans les Couches Nanométriques Epitaxiées »* (GdR 0687 of the Centre National de la Recherche Scientifique).

The aim of this school was to review the main concepts necessary to understand the role of interfacial stress, strain and relaxation in crystal growth by heteroepitaxy. By bringing together scientists from various fields (physics, chemistry, materials science and engineering) which daily use complementary methodological approaches (experiment, theory, modelization), the school allowed to offer them 11 multidisciplinary courses. Thus it has been possible to address the state of art of stress in epitaxial materials, to describe the various methods to measure the atomic displacement and stress fields, to review the spectroscopic methods necessary to map the interface chemistry, to detail the theoretical methods and concepts which are needed to predict them and to question the fact that stress and relaxation can induce specific properties in magnetism, catalysis, electron transport.

The field of *stress and strain in heteroepitaxy* has indeed known large developments during the last ten years. New techniques have been used to set up new devices in which functionalities are obtained through structuration at a nanometer scale. Large scale integration and reduced dimensions are the key factors to optimize the achievements of these devices. Already used in industry (quantum wells, magnetic sensors), these devices are obtained by molecular beam epitaxy, sputtering or pulsed laser deposition. Their reduced dimensionality increases the number of surfaces and interfaces, the role of which has to be precised. Experimentalists try now to associate materials having very different crystal structure and chemical composition. The elastic stress stored in the device can induce various phenomena which have to be evaluated, understood and predicted as much as possible.

The lectures have been followed by round tables, specific workshops and poster presentations. Most of the papers outcoming from these poster presentation will be published in a special issue of Applied Surface Science (Appl. Surf. Sci. 177/4, 2001). They show that many questions are still in debate. We hope that the discussion which occured around them will allow to progress rapidly in the domain of heteroepitaxy and new related properties.

Very special thanks go to all colleagues who participated very actively in this school by lecturing and by preparing the present keynotes. Finally the organizers would like to thank the *Centre National de la Recherche Scientifique* for its regular budget given to the Porquerolles' School, allowing to organize annual meetings on special topics in Surface Science. They are also happy to thank all other institutions and companies which are listed below for their repeated and unfailing help.

Margrit Hanbücken, CRMC2, Marseille
Jean-Paul Deville, IPCMS, Strasbourg

Conference Organization

Scientific Committee

P. Alnot (Nancy)
S. Bourgeois (Dijon)
J.C. Bertolini (Lyon)
B. Carrière (Strasbourg)
M.J. Casanove (Toulouse)
J.P. Deville (Strasbourg)
H.-J. Ernst (Saclay)
M. Hanbücken (Marseille)

J. Massies (Sophia-Antipolis)
A. Marty (Grenoble)
P. Müller (Marseille)
A. Rocher (Toulouse)
D. Schmaus (Paris)
F. Solal (Rennes)
A. Taleb-Ibrahimi (Orsay)
O. Thomas (Marseille)

Organizing Committee

Chairperson: Margrit Hanbücken, CRMC2, Marseille
Co-Chairperson: Jean-Paul Deville, IPCMS, Strasbourg

with

J. Massies (CRHEA, Sophia-Antipolis)
A. Piednoir (CRMC2, Marseille)

P. Müller (CRMC2, Marseille)
O. Thomas (TECSEN, Marseille)

This conference was held under the auspices and financial help of:
- Centre National de la Recherche Scientifique (CNRS)
- Groupe Français de Croissance Cristalline (GFCC)

It is our pleasure to acknowledge the financial assistance provided by:

Commissariat à l'Energie Atomique - Département de Recherche sur L'Etat Condensé, les Atomes et les Molécules (CEA - DRECAM)

Ministère de la Défense - sous-direction scientifique du service de la recherche et des études - DGA

The University Aix-Marseille II, Faculté des Sciences de Luminy

The University Aix-Marseille III, Faculté des Sciences et Techniques de Saint-Jérôme

Centre de Recherche sur les Mécanismes de la Croissance Cristalline (CRMC2), Marseille

Conseil Régional Provence Alpes Côte d'Azur

Karl-Süss-Technique, Saint-Jéoire (France)

Omicron EURL - France, Saint Cannat (France)

Contents

Foreword v
Organizers and Sponsors vii

1. **Introduction** 1
   Raymond KERN
   CRMC2, Marseille

2. **Some elastic effects in crystal growth** 3
   Pierre MÜLLER and Raymond KERN,
   CRMC2, Marseille

3. **Introduction to the atomic structure of surfaces: a theoretical point of view** 63
   Marie-Catherine DESJONQUERES, CEA Saclay and
   Daniel SPANJAARD, LPS, Orsay

4. **Dislocations and stress relaxation in heteroepitaxial films** 99
   Ladislas KUBIN,
   ONERA, Chatillon

5. **An atomistic approach for stress relaxation in materials** 119
   Guy TREGLIA,
   CRMC2, Marseille

6. **Ab initio study of the structural stability of thin films** 151
   (only abstract and references in this book)
   Alain PASTUREL,
   LPM2C, Grenoble

7. **Stress, strain and chemical reactivity: a theoretical analysis** 155
   Philippe SAUTET,
   IRC, Lyon

8. **Strain measurements in ultra-thin films using RHEED and X-ray techniques** 173
   Bruno GILLES,
   ENSEEG-INPG, Saint Martin d'Hères

9. **Measurements of displacement and strain by high-resolution transmission electron microscopy** 201
Martin HYTCH,
CECM, Vitry

10. **Stress measurements of atomic layers and at surfaces** 221
Dirk SANDER,
Max-Planck-Institut für Mikrostrukturphysik, Halle, Germany

11. **STM spectroscopy on semiconductors** 243
Didier STIEVENARD,
IEMN-ISEN, Lille

12. **Spatially resolved surface spectroscopy** 287
Jacques CAZAUX. Université de Reims and
Jean OLIVIER, LCR, Thomson-CSF, Orsay

Introduction

R. Kern

CRMC2 – CNRS, Campus Luminy, Case 913, 13288 Marseille

Forty years ago surface science was devoted to some fewhappy, developing new concepts as well as new instruments. The classical disciplines such as physics, chemistry, metallurgy and more recently materials science became interested by the new perspectives surface science offered. Journals appeared, conferences flourished but textbooks are still largely missing. This is due to the difficult task to transmit multidisciplinary knowledge. Schools either in summer or in winter are good places to promote with some success this new knowledge among young and less younger researchers. There have been organized in the past many schools in surface science, the first were of general nature. The field grew however so fast and large that special topics had to be treated, each time a new shoot appeared. This was the case in the past twenty years for crystal growth and especially for Epitaxial Growth called shortly Epitaxy. The name epitaxy appeared first time for "regular overgrowth of two crystalline species" in the seminal thesis of L. Royer, Université de Strasbourg, 1928, published in Bull. Soc. Fr. Min. 51, 7 (1928). Today technicians use sometimes the nickname epi. Due to the many, widespread industrial applications in electronics, optics, magnetism, this topic was strongly boosted so that it was divided again. On one side there are treated the physical-chemical problems concerning epitaxial growth, on the other side the physics and engineering problems. Fortunately some topical schools and workshops are devoted to bring again these various scientists together. This was the case for the last school in October 1999 at Porquerolles with the thematic "micro and nanotechnology" (published in Applied Surface Science, Vol. 164, September 2000, p. 1 – 286).

During the last ten years, new epitaxial systems were developed where mechanical stress and strain are intentionally controlled during epitaxial growth. Sometimes stress release is wanted, sometimes stress is preserved depending upon the applications. There was the need this time at the Porquerolles school, in October 2000, to offer to the mostly young assistance, lectures, workshops, round tables and poster evenings with their own scientific production. The title was "Stress and strain in epitaxy". This book contains the tutorial lectures (two or three hours each) given at the school. They are numbered in the content list. Some of these lectures concern general knowledge in surface science. Lecture 3 connects theoretical solid state physics to surface physics, lectures 11, 12, 13 describe experimental analytical surface tools, becoming ready in the next future to approach the proper thematic of the school. Then the proper thematic of stress and strain is approached. In lectures 2 and 4 the macroscopic point of view is treated. In 2 bulk and surface elasticity are defined, in 4 stress relaxation by dislocations is considered. In both lectures epitaxial layers and dots are the specific matter to be illustrated. The symmetric part to these lectures is lecture 5 where the atomistic approach is preferred but making the necessary connection with the above macroscopic treatments. The atomistic approach brings new insights in alloying, new ways to relax stress, structurally and chemically. All these three theoretical lectures not only describe the present state but also the kinetic change or trend, a given state is generated or relaxed. There is also the important thematic: what kind of elastic data can one use and find for finite size materials? In lecture 6 the answer was: calculate them ab initio for specific materials under specific thin film conditions (finite size). This is especially true since thin films appear very often as metastable phases, unknown in bulk materials. Unfortunately this book does not contain this full lecture, its abstract sends back to original literature without the precious selection and advice of the teacher.

The counterparts to these theoretical lectures are the experimental ones numbered 8, 9 and 10. Their concern is strain and displacement measurements using diffraction techniques. Grazing incidence methods (8), electrons and X-rays may operate in-situ during deposition, the HRTEM electron microscope studies (9) operating mostly ex-situ but having the advantage of imaging what becomes crucial when defects or composition changes relax strain. The chemist brings his point of view in lecture 7 about stress and strain in connection with reactivity.

Stress measurements are treated in 10 where the cantilever method is studied in details. Due to its sensitivity to adsorption it has to be used under UHV conditions. This allows to use LEED and RHEED, so that structural changes and strain can be evaluated. It concentrates more on sub- and monolayers than thin films. The method detects phase transitions and identifies stress induced by chemisorption as the driving force of these transitions. Stress – strain relaxation by alloying or exsolution followed experimentally could not really be covered. In the future I hope we will see that in-situ control of stress, strain, composition and structure can be done simultaneously. Theoretical knowledge is still ready.

Many uncovered subjects have been presented as original oral contributions and discussed during the poster evenings and round tables. Some of them are published in a special issue of Applied Surface Science 2001.

Some elastic effects in crystal growth

P. Müller, R. Kern

Centre de Recherche sur les Mécanismes de la Croissance Cristalline[*]
Campus de Luminy, case 913, F-13288 Marseille Cedex 9, France

ABSTRACT

These lectures deal with some elastic effects in crystal growth. We recall some basics results about the elastic description of a bulk solid and its surface, then we emphasize on surface stress and surface strain quantities and on the description of surface defects in terms of point forces. Then we focus on the morphological stability of a stressed surface and epitaxially strained crystal as well. We will show how surface stress modifies wetting conditions and how bulk stress modifies the equilibrium state. For 2D growth (perfect wetting) bulk strain modifies the chemical potential of each layer and due to finite size wetting we introduce, it results a number of equilibrium layers for each imposed undersaturation.

For 3D growth (no perfect wetting) the epitaxial stress acts against wetting and leads to a global thickening of the equilibrium shape. We also show how elastic relaxation is a prerequisite for the simultaneous existence of 2D layers and 3D crystals (Stranski Krastanov or SK growth). In the three cases of 2D, 3D or SK mixed mode, beyond some critical size, plastic relaxation may occur. In a last part we consider elastic effects on growth mechanisms. We show that, except for Stranski Krastanov growth, the activation barrier for nucleation is not significantly influenced by strain. In contrast strain plays a role on the detachment rate of atoms (strain lowers the barrier to detachment of atoms from laterally large islands in respect to laterally small islands) and then on kinetics. Then we focus on strain effects on step flow growth and show how step-step and/or adatom-step elastic interactions may give birth to a supplementary net force on each adatom. This force modifies the net current of adatoms and thus leads to some new growth instabilities. The surface diffusion coefficient itself may also be modified by strain but without noticeable modification of growth mechanisms. At last we mention some collective effects.

---

[*]Associé aux Universités Aix-Marseille II et III.

INTRODUCTION

Since Royer [1] the regular oriented over-growth of a crystalline material A onto a single crystal surface B is called epitaxy. Two lattice planes of A and B and at least two lattice rows come in contact and in case of *coherent epitaxy* accommodate their two dimensional (2D) misfit. By this means the couple A/B stores a certain amount of elastic energy. The so-stored elastic energy has been recognised so far as a source of mechanical problems such as cracking, blistering, peeling… Then for many years the main problem of crystal growers was to avoid strain by choosing very low-mismatched systems. Nevertheless it has also been recognised that stress can modify some crystal properties. This is the case of the functional performance of devices such as the possibility of band-gap engineering involving strained structures [2] or the correlation between mechanical stress and magnetic anisotropy in ultra thin films [3]. These technological considerations have stimulated crystal growers to consider also crystal growth properties induced by stress. Nevertheless the problem of formation of a strained crystal on a single crystal is complex. The difficulties basically have three origins.

The first difficulty arises from the fact that since the equilibrium shape of a crystal essentially depends upon surface energy considerations [4,5], a good description of the thermodynamic state of a strained crystal needs to accurately define the role of stress and strain on specific surface energies. This can be done by properly defining surface stress and strain quantities as partially done by Gibbs [6], Shuttleworth [7], Herring [5] and others [8,9,10].

The second difficulty arises from the fact that most strains are anisotropic and inhomogeneous. Indeed on one hand because of the Poisson effect the in-plane strain due to misfit accommodation is accompanied by a vertical opposite strain. Furthermore on the other hand, islands or nuclei can relax by their edges. Obviously this elastic relaxation depends on the shape of the island and therefore cannot be homogeneous! Thus a good description of the bulk elastic energy needs to calculate accurately elastic relaxation.

The third difficulty arises from the fact that, even weak, the elastic effects dominate at long range. Thus elasticity may also affect long-range behaviour usually driven by surface diffusion considerations. In other words not only the energetics of crystal growth may be altered by elasticity but kinetics behaviour may also be altered.

Our purpose in these lectures is to describe some elastic effects on crystal growth. For the sake of simplicity we will only consider pure cubic materials A over B and furthermore do not consider alloy composition, especially changes induced by strain (except briefly in section 3.3.4.). Furthermore as in most of the analytical formulations we will use macroscopic and linear elasticity. For too high misfits (>1%) linearity may fail, for studying the first stages of growth as nuclei the macroscopic treatment may be questionable too. The main advantage of linear elasticity is the possible analytic form of the results (even when complex) that gives the basic tendencies. Obviously in a specific treatment theses results have to be compared with atomistic (if better) calculations.

Last but not least, though kinetics may more or less slow down the realization of the final state, thermodynamics remains the primer way to tell what is possible so that in the framework of these lectures we first focus on thermodynamic properties then on kinetics. The lectures are divided into three parts corresponding approximately to the three above-mentioned difficulties.

In the first part (section 1.) we will recall some basic results of the classical elasticity theory. Since considering crystal growth and thus surface phenomena, we will focus on the accurate description of elastic properties of surfaces. For this purpose we will introduce surface stress and surface strain as surface excess quantities. Furthermore since crystal growth often starts on surface defects (such as steps) we will also describe elastic fields induced by such surface defects (adatoms, steps, domains…)

In the second part (section 2.) we will focus on elasticity effects on the macroscopic thermodynamic state. For this purpose we will revisit Bauer's thermodynamic analysis of epitaxial growth [11,12] by taking into account elastic energy of bulk and surface as well. More precisely we will consider elastic effects on two dimensional (2D), three dimensional (3D) and mixed 3D/2D (or Stranski-Krastanov) growth modes. We will see that, even weak, elasticity may play a major role on the equilibrium properties such as the number of equilibrium layers of 2D film or the equilibrium aspect ratio of 3D crystals. More exciting is the fact that, if in absence of elastic relaxation there is no place for the Stranski-Krastanov growth mode in equilibrium conditions (except in case of some exotic structure change or for kinetics reasons) elasticity considerations open a place for equilibrium Stranski-Krastanov transition.

The third part (section 3.) concerns elasticity effects on microscopic or elementary growth mechanisms. For this purpose we will see how elastic interactions may influence nucleation and step flow. In the first case we only have to consider the role of the elastic interactions on the nucleation activation energy. In fact it will be very weak. In the second case we will reconsider usual kinetics formulation of step flow with as a new ingredient an elastic contribution to the net current of adatoms due to step-step or adatom-step elastic interactions. The main effect of elasticity is then to favour the appearance of new kind of surface instabilities

At last in a short conclusion we will mention some elastic effects we do not take up in detail in these lectures.

1. ELASTIC DESCRIPTION OF A SOLID AND ITS SURFACE

1.1. Elastic description of bulk phases

When a bulk material is stressed (resp. strained) it becomes strained (resp. stressed). Stress and strain are connected by the elastic constants of the material. Many textbooks deal with elastic properties of solids, fundamental aspects are given in [13,14] whereas [15] essentially focus on anisotropic properties.

In this first section we only recall some fundamental concepts.

1.1.1. Bulk stress tensor

Let us consider an elementary parallelepiped (volume $dV=dx_i dx_j dx_k$) centred on a point x_i in a stressed solid (see figure 1a). Each of its faces (area $dx_i dx_j$) normal to the x_k (i,j,k=1,2,3) direction is submitted to a force per unit area (a pressure when negative) whose i^{th} component reads σ_{ik} in the homogeneous case or $\sigma_{ik} + 1/2(\partial \sigma_{ik}/\partial x_k)dx_k$ up to the first order in the inhomogeneous case (independent of x_i). The bulk stress thus is defined by a third order tensor of rank two [σ]. The three σ_{ii} components describe normal stress whereas σ_{ik} components with i≠k define shearing stresses. The components σ_{ij} of [σ] are not invariant under axis rotation (only the trace of the tensor whose mean value equals the mean negative pressure is invariant). As for all symmetric second rank tensors, the components transform as $\sigma'_{mn} = a_{mi} a_{nk} \sigma_{ik}$ where[†] a_{ij} are the components of the matrix of axis transformation [15].

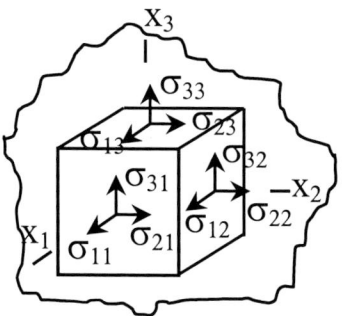

Figure 1a: Action of the components σ_{ij} of the bulk stress tensor applied on the three front faces of the elementary cube. Each face normal to x_j bears a triplet $\sigma_{1j}, \sigma_{2j}, \sigma_{3j}$, the first index $i=1,2,3$ giving the direction x_i where the stress acts. When $i=j$ they are normal stresses, when $i \neq j$ they act in the face. On the back faces of that cube are acting identical stresses of opposite signs or slightly different ones for inhomogeneous stresses.

When the elementary parallelepiped is in mechanical equilibrium that means when no resultant force or torque displaces or rotates it, the bulk stress tensor fulfils the following conditions [15] (see foot note)

$$\partial \sigma_{ik} / \partial x_k = f_i \text{ and } \sigma_{ik} = \sigma_{ki} \qquad (1)$$

where f_i is the ith component of the bulk density of forces. Bulk density of forces generally comes from gravity and can often be neglected when considering nano-crystals. Owing to its diagonal property (1), the bulk stress tensor [σ] can also be written as a 6 dimensional vector with $\sigma_i \equiv \sigma_{ii}$, $\sigma_4 \equiv \sigma_{23}$; $\sigma_5 \equiv \sigma_{13}$ and $\sigma_6 \equiv \sigma_{12}$ (Voigt notation). However σ_m components transform differently than σ_{ij} components under axis transformation [16].

1.1.2. Bulk strain tensor

The symmetric bulk strain tensor components are defined by $\varepsilon_{ik} = \frac{1}{2}\left(\frac{\partial u_i}{\partial x_k} + \frac{\partial u_k}{\partial x_i}\right)$ where u_i are the components of the displacement field [13,15]. The artificial symmetrisation of the strain tensor avoids considering a simple rotation as a deformation [15]. The ε_{ii} components describe the relative elongation of an infinitesimal length parallel to axis x_i, whereas $\pi/2 - 2\varepsilon_{ik}$ with $i \neq k$ is the deformation angle measured between two straight lines initially parallel to axis x_i and x_k respectively. As for σ_{ii}, the trace ε_{ii} is rotation invariant as it should be obviously since it represents the bulk dilatation.

As for the bulk stress tensor one can define a 6 dimensional strain vector as $\varepsilon_i \equiv \varepsilon_{ii}$, $\varepsilon_4/2 \equiv \varepsilon_{23}$, $\varepsilon_5/2 \equiv \varepsilon_{31}$ and $\varepsilon_6/2 \equiv \varepsilon_{12}$ where the factor 2 is introduced for further simplifications. Let us note that ε_{ij} (resp ε_m) components transform as σ_{ij} (resp σ_m) components [15,16].

[†] We use Einstein notation, thus summation has to be performed on repeated indices.

1.1.3. Hooke's law

In the framework of linear elasticity, relationships between stress and strain can be written up to the first order as

$$\begin{cases} \sigma_{ik} = C_{ikmn}\varepsilon_{mn} \\ \varepsilon_{ik} = S_{ikmn}\sigma_{mn} \end{cases} \quad (2)$$

where C_{ikmn} and S_{ikmn} are called stiffness and compliance coefficients respectively. These coefficients describe the elastic properties of the material. Since stiffness [C] and compliances [S] are fourth rank tensors, they contain in 3D, 81 components which transform under an axis transformation as $C'_{ikmn} = a_{io}a_{kp}a_{mq}a_{nr}C_{opqr}$. In fact owing to stress and strain tensor intrinsic-symmetry and energy invariance as well, these tensors only contain 21 independent components. Furthermore crystalline symmetries (extrinsic) reduce the number of independent components from 21 for triclinic crystals to 3 for cubic crystals [15]. Isotropic material (such as glass) are simply described by two elastic constants.

Obviously, using the vectorial (or Voigt) notation of stress and strain tensors, Hooke's law (2) can also be written

$$\begin{cases} \sigma_i = C_{ik}\varepsilon_k \\ \varepsilon_i = S_{ik}\sigma_k \end{cases} \quad (3)$$

with i,k=1,2,3,4,5,6. C_{ik} and S_{ik} thus are 6x6 matrices inverse each other. They are not tensors. Relationships between tensorial and matrix components are $C_{ijkl} = C_{mn}$ with ii=m whatever i≠j, m=4 for ij=23, m=5 for ij=13 and m=6 for ij=12. In contrast $S_{ijkl} = S_{mn}$ for m and n =1,2,3; $2S_{ijkl} = S_{mn}$ for m or n =4,5,6; $4S_{ijkl} = S_{mn}$ for m and n =4,5,6 with as in section 2.1. ii=m whatever i, m=4 for ij=23, m=5 for ij=13 and m=6 for ij=12[‡]. Elastic constants values are generally given in this Voigt notation for some particular crystallographic orientation [17,18]. For other orientations the elastic constants have to be recalculated by the very lengthy transformation of the components of the fourth rank tensors [C] or [S]. Once the transformation has been performed the elastic tensors can be again written in Voigt's notation. Some usual transformations can be found in [19,20], but for a very efficient and general matrix method see Angot [16].

1.1.4. Bulk elastic energy

The elastic energy can be defined as the work of the forces per unit area (σ_{ik}) against the bulk deformation (ε_{ik}) and thus reads for a material of volume V [13]

$$W_{el}^{Bulk} = \frac{1}{2}\int \sigma_{ik}\varepsilon_{ik}dV \quad (4)$$

[‡] Let us warn that some authors use different definitions.

For homogeneous stress and strain one thus can obtain with (2) or (3) the energy density (4) under the following equivalent forms

$$w_{el.} = \frac{dW_{el}^{Bulk}}{dV} = \frac{1}{2}C_{ikmn}\varepsilon_{ik}\varepsilon_{mn} = \frac{1}{2}S_{ikmn}\sigma_{ik}\sigma_{mn} = \frac{1}{2}C_{ik}\varepsilon_i\varepsilon_k = \frac{1}{2}S_{ik}\sigma_i\sigma_k \tag{5}$$

In most of the practical cases elastic energy may be roughly estimated by assuming the material to be isotropic and thus only described by two elastic constants C_{11} (S_{11}) and C_{12} (S_{12}) in matrix notation or more currently the Young modulus E_{is}^{3D} and Poisson ratio v_{is}^{3D} defined by [15]

$$E_{is}^{3D} = \frac{(C_{11}-C_{12})(C_{11}+2C_{12})}{C_{11}+C_{12}} = \frac{1}{S_{11}} \quad \text{and} \quad v_{is}^{3D} = \frac{C_{12}}{C_{11}+C_{12}} = -\frac{S_{12}}{S_{11}} \tag{6}$$

For an isotropic material the elastic energy density (5) when isotropically strained ($\varepsilon_{ii} = \varepsilon$) thus reads after development of (5) with (6)

$$w_{el.} = \frac{E_{is}^{3D}}{(1-v_{is}^{3D})}\varepsilon^2 \tag{7}$$

For anisotropic crystals, relation (5) and corresponding elastic constants have to be used.

For biaxially strained films things can nevertheless be simplified by introducing two dimensional Young modulus and Poisson ratio. For instance let us consider a (001) biaxially strained layer ($\varepsilon_{11} = \varepsilon_{22} = m_o$) of a *cubic crystal*. Thus from appendix A there is:

$$w_{el}^{(001)} = E_{(001)}^{2D} m_0^2$$

with $E_{(001)}^{2D} = C_{11} + C_{12} - 2\frac{C_{12}^2}{C_{11}}$ and $v_{(001)}^{2D} = -2\frac{C_{12}}{C_{11}}$

where $E_{100}^{2D} = C_{11} + C_{12} - 2C_{12}^2/C_{11}$ is the two dimensional Young modulus in the (100) face plane (see appendix A). Notice that in all the cases the elastic energy density is quadratic in respect to strain. For other crystallographic orientations elastic constants have to be transformed under axis rotation. However *the elastic energy density of epitaxially strained layers can always be written* (in absence of relaxation) *under the form*

$$w_{el} = Ym_0^2 \tag{8}$$

where Y is a combination of elastic constants C_{ij} or S_{ij}. Usually $Y \approx 10^{12}$ erg.cm^{-3} = 10^2 GJm^{-3} the elastic energy density is of the order of the chemical bonding (2 eV/at.≈0.25GJm^{-3}) for strain m_o of roughly 5%. Obviously such important energy density cannot be neglected when formulating thermodynamic description of crystal growth, as we will see in section 3.

1.2. Elastic description of ideal planar interfaces

The elastic properties of surfaces can be described by surface stress and strain as excess quantities as first described by Gibbs [6], Shuttleworth [7] and Herring [5] (for surface stress) then Andreev and Kosevitch [8] and Nozières and Wolf [9,10] who furthermore introduced surface strain. In the following we will follow Nozières approach.

1.2.1. Surface stress

Let us consider an infinite coherent planar interface (*whose normal is the axis x_3*) in between two bulk phases I and II both characterised by their own homogeneous stress tensor σ_{ik}^I and σ_{ik}^{II}. Since in the infinite interface the stresses cannot depend on in-plane x_1 and x_2 coordinates, the mechanical equilibrium condition (1), written in absence of bulk forces f_i gives at the interface (x_3=0); $\partial \sigma_{i3}/\partial x_3 = 0$.

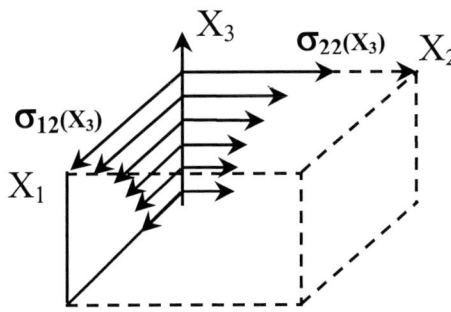

Figure 1b: Surface stress as an excess of bulk stress second rank tensor. Schematically for the surface X_3=0 as a perspective view. It contains the σ_{22} and σ_{12} (or σ_{21} and σ_{11}) components at different levels X_3. At surface equilibrium $\sigma_{33}=\sigma_{13}=\sigma_{23}=0$ and $\sigma_{12}(X_3)=\sigma_{21}(X_3)$. The surface stress is thus the excess

$$s_{ij}^{(3)} = \int \sigma_{ij}(X_3)dX_3 / \int dX_3 - \sigma_{ij}(X_3 \to -\infty)$$

(i,j=1,2). Obviously it is an intrinsic material property independent of the external stresses applied on the body.

Thus there is no interfacial excess of the σ_{i3} components. In contrast, at the interface (x_3=0) $\sigma_{ij}^I \neq \sigma_{ij}^{II}$ (i,j=1,2). The interfacial stress components s_{ij} are thus defined as the surface excess quantity of the components σ_{ij} (i,j=1,2) of bulk stress tensor. Obviously s_{ij} (i,j=1,2,3) with s_{i3}=0 when relaxed in respect to surface x_3=0. In figure 1c it can be seen that s_{ij} at equilibrium is a degenerated 3D, second rank tensor or 2D second rank tensor. On figure 1b we sketch and precise surface stress excess at a vacuum-crystal interface.

1.2.2. Surface strain

Let us suppose that there is forced lattice coherence from bulk phase I to bulk phase II along x_3=0. Then by definition of coherence at the interface there is $\varepsilon_{i3}^I = \varepsilon_3^{II}$ with i,j=1,2. Thus one only can define the interfacial strain tensor e_{ij} as the excess of the components $\varepsilon_{ij}^I, \varepsilon_{ij}^{II}$ of the bulk strain tensors. When one bulk phase is a fluid or vacuum, the solid near x_3=0 presents an intrinsic excess of ε_{i3} depending on the symmetry. It is called: surface strain. As surface stress, surface strain also is a symmetric second rank tensor e_{ij} (i,j=1,2,3) with e_{ij}=0 (if i,j=1,2) (for the surface x_3=0). In figure 1c we illustrate the complementary nature of surface stress and surface strain. Let us note as underlined by Nozières and Wolf [9,10] that if generally the e_{33} component is an elastic relaxation it can also correspond to a mass transfer across the interface. It is important to notice that surface stress and surface strain being surface excesses, in the framework of linear bulk elasticity and at mechanical equilibrium, *surface stress and* surface strain are independent quantities. In other words there is no Hooke law for the surface [8]. Furthermore they are bulk stress and strain independent.

$$[s_{ij}] = \begin{bmatrix} s_{11} & s_{12} & 0 \\ s_{12} & s_{22} & 0 \\ 0 & 0 & 0 \end{bmatrix} \quad ; \quad [e_{i3}] = \begin{bmatrix} 0 & 0 & e_{13} \\ 0 & 0 & e_{23} \\ e_{13} & e_{23} & e_{33} \end{bmatrix}$$

Figure 1c: Illustration of the complementary nature for a given face (3) of the two intrinsic surface properties, sur(inter)face stress and strain at mechanical equilibrium.

1.2.3. Surface elastic energy

The surface (resp. interface) elastic energy can be defined as the work done to deform the surface n (resp. interface). It simply reads for the interface n=3 [9,10],

$$W_{el}^{Surf.} = \int (\sigma_{i3} e_{i3}^n + s_{ij}^n \varepsilon_{ij}) dS_n \text{ with i,j=1,2,3}.$$

When the bulk phase I is vacuum and the phase II a solid, the mechanical equilibrium (1) gives at the surface $x_3=0$, $\sigma_{i3}^{II} = \sigma_{i3}^{I} = 0$. Then the surface elastic energy reduces to

$$W_{el.}^{Surf.} = \int s_{ij}^n \varepsilon_{ij} dS_n \qquad (9)$$

The integral is performed over the non-deformed surface.

In fact (9) is the original definition of surface stress of Gibbs [6]. *The surface stress s_{ij}^n of a face n is the work done to deform the surface n at a constant number of atoms.* Surface stress thus must not be confused with the surface energy γ_n (the energy of creating a surface without deformation) of the face n, which is the work done to create this surface at constant strain. In fact for a free crystalline face there is [7]

$$s_{ij}^n = \frac{\partial(\gamma_n S_n)}{\partial \varepsilon_{ij}} = \gamma_n \frac{\partial S_n}{\partial \varepsilon_{ij}} + S_n \frac{\partial \gamma_n}{\partial \varepsilon_{ij}} = \gamma_n \delta_{ij} + \frac{\partial \gamma_n}{\partial \varepsilon_{ij}} \qquad (10)$$

where δ_{ij} is the Kronecker symbol. Relation (10) known as Shuttleworth relation [7] shows that for a liquid (where a surface cannot be deformed at a constant number of atoms so that $\partial \gamma_n / \partial \varepsilon_{ij} = 0$) surface stress and surface energy are numerically equal (see more complete discussion in [104]). In order to recall the physical difference between surface stress and surface energy it is common to express s_{ij}^n as a force per unit length and γ_n as energy per unit area. Notice that these two quantities have the same magnitude ($0.5 Jm^{-2}$). However γ_n is always a positive quantity s_{ij}^n can be positive (tensile component) or negative (compressive component).

In the following for the sake of simplicity we will mostly consider isotropic surfaces (symmetry greater than binary axis) where the surface stress tensor [s^n] reduces to a scalar s^n for a given orientation n but nevertheless varies with orientation in a cubic crystal. Only in amorphous materials s and γ are direction n-independent.

1.3. Elastic description of real surfaces

It is well known that surface defects such as adatoms or steps change the surface energy. For instance, foreign adsorption decreases the surface energy γ by a quantity $\partial\gamma/\partial\mu = -\Gamma$ where μ is the chemical potential of the adsorbed species and Γ the adsorption density [6]. On the other hand the surface energy of a vicinal surface (angle α) can be written $\gamma = \gamma_o + \beta_1 p + \beta_2 p^3$ where γ$_o$ is the surface energy of the terrace (reference face with a cusp in the γ-plot), p=tgα the macroscopic slope of the vicinal face and β$_i$ some coefficients depending on step energy (β$_1$) and step-step interaction (β$_2$) (for a review see [21]).

In fact adatoms and steps also create a strain field in the otherwise flat surface and the underlying bulk substrate as well. Thus a complete elastic description of crystalline surfaces must include an elastic description of the defects. Indeed we will see (section 3.) that such elastic defects may have dramatic effects on crystal growth mechanisms.

1.3.1. Elastic description of adatoms

In 3D elasticity it is well known [22,13,14] that the field of elastic strain far away a bulk defect located at x=x$^{(i)}$ can in the solid be calculated by considering the action at x>x$^{(i)}$ of a point force F(x-x$^{(i)}$) located at x=x$^{(i)}$. At mechanical equilibrium these forces reduce (up to the first order) to a localised force doublet. Its components can thus be written:

$$F_\alpha(x - x^{(i)}) = A_{\alpha\beta} \frac{\partial}{\partial x_\beta} \delta(x - x^{(i)}) \tag{11}$$

where δ(x) is the 3D dirac function, (α,β=1,2,3) and $A_{\alpha\beta} = \sum_j f_{\alpha,j} x^{(i)}_{\beta,j}$ a 3D tensor with $f_{\alpha,j}$ the αth force component in between the defect i and atoms j $\vec{x}_j^{(i)}$ (components: $x_{\beta j}^{(i)}$) apart. The point force at (i) (11) produces in the planar solid a displacement field whose components $u_\alpha^{(i)}$ are [9,13,23]:

$$u_\alpha^{(i)} = -\sum_\beta \int D_{\alpha\beta}(x',x) F_\beta(x - x^{(i)}) dx^3 \tag{12}$$

where D$_{\alpha\beta}$(x',x) is the 3D Green tensor given for isotropic elastic bodies in classical papers [13,22] when (i) is located at the surface [13] and given in [22] when located in the bulk.

In a seminal paper Marchenko and Parshin [24] following Lau and Kohn [25] consider adatoms as true 2D elastic surface defects and thus model them by 2D point forces doublets. In other words they use relations (11) and (12) but only with α,β=1,2 [24,25] and use the 2D Green tensor [13] to calculate elastic displacements in the plane of the surface. In fact, as depicted by Kern and Krohn [26], adatoms produce a force distribution around them and thus distort also the underlying bulk substrate. Adatoms thus must be described as 3D defects (α,β=1,2,3 in (11) and (12)) and the displacement field calculated by means of 3D Green tensor [22]. The 3D dipole can only be reduced to a 2D one when the normal component of the force exerted by the adatom onto its underlying substrate can be neglected (see such a specific case in appendix I).

1.3.2. Elastic description of steps
* Step on a stress free body:*

As an adatom, a step is not a true surface defect and thus can be described as a row of 3D (and not 2D) dipoles. In the following we assume the steps (located at $x_1=0$) parallel to the direction x_2, furthermore we assume that the steps are infinite in this direction. Thus the point forces describing the displacement field due to a step can be generally written

$$\begin{cases} F_1 = A_{11} \frac{\partial}{\partial x_1} \delta(r) + A_{13} \frac{\partial}{\partial x_3} \delta(r) \\ F_3 = A_{31} \frac{\partial}{\partial x_1} \delta(r) + A_{33} \frac{\partial}{\partial x_3} \delta(r) \end{cases} \quad (13)$$

Nevertheless as shown by Marchenko and Parshin [24], Andreev and Kosevich [8] and Nozières [27] a step has a non-vanishing moment. Indeed since a step divides the surface in two *equivalent* terraces, it is submitted to the surface stress s of the two neighbouring terraces which exert a torque of moment sh (h being the height of the step) and thus tends to twist the crystal.

Obviously the force distribution (13) must restore the torque. For a high symmetry surface ($s_{ij}=s\delta_{ij}$), mechanical equilibrium conditions then lead to $A_{31} - A_{13} = sh$ where h is the height of the step.

Notice that usually, following Marchenko et al. [24] and Andreev et al. [8] steps are described in a simpler way by defining a single force dipole F_1 in the direction x_1 completed by a vertical force dipole F_3 restoring the moment sh. The point force distribution describing the step thus reads:

$$\begin{cases} F_1 = A_{11} \frac{\partial}{\partial x_1} \delta(r) \\ F_3 = A_{31} \frac{\partial}{\partial x_1} \delta(r)) \end{cases} \quad (14)$$

with $A_{31} = sh$. Such a notation is justified in [26] (see section 3.1. of [26])

It is important to notice that when a step separates two domains of different surface stress s_1 and s_2 (it is for example the case of Si(100) face where monoatomic steps separate (2x1) and (1x2) reconstructed surfaces) the point forces distribution that describes the step must include a non-vanishing resulting force (s_1-s_2) near the step before relaxation. In this case an elastic monopole has to be added and thus (14) becomes

$$\begin{cases} F_1 = (s_1 - s_2)\delta(x) + A_{11} \frac{\partial}{\partial x_1} \delta(r) \\ F_3 = A_{31} \frac{\partial}{\partial x_1} \delta(r)) \end{cases} \quad (15)$$

Obviously A_{11} and A_{31} values can only be obtained from atomistic calculations.

* *Step on a stressed body*:

Last but not least let us note that until now we have only considered elastic description of a step on a crystal free of any bulk stress. However there is an additional effect at the surface of a bulk stressed solid (or a non relaxed epitaxial layer). Indeed in this case there is a discontinuity of bulk stress in the surface height and the lateral force on one side of the solid is not compensated by an equal force on the other side. The result once more is a supplementary force monopole located at the steps. The intensity of this force naturally is the bulk stress σ (see 1.1.) times the step height h.

Generally for such stressed solids the dipolar contribution can be neglected (upper order) and thus the step before relaxation is described by the following point force distribution:

$$F_1 = \sigma h \delta(x) \tag{16}$$

1.3.3. Elastic interaction of elastic defects

Since surface defects create a displacement field in the underlying substrate, surface defects interact by way of the underlying deformation. The interaction energy between two defects is simply the work done by the force distribution $f_\alpha^{(1)}$ of the first defect due to the displacement field $u_\alpha^{(2)}$ generated by the other defect. It reads [13,14,23]

$$U^{int.} = \frac{1}{2}\int f_\alpha^{(1)}(x-x^{(i)})u_\alpha^{(2)}(x)dV + \frac{1}{2}\int f_\alpha^{(2)}(x-x^{(i)})u_\alpha^{(1)}(x)dV \tag{17}$$

A lot of literature on elastic interactions between point defects exists for many situations. For defects at the surface Rickman and Srolovitz have proposed a generalised approach [28]. Their results are summarized on table I (in absence of normal forces F_3) where each surface defect is characterized by its dimension (D) and its multipole order (m). An adatom thus is characterized by D=0 and m=1(dipole), whereas a step (D=1) can be characterized by m=1 (dipoles) or m=0 (monopoles) according to the nature of the step (step on a stress free surface or boundary in between two stressed domains) (see section 1.3.2. and figure 2).

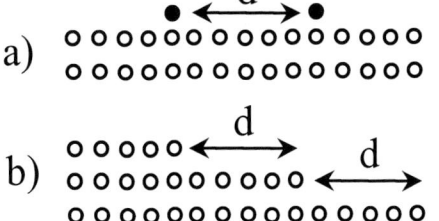

Figure 2: Elastic interaction in between:
a/ 2 adatoms considered as elastic dipoles (D=0,m=1),
b/ 2 steps bearing dipoles (D=1,m=1) of same sign or (and) monopoles (D=1,m=0) of same sign when the solid is stressed;
c/ 2 steps of opposite sign.

	Adatom *dipole* (D=0, m=1)	Step on a stress free body *line of dipoles* (D=1,m=1)	Step **on a stressed body** *line of monopoles or (semi infinite sheet of dipoles)* (D=1,m=0)
Adatom *dipole* (D=0, m=1)	$+d^{-3}$ (repulsion if same sign)	$+d^{-2}$ (repulsion if same sign)	$+x^{-1}$
Step on a stress free body *line of dipoles* (D=1,m=1)	d^{-2}	d^{-2} (repulsion if same sign)	------------
Step **on a stressed body** *line of monopoles or (semi infinite sheet of dipoles)* (D=1,m=0)	$+x^{-1}$	------------------	$\ln(d)$ (attraction if same sign)

Table I: Elastic interaction dependence versus the distance d>0 or the abcissa $-\infty<x<+\infty$ in between some elastic defects when the normal force F_3 is neglected.

The so obtained results are easy to memorize if one admits the repulsive law d^{-3} between similar adatoms (see figure 2a). The interaction between a line of dipoles (step) and a dipole is thus simply obtained by integration along the line and leads to an interaction in d^{-2}. (The elastic interaction between two steps on a stress free body (see figures 2b and 2c) is obtained by integration over the step of infinite length of the d^{-3} law then multiplied by the number of dipoles in the other step giving again a d^{-2} law). The interaction of adatoms (dipoles) with step of a stressed body (the step bears monopoles) gives an x^{-1} law interaction that means interaction changes sign with abscissa x. (Nevertheless the definite sign of the interaction depends upon the respective signs of monopoles and dipoles). This result is equivalent to the interaction of a dipole and a semi-infinite sheet of dipoles. Indeed as in electrostatics a semi-infinite sheet of dipoles is equivalent to a distribution of monopoles located at the border of the sheet. Thus the interaction between a dipole and a step on a stressed body can thus be obtained by a supplementary integration of the d^{-2} law over the various rows of dipoles constituting the domain and thus give a x^{-1} interaction law.

At last the interaction between two parallel steps on a stressed body (two lines of monopoles of same sign. (Fig 2b)) gives an interaction law $+\ln(d)$. This *attractive interaction* can be also obtained from the electrostatic equivalence from a new integration of the x^{-1} law over the various rows of dipoles. Such steps of opposite sign (fig 2c) give a repulsive interaction $-\ln(d)$.

From these results it is easy to see that the interaction between an adatom and a stressed island of lateral size L must vary from d^{-1} for large L (interaction between an adatom and a sheet of dipoles) to d^{-2} for weak L (interaction in between adatoms and a row of dipoles). Obviously the intensity of the interaction depends upon the detail of the calculation of the coefficients $A_{\alpha\beta}$ but does not change nature and sign.

In fact, for completeness we stress the fact that all these classical works [8,24,26,28] summarized in table I only hold for semi-infinite isotropic substrates where Green's function formalism applies. This is no more the case for cubic crystals [29,30]. On the (001) or (111) faces of these crystals the interaction energy of identical adatoms become very anisotropic even with change of sign, however interaction of dipole rows remain with the same sign whatever their azimuth. More recently Peyla et al [31,32] showed that for very thin substrates things change too. For example identical adatoms deposited onto a true 2D isotropic layer may attract or repel each other according to the in-plane direction. The local force distribution seems to be responsible. For thicker sheets this effect goes backwards to usual d^{-3} repulsion valid for thick isotropic substrates.

1.4. Morphology and surface stability of a stressed body

A stressed body may become unstable against undulations or spontaneous formation of stressed domains. In this section we will only give some semi quantitative arguments for a better understanding of basic phenomena leading to such instabilities. In connection with growth mechanism we will examine another instability, the step bunching in section 4.3.

1.4.1. Asaro-Tiller-Grienfeld (ATG) instability

Since Asaro and Tiller [33] then very later Grienfeld [34,35] it is known that a planar surface of a stressed solid is unstable against undulation. There is an abundant theoretical [33-41] and some experimental facts [42,43] on this instability. In the framework of these lectures we will follow [33,36,39] most simplest arguments. For this purpose we consider the free energy change ΔF *per unit area* induced by a periodic one-dimensional surface *undulation* $z(x) = h\cos(\omega x)$ of the surface of a uniaxially in-plane stressed solid $\sigma = \sigma_{xx}$ (ω is the wave vector $\omega = 2\pi/\lambda$, $h>0$ the amplitude of the undulation, Oz pointing in the solid) (see figure 3).

This free energy change per unit area contains two terms. The first one ΔF_1 is the capillary energy change due to surface area increase (surface energy γ being isotropic):

$$\Delta F_1 = F_1^{und.} - F_1^{flat} = \gamma \frac{\iint \left\{ \sqrt{1+\left(\frac{\partial z}{\partial x}\right)^2} - 1 \right\} dxdy}{\iint dxdy} \approx \frac{\gamma}{2} \frac{\iint \left(\frac{\partial z}{\partial x}\right)^2 dxdy}{\iint dxdy} = \frac{\gamma}{4} h^2 \omega^2 \quad (18)$$

where the last expression is valid for small slope $\partial z/\partial x < 1$ or $h/\lambda << 1$

The second term ΔF_2 is the elastic energy change per unit area induced by the undulation. Even for isotropic elasticity it is difficult to calculate since it depends upon the detail of the elastic relaxation. We follow the original paper [36] with a simplified version of first order in h/λ where the approximation (trick) is to apply sinusoidal forces along a flat surface (The undulation is treated as a surface defect on a planar semi-infinite surface). In fact along a flat surface at z=0, as well as in the bulk (z>0), there is in the solid an in-plane constant stress $\sigma_{xx}(x,z) = \sigma$ so that nowhere f_x forces apply, gravity or other so-called body forces being excluded. But when some 1D undulation z(x) of some arbitrary amplitude h>0 at the surface and $\omega=2\pi/\lambda$ is created (fig.3), there develops at the surface some excess forces. When the amplitude is small in respect to λ, ($h/\lambda <<1$) there may tentatively be a force density

$f_x(x,0) = \frac{d}{dx}(\sigma z) \approx \sigma \frac{dz}{dx}$, with $f_y(x,0) = 0$ of course and a normal force $f_z(x,0)$ close to zero at second order of the slope dz/dx.

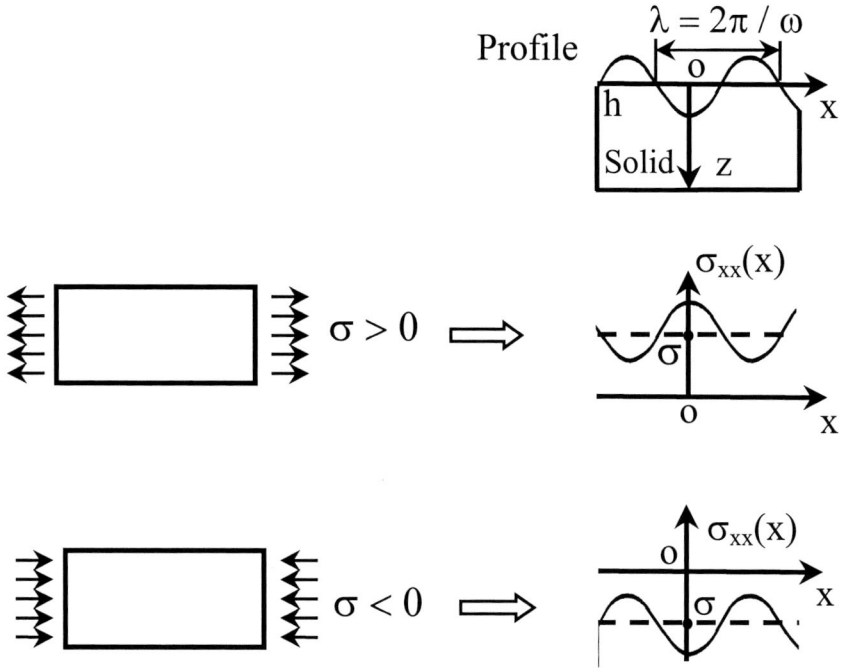

Figure 3: *Azaro-Tiller-Grienfeld instability of a planar surface under an external uniaxially stress $\sigma_{//} = \sigma$. When the surface develops a 1D sinusoidal undulation $h\cos\omega x$, $h>0$, Oz pointing in the bulk, the surface at first order develops a stress concentration either tensile ($\sigma>0$) or compressive ($\sigma<0$) in the valleys.*

Let us suppose that such a line of forces $f_x(x,0)$ is applied on the surface of a planar semi-infinite isotropic solid, according to Love [14] the stress in excess of σ is given by:

$$\sigma_{xx}(x,0) - \sigma = -\frac{2}{\pi}\int_{-\infty}^{+\infty}\frac{f_x(x',0)}{x'-x}dx' = \frac{2\sigma h\omega}{\pi}\int\frac{\sin\omega x'}{x'-x}dx'$$

According to [36] and appendix B the principal value of this Cauchy integral is $\pi\cos\omega x$ so that

$$(\sigma_{xx}(x,0) - \sigma)/\sigma = 2\omega z(x) + O\left[(h/\lambda)^2\right]; \quad \sigma_{yy}(x,0) = 0 \tag{19}$$

Thus at the surface an excess modulation of the in-plane stress appears. In figure 3 we observe for $\sigma>0$ that valleys bear maximum tensile stress, in figure 3 for $\sigma<0$ they bear maximum compressive stress. Crest at contrary, have smaller stress than the mean value σ, which is also the bulk value whatever tensile or compressive.

Hooke's law under plane strain ($\varepsilon_{yy} = 0$ when the solid is infinite along Oy), at the surface z=0, associates to the in-plane stress xx the only in-plane strain xx modulation from (19):

$$\varepsilon_{xx}(x) = \frac{1-v^2}{E}\sigma_{xx}(x) = \varepsilon_{//}(1+2\omega z(x)) \tag{20}$$

It follows the strain modulation around the mean in-plane strain $\varepsilon_{//} = \frac{1-v^2}{E}\sigma$. Due to Poisson's effect the normal strain modulates with opposite phase $\varepsilon_{zz}(x) = -\frac{v}{1-v}\varepsilon_{xx}(x)$. Along the undulated surface the lattice parameters vary as $a_{//}(x) = a_o(1+\varepsilon_{xx}(x))$, $a_{\perp}(x) = a_o(1+\varepsilon_{zz}(x))$ where a_o is the parameter of the stress free crystal. In the valleys x=nλ (fig 3): $[a_{//}(n\lambda) - a_o]/a_o = \varepsilon_{//}(1+4\pi h/\lambda)$ there is strain concentration whereas on the crests x=(2n+1)λ: $[a_{//}((2n+1)\lambda) - a_o]/a_o = \varepsilon_{//}(1-4\pi h/\lambda)$ there is strain deficit. One says that the crests relax.

Let us note that the strain-stress modulations penetrate in the bulk z>0. The more general Green's kernel $Re\left[\frac{x'-x}{(x'-z)^2}\right]$ has to be used. The result for z>0 of (19) is then valid when its right side is multiplied by $(1-\pi z/\lambda)\exp(-2\pi z/\lambda)$ showing that the in-plane stress damps exponentially with a decay distance λ/2π independent of the amplitude h (at this degree of approximation). This result and others for σ_{yy}, σ_{xz} have been successively obtained by different methods by [33,34,36,39].

The matter is now to calculate the strain energy per unit area, of the undulated surface, that means the integral (4) $F_2^{und} = \frac{1}{2}\int_V \sigma_{ik}\varepsilon_{ik}dV / \int_S dS$ extended over all the semi-infinite solid z>0. This would be a cumbersome task since in the bulk Hookes'law gives several ε_{ik} components for each σ_{ik}. Fortunately Marchenko [44] showed how to reduce this volume integral in a surface integral where then the ε_{xx} component is connected to one σ_{xx} component (see appendix C) so that

$$F_2^{und} = \frac{1}{2}\int_S f_i u_i dS / \int_S dS \tag{21}$$

where f_i are the surface force density components inducing the surface displacement components u_i. Since on the slightly undulated surface exists only, at first order, the $f_x(x,0) = 2\sigma\omega(dz/dx)$ component and the corresponding in-plane displacement from (20) $u_x(x,0) = \int \varepsilon_{xx}(x,0)dx = \frac{1-v^2}{E}\sigma[x+2h\sin\omega x]$ where we put $u_x(x,0) = 0$ by convention. From (21) with these functions:
$F_2^{und} = -4(1-v^2)\sigma^2 h^2\omega/E$ so that for the flat surface $F_2^{flat} = 0$ and

$$\Delta F_2 = F_2^{und} - F_2^{flat} = -4(1-v^2)\sigma^2 h^2 \omega / E \tag{22}$$

We learn that the elastic energy density drops when an undulation develop on a in-plane $\sigma_{xx} = \sigma$ stressed solid irrespective stress is tensile or compressive. To this opposes the surface energy change (18) so that the total free energy change is

$$\Delta F = \Delta F_1 + \Delta F_2 = h^2 \omega \left[\frac{\gamma}{2} \omega - \frac{4(1-v^2)}{E} \sigma^2 \right] \tag{23}$$

Instability occurs when $\Delta F<0$ that means the applied stress, tensile or compressive, overpasses the critical value

$$|\sigma| \geq \left[\frac{E\gamma\omega}{8(1-v^2)} \right]^{1/2} \tag{24}$$

In terms of critical strain (20) and wavelength (24) gives $\varepsilon_{//} \geq \left[\frac{\pi(1-v^2)\gamma}{4E\lambda} \right]^{1/2}$. Since γ/E values of a wide variety of materials [45] are among 10^{-2} nm, for usual strain $\varepsilon_{//}=10^{-2}$ as e.g. epitaxial strain, $\lambda \approx 10^2$ nm so that nanoscopic undulations may develop. The early time evolution of the surface profile is driven by the chemical potential gradient along the surface [33,38].

Obviously a more complete treatment must contain first real anisotropy γ behaviour, then surface stress and gravity as well. The surface stress effect on the instability has been studied by Grilhe [46] and Wu [44]. An interesting result [46] is that the symmetry between compression and tension is broken by surface stress since the σ^2 dependence adds a σ contribution. If $(s-\gamma) = \partial\gamma / \partial\varepsilon$ has same sign as the bulk stress σ the critical wavelength is reduced. When introducing gravity [27], a new stabilizing term may appear in ΔF when acting in a proper direction. The effective contribution to instability would start at sizes h of the order of the millimeter. More precisely the ATG instability is thus in between two critical wavelengths. Only a non-linear analysis can give the true final shape [48]

Let us note that Spencer et al. [40,41] have studied the case of an epitaxial stressed layer on a lattice mismatched substrate but the same limitation of constant γ are used. The free surface also is unstable in respect to a sufficiently long wavelength but the critical wavelength now depends upon the thickness d of the strained film in the very special case the substrate is infinitely stiff. Such undulations have been interpreted as a possible origin of the Stranski Krastanov transition on kinked faces (diffuse faces) [35]. For F faces (singular faces) the origin of Stranski-Krastanov transition will be discussed in section 3.2.3.3.

1.4.2. Spontaneous formation of stressed domains

Marchenko [44] was the first to propose that elastic interaction between surface antiphase domains may lead to periodic patterns due to alternating surface stress discontinuities $\pm\Delta s$ (see figure 4a). A special case is for example the Si(100) surface that exhibits reconstructed (1x2) or (2x1) domains of surface stress s_1 and s_2 according to the level of the reconstructed terraces. Due to the anisotropy of the flat Si(100) surface, a vicinal Si(100) surface can be represented by a parallel line of alternating monopoles bearing the stress discontinuities $\Delta s = s_1 - s_2$ (see (15) and figure 4b). Alerhand et al [49] took this over for this case and made the balance between the domain boundary creation energy and the elastic interaction that leads to a selection of the size of the domains.

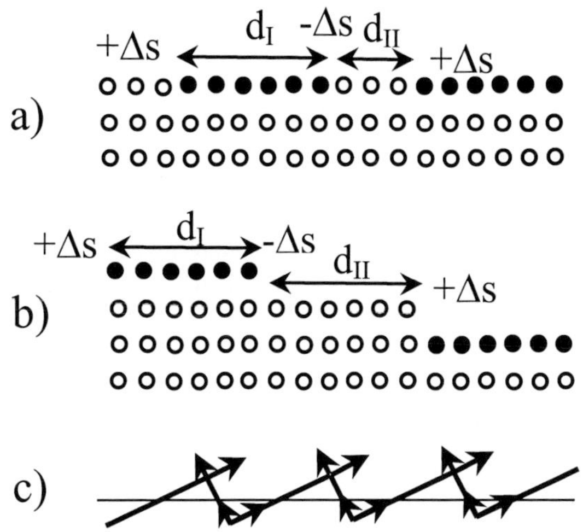

Figure 4:
a/ antiphase surface domains (D=1,m=0) bearing surface stress discontinuities($-\Delta s = s_1 - s_2$, $+\Delta s = s_1 + s_2$);
*b/ case of the (001) Si vicinal face similar to **a**;*
c/ unstable flat face developing a vicinal of macrofacets, the arrows representing the non-compensated surface stress of the macro-facets.

More precisely for a surface with periodic (period $d = d_I + d_{II}$) alternating domains of size d_I and d_{II}, the total energy per unit length reads [44,49]
$$U = \frac{\chi h}{d} - \frac{1-v^2}{\pi E} \Delta s^2 \ln\left(\frac{d}{2\pi a} \sin \pi \frac{d_I}{d}\right)$$
where the first term is the boundary creation energy per unit length (χ) and the second term the elastic energy where one recognises the ln(d) dependence (see table I) and where the sinus comes from the periodicity. The minimisation of U in respect to d_I gives the equilibrium size of each domain. Furthermore, stretching or compressing the crystal parallel to the surface and normally to the steps favours one type of domain over the other as found experimentally by Webb et al. [50]. More recently, B.Croset et al. [51] have completed Marchenko-Alerhand's theory by taking into account the elastic self-energy of each domains so that epitaxial strain somewhat relaxed by dislocations of Frenkel-Kontorova type could be considered. For completeness notice that the Marchenko elastic interaction is also at the basis of spontaneous periodic faceting of unstable crystal surfaces (see figure 4c). For a review see [52,53].

2. MACROSCOPIC THERMODYNAMIC TREATMENT OF EPITAXIALLY STRAINED CRYSTALS

2.1. Wetting conditions versus elasticity

Let us recall that three possible mechanisms of epitaxial growth of a crystal A onto a crystal B have been recognized: the 3D (or Volmer-Weber growth), the layer by layer (or Frank-van der Merwe growth) and the layer by layer growth followed by 3D growth (or Stranski

Krastanov growth). In absence of misfit Bauer [11,12,54] rationalised these growth modes by defining the so-called wetting factor

$$\Phi_\infty = 2\gamma_A - \beta = \gamma_A + \gamma_{AB} - \gamma_B \qquad (25)$$

In (25) we also use the Dupré relation [55] $\beta = \gamma_A + \gamma_B - \gamma_{AB}$ with γ_B the surface energy of B, γ_{AB} the interfacial energy and β the adhesion energy of A on B. The wetting factor (25) in fact is connected to the capillary energy change ΔW_{cap} per unit area during the thermodynamical process in which a crystal A is created (volume $h\ell^2$ with 2 basal faces of energy γ_A and 4 lateral faces of energy γ_A for a parallelepiped crystal) then stuck on a substrate B (adhesion energy $-\beta$) (see figure 5a): $\Delta W_{cap} = \Phi_\infty \ell^2 + 4\Phi'h\ell$ where $\Phi' = \gamma'_A$

When $\Phi_\infty < 0$ (more than perfect wetting) $\gamma_A + \gamma_{AB} < \gamma_B$ (or using Dupré relation $2\gamma_A < \beta$) so that the minimum state of energy of the system is reached for an increasing interface that means a 2D film is more stable than a 3D crystal. When $\Phi_\infty > 0$ (no perfect wetting) $\gamma_A + \gamma_{AB} > \gamma_B$ (or $2\gamma_A > \beta$) so that the minimum state of energy is reached for a decreasing interface. A 3D crystal thus is more stable than a 2D film.

Notice that from this thermodynamical point of view 2D and 3D growth are thus well differentiated and cannot occur simultaneously. The case of Stranski Krastanov growth mode is more complex since obviously[§] the wetting energy must vary during the growth from negative values to positive values in such a manner 2D growth is relayed by 3D growth. Such dependence of wetting energy upon the number of deposited layers z can be justified by introducing long-range effect [56]. The wetting factor thus becomes $\Phi(z) = \Phi_\infty^0 f(z)$ where $f(z)$ is a decreasing function[**] of z. We will come back to this point in section 2.2.2.

Naturally arises the question of the modification of the wetting factor (25) when there is a misfit and thus incidentally the modification of the growth mode by elasticity. For this purpose we just consider a new thermodynamical process analogous to the one we just described (figure 5a) but where now the crystal A is homogeneously strained before coherent accommodation onto a lattice mismatched substrate B (figure 5b). For the sake of simplicity we will consider a semi-infinite planar substrate B. The crystals are cubic of crystallographic parameter a (for A) and b (for B) with a (100) as contact plane both having parallel axis. We define the misfit strain as

$$\varepsilon_{11} = \varepsilon_{22} = m_o = (b - a)/a \qquad (26)$$

[§] In this paper we exclude growth mode change due to structural change during crystal growth.
[**] Let's mention that in the original paper of Stranski and Krastanov [57], ionic NaCl crystals have been considered so that normal to a (001) face there is an oscillating potential changing sign from a simple to a double layer.

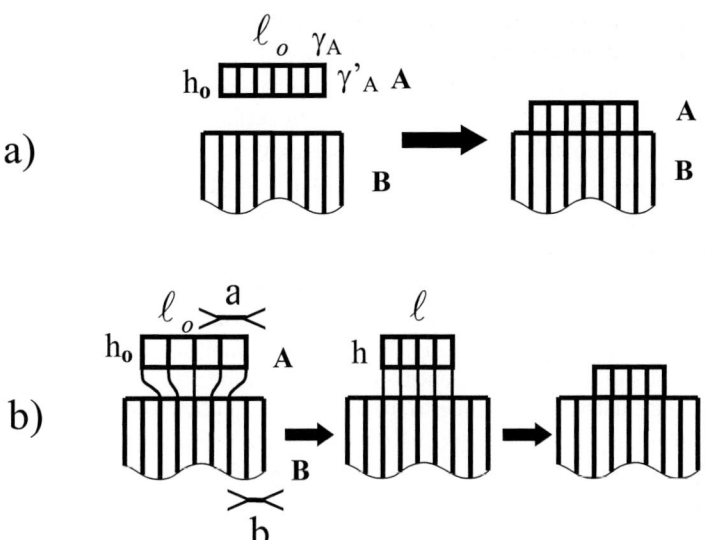

Figure 5: Schematic thermodynamic process of formation: *a/* in absence of misfit the 3D crystal A is created then accommodated onto a isomorphous substrate B. *b/* In the presence of misfit the crystal A (parameter a) is homogeneously strained (from $V_o = h_o \ell_o^2$ to $V = h\ell^2$) before accommodation and adhesion onto the lattice mismatched substrate B (parameter b).

The free energy change following the sequence (figure 5b) reads
$$\Delta W = \left[2\gamma_A \ell_o^2 + 4\gamma'_A \ell_o h_o\right] + \left[2s_A(\ell^2 - \ell_o^2) + 4s'_A(\ell h - \ell_o h_o)\right] - \beta \ell^2$$
where the first term is the energy of formation of the surfaces of the *non-deformed* crystal ($V_o = h_o \ell_o^2$), the second term is the surface elastic work against surfaces deformation (from V_o to $V = h\ell^2$) (see (9)) and the last term the adhesion energy of the *deformed* crystal on the substrate. Then using $\ell = \ell_o(1+m_o)$ and $h = h_o\left(1 - \frac{2\nu}{1-\nu}m_o\right)$ one obtains

$$\Delta W = \Phi_\infty^{m_o} \ell^2 + 4\Phi'_\infty h\ell$$

with up to the second order in m_o:

$$\Phi_\infty^{m_o} = 2\gamma_A - \beta + 4m_o(s_A - \gamma_A) \quad \text{and} \quad \Phi'_\infty = \gamma'_A + m_o(s'_A - \gamma'_A)\frac{1-3\nu}{1-\nu}. \tag{27}$$

$\Phi_\infty^{m_o}$ thus is a generalised wetting factor that replaces the usual wetting factor Φ_∞ (25) when surface stress acts [58].

Owing to the Shuttleworth relation (10), relation (27) is nothing other than the expansion of the wetting factor (25) up to the first order in strain when β is the adhesion energy of the accommodated material A over B.

For usual epitaxial material $m_o=10^{-2}$ and for clean surfaces s_A and γ_A roughly have the same order of magnitude so that the corrective term $4m_0(s_A-\gamma_A)$ to the wetting energy remains weak. However surface stress contributes more to the wetting condition the smaller the wetting factor (25) is and surface stress differs from surface energy. It is the case of Si/Ge system ($m_o\approx+4\%$) where $\Phi_\infty^0 = 150\ erg\ cm^{-2}$ whereas $\Phi_\infty^m = 380\ erg\ cm^{-2}$ [58]. However for such a case of 3D growth (since $\Phi_\infty^0 > 0$) the surface stress effect is overestimated. Indeed owing to the condition $\sigma_{iz}=0$, 3D crystals must relax by their free faces and thus the bulk strain is lowered from homogeneous misfit m_o to non-homogeneous residual values $\varepsilon(x_1,x_2,x_3)$. Furthermore if both crystals remain coherent, the interfacial stress s_{AB} also works during the relaxation from m_o to $\varepsilon(x_1,x_2,x_3=0)$. The new wetting factor in presence of elastic relaxation Φ_m is easy to write formally. However it cannot be calculated in a simple way since the residual strain tensor components depend upon the precise shape of the 3D crystal and have to be calculated using some mechanical model. Furthermore interfacial stress values are poorly known.

In fact we will only remember that in most usual cases surface and interfacial stress corrections to the wetting conditions remain weak.

2.2. Equilibrium state versus elasticity

2.2.1. Description of an epitaxial model system

Our purpose in this section is to seek for the thermodynamical and mechanical equilibrium state of A/B.

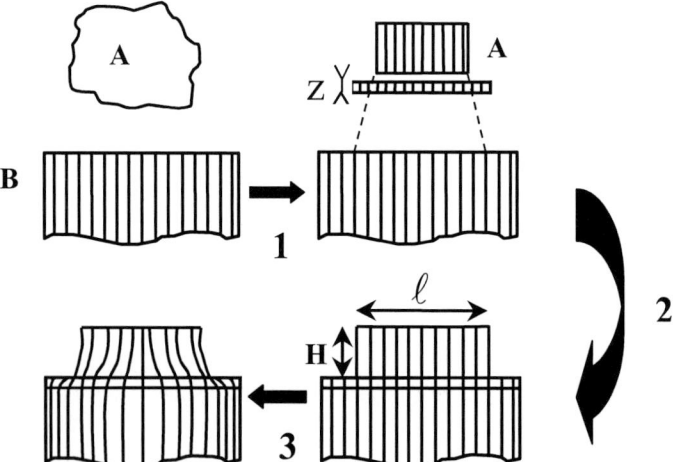

Figure 6: *Schematic thermodynamic process of formation in the case of Stranski Krastanov growth. 1. formation, 2. homogeneous deformation for accommodation then adhesion, 3. inhomogeneous elastic relaxation.*

Therefore we will only consider a thermodynamic process where the deposit A is obtained from the condensation of a perfect vapour onto a lattice mismatched semi-infinite crystal. Furthermore both crystals A and B are cubic and supposed not to mix. The epitaxy is with parallel axis on a (001) plane. The (001) surfaces of A and B are supposed to be stable, that means each having a cusp in their surface energy plot γ_A and γ_B. Furthermore the interface (001) has to be stable that means $\gamma_{AB} > 0$ and having a cusp. The crystallographic parameters are a and b respectively, the in-plane misfit being given by (26).

We will consider the final state as a 3D crystal of volume V sitting on z pseudomorphic layers over B that means Stranski Krastanov case (see figure 6). Indeed such formulation will allow discussing the two other cases of Volmer-Weber and Frank-van der Merwe growth as limiting cases for z=0 and V=0 respectively. Furthermore for the purpose of this lecture we will only consider box shaped 3D crystals. Other shapes will be discussed on the basis of some other papers [59-65]. Since as abovementioned surface stress generally plays a minor role on the wetting condition, in the following we will neglect surface stress and consider bulk elasticity only. We will come back to surface stress effects in section 3.2.3.4.

2.2.2. Free energy change of the SK condensation

Considering the condensation process described in figure 6 where 3D islands and 2D layers are formed from a vapour, the free energy change is composed of three terms:

* ΔF_1 is the chemical work to form (on an area L^2) a 2D film of z layers and an island (volume $V = h\ell^2$) from an infinite reservoir of matter A defining the saturation pressure P_∞. It reads

$$\Delta F_1 = -\Delta\mu\left(zaL^2 + h\ell^2\right) \tag{28}$$

where $\Delta\mu = \dfrac{kT}{a^3}\ln(P/P_\infty)$ is called the supersaturation per unit volume of vapour A at pressure $P > P_\infty$ (supposed to be perfect) and a an atomic linear size.

* ΔF_2 corresponds to the formation of the surfaces of the crystal A followed by its adhesion on the bare substrate B. It reads:

$$\Delta F_2 = \Phi_\infty\left[\left(L^2 - \ell^2\right)f(z) + \ell^2 f(z+h)\right] + 4\gamma'_A h\ell$$

where we consider the wetting energy Φ_∞ of (25) as being size dependent as mentioned in section 2.1. More precisely since surface and adhesion energies are excess quantities they are only well defined for semi-infinite solids. For a finite solid that only contains a few layers, surface and adhesion energy must depend on the number of layers z and thus read $\gamma_A(z)$ and $\beta(z)$. The evolution with z by a decreasing function f(z) is quite usual in 2D multiplayer condensation (see [66]) or surface melting [67,68]. In the following we choose an exponential behaviour as used and justified for semi-conductors. The wetting energy now reads [69,70]:

$$\Phi(z) = \Phi_\infty\left(1 - \exp(-z/\zeta)\right) \tag{29}$$

where ζ is a screening factor close to unity so that in the following we put $\zeta=1$.

Thus ΔF_2 reads:

$$\Delta F_2 = \Phi_\infty \left[\left(L^2 - \ell^2\right)\left(1 - \exp(-z)\right) + \ell^2 \left(1 - \exp(-(z + h/a))\right) \right] + 4\gamma'_A h\ell \qquad (30)$$

In the following since z+h/a>h/a and owing to the quick variation of the exponential we will neglect $\exp(-(z+h/a))$ against $\exp(-z)$. Notice again that we have neglected surface stress work.

* ΔF_3 is the total elastic energy stored by the composite system A/B. It can be written as the sum of the elastic energies stored in the crystal A and in the substrate B respectively and thus reads:

$$\Delta F_3 = \mathcal{E}_o \left[zaL^2 + h\ell^2 R(h,\ell) \right] \qquad (31)$$

where $\mathcal{E}_o = Y m_o^2$ where Y is a combination of elastic constants (see section 1.1.4.). For micro or nano crystals naturally arises the question of the validity of bulk elastic constants. Nevertheless it seems [71] that surface stress considerations avoid using size dependent elastic constants (see appendix D). The first term of relation (31), is the homogeneous energy stored by z pseudomorphous layers of thickness a and lateral size L. The second term is the elastic energy originating from the 3D upperlying crystal of volume $V = h\ell^2$. The factor $0 < R(h,\ell) < 1$ is a relaxation factor. Obviously in absence of 3D crystals V=0, one recovers $\Delta F_3 = \mathcal{E}_o z a L^2$ the elastic energy of the pseudomorphous film. Let us note that the relaxation factor $R(h,\ell)$ has to be calculated for each specific case. It originates from the fact that the normal stress components σ_{iz} along the free surface has to vanish at mechanical equilibrium. It has not the same expression in case of 3D growth (z=0), SK growth (z≠0,V≠0) or 2D growth where it does not play any role (V=0) or more exactly a minor role since $h = a$ and $r = a/\ell$.

For a 3D crystal sitting on a bare substrate B (Volmer-Weber growth) the epitaxial contact in between the 3D deposit and its lattice mismatched substrate is supposed to be coherent and to remain coherent during the elastic relaxation of the 3D crystal. In this case, during the relaxation the crystal A drags the atoms of the contact area and produces a strain field in the substrate B which was initially strain free. This created strain field may be calculated by using point forces [59,72] or more properly by using a self-consistent approach [73,74]. After relaxation the 3D crystal and its substrate are inhomogeneously strained (see figure 6). This means that although the total energy density has been lowered by elastic relaxation, the elastic energy density in the substrate has increased.

The Stranski-Krastanov case is more complicate. Indeed before relaxation, deposit A (3D crystal A and the z pseudomorphous layers) is homogeneously strained whereas the substrate B is stress free. After relaxation the elastic energy density in the 3D crystal as well as the elastic energy density in the underlying film have been lowered in respect to the initial pseudomorphous strained layers (see figure 6). If the number of 2D layers is weak enough the inhomogeneous strain field induced in the 2D layers by the relaxation of the 3D crystal penetrates into the underlying foreign substrate B. The elastic energy density in the substrate B thus increases in respect to the initial stress free lattice mismatched substrate (see figure 7). If on contrary the number of layers increases the positive contribution of B vanishes. In other words the strain fields, induced by the elastic relaxation of 3D islands are not the same for a bare substrate and a composite (2D A layers + B) substrate. In the former VW case, the elastic strain density can be analytically evaluated in a self-consistent fashion (see figure 7), but to our best knowledge it is not the case up to now for a composite substrate. So in the discussion we will distinguish the relaxation factor in case of 3D growth $R^{3D}(h,\ell)$ and in case of Stranski Krastanov growth $R^{SK}(h,\ell,z)$.

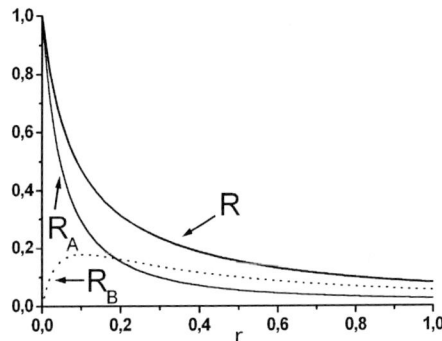

Figure 7: *Elastic energy relaxation factor $R^{3D}(r)$ versus shape ratio $r=h/\ell$ for a 3D box shaped crystal calculated from [74,87] when the deposit and the substrate have the same elastic properties. R_A and R_B are the relaxation factors of the deposit A and substrate B respectively. $R=R_A+R_B$. Multiplied by $\mathcal{E}_o = Ym_o^2$, $Y=E/(1-v)$ each gives the corresponding elastic energy density.*

Finally it is very convenient to write the total energy change $\Delta F = \Delta F_1 + \Delta F_2 + \Delta F_3$ as a function of the volume of the 3D island $V = h\ell^2$, of the aspect ratio $r = h/\ell$, the number of underlying layers z, and the film area L^2 containing one island, using equations (28) to (31) ΔF can be written

$$\Delta F = -\Delta\mu(V+L^2za)+\Phi_\infty\left[\left(L^2-\left(\frac{V}{r}\right)^{2/3}\right)(1-e^{-z})+\left(\frac{V}{r}\right)^{2/3}\right]+4\gamma_A'V^{2/3}r^{1/3}+ \mathcal{E}_o(VR(r)+zaL^2) \qquad (32)$$

where a is an atomic size and where for a box shaped crystal the relaxation factor only depends upon the aspect ratio $r = h/\ell$ and upon z for Stranski-Krastanov case.

Notice that if 2D layers have to be formed A must wet B and thus Φ_∞ must be negative (see (25)) Thus if $\Phi_\infty<0$ and $V>0$, ΔF given by (32) is the free energy change due to SK condensation. The free energy change for single 2D condensation (Frank van der Merwe growth) is thus simply obtained by taking V=0 with $\Phi_\infty <0$ in (32). On the contrary the free energy change due to 3D growth onto a bare substrate (Volmer Weber growth) is obtained by taking $\Phi_\infty >0$ in (32) with z=0.

Globally in all cases condensation takes place for $\Delta F<0$. To this can contribute the first two terms of (32): $\Delta\mu$ when positive and wetting Φ_∞ when negative. In the following, we will distinguish the growth of 2D layers and the subsequent growth of the 3D crystals.

2.2.3. Equilibrium state

The equilibrium state is found by minimisation of the total energy change ΔF, the zeros of the partial derivatives $\left.\frac{\partial \Delta F}{\partial z}\right|_{V,r}$; $\left.\frac{\partial \Delta F}{\partial V}\right|_{z,r}$ and $\left.\frac{\partial \Delta F}{\partial r}\right|_{z,V}$ of (32) giving the equilibrium values of z, V and r noted z*, V* and r* respectively. In the following, supersaturation $\Delta\mu$ as well as

$$\theta = (\ell/L)^2 \tag{33}$$

the fraction of the film surface covered by 3D crystals are considered as constant parameters. According to the sign of the wetting energy Φ_∞ the three previous partial derivatives give the following relations given in table II where we distinguish $R^{3D}(r)$ from $R^{SK}(r,z)$. We also put:

$$r_o = |\Phi_\infty|/2\gamma'_A \tag{34}$$

We will show in section 3.2.3.2. that r_o is the aspect ratio of a deposited crystal in absence of misfit.

	$\Phi_\infty < 0$ (2D or SK growth)	$\Phi_\infty > 0$ (3D growth)
$\left.\dfrac{\partial \Delta F}{\partial z}\right\|_{V,r}$	$z^* = \ln\left[\dfrac{|\Phi_\infty|}{(\mathcal{E}_o - \Delta\mu)a}(1-\theta)\right]$ (a)	
$\left.\dfrac{\partial \Delta F}{\partial V}\right\|_{z,r}$	$V^*(r) = \left[\dfrac{\frac{8}{3}\gamma'_A - \frac{2}{3}\dfrac{|\Phi_\infty|}{r}e^{-z}}{\Delta\mu - \mathcal{E}_o R^{SK}(r,z)}\right]^3 r$ (b)	$V^*(r) = \left[\dfrac{\frac{8}{3}\gamma'_A - \frac{2}{3}\dfrac{|\Phi_\infty|}{r}}{\Delta\mu - \mathcal{E}_o R^{3D}(r)}\right]^3 r$ (d)
$\left.\dfrac{\partial \Delta F}{\partial r}\right\|_{z,V}$	$r^{*2/3} = -\dfrac{4\gamma'_A}{3\mathcal{E}_o}V^{-1/3}\left[1 - \dfrac{r_o}{r}\exp(-z)\right]\left(\dfrac{dR^{SK}}{dr}\right)^{-1}$ (c)	$r^{*2/3} = -\dfrac{4\gamma'_A}{3\mathcal{E}_o}V^{-1/3}\left[1 + \dfrac{r_o}{r}\right]\left(\dfrac{dR^{3D}}{dr}\right)^{-1}$ (e)

Table II: *equilibrium values z^*, V^*, r^* according to growth conditions*

On the basis of these results, let us discuss the elasticity effects on the equilibrium state. In the three cases under study (2D, 3D, Stranski-Krastanov growth mode) we will consider growth conditions, then equilibrium properties and at last plastic-elastic interplay.

2.2.3.1. Layer by layer growth

* Here we are only concerned with Frank-van der Merwe growth that means $\Phi_\infty < 0$ for having 2D condensation (see at the end of section 3.2.2.) with V=0 and thus $\theta = 0$ so that from formula (a) of table II there is

$$z^* = \ln\left[\dfrac{|\Phi_\infty|}{(\mathcal{E}_o - \Delta\mu)a}\right] \tag{35}$$

From (35), since z must be positive, the z layers can only exist for

$$-\infty < \Delta\mu < \mathcal{E}_o \qquad (36)$$

Thus *for having 2D growth, the supersaturation cannot overpass the bulk elastic energy density stored in the strained layers*. In absence of misfit ($m_o=0$) the usual condition for 2D growth $\Delta\mu<0$ is recovered.

For such layer by layer growth the free energy density of relation (32) reads with V=0, r=0, $L^2 \to \infty$

$$\Delta F / L^2 = -(\Delta\mu - \mathcal{E}_o)za - |\Phi_\infty|(1-\exp(-z)) \qquad (37)$$

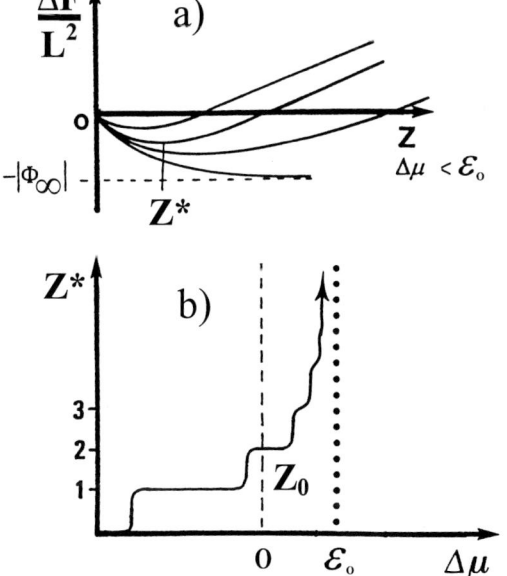

Figure 8:
a/ Free energy density change $\Delta F/L^2$ for layer growth as a function of the number of layers z for different chemical potentials $\Delta\mu < \mathcal{E}_o$. $\Delta F/L^2$ only has minimum for $\Delta\mu < \mathcal{E}_o$.

b/ number of equilibrium layers z versus the chemical potential $\Delta\mu$. For $\Delta\mu = \mathcal{E}_o$, z* tends towards infinity.*

In figure 8a we plot the free energy density $\Delta F/L^2$ as a function of z for different chemical potentials $\Delta\mu$. $\Delta F/L^2$ shows minima, at z=z*, for $\Delta\mu < \mathcal{E}_o$. In this case since $(\Delta F/L^2)_{z^*} < 0$ 2D layers form spontaneously. Notice again that this spontaneous layer formation is precluded for $\Phi_\infty > 0$.

* When relation (36) is fulfilled provided $\Phi_\infty<0$, each layer z is a 2D phase, built at a given undersaturation $\Delta\mu_z = \mathcal{E}_o - |\Phi_\infty/a|\exp(-z)$ obtained from (35). Up to saturation $\Delta\mu=0$ there builds up a finite number of layers z_o (see fig 8b) which for a non covered film $\theta=0$ is given by:

$$z_o = \ln\left[\frac{|\Phi_\infty|}{a\,\mathcal{E}_o}\right] \qquad (38)$$

This number only depends on the wetting over strain energy ratio $|\Phi_\infty|/a\,\mathcal{E}_o = |2\gamma_A - \beta|/[Ym_o^2]$ (see (25) and (8)). This result is largely experimentally supported on very different pairs A/B: reversible multilayers adsorption measurements (see [76-79]). The result is remarkable so that it has to be illustrated with a striking picture: provided $2\gamma_A - \beta < 0$ *a piece of material A put in the vicinity of a crystal B exposing a (001) face, placed in a box of uniform temperature sublimates spontaneously on B up to z_o epitaxial layers*. The only driving force to oppose to strain energy of the created A layers is due to $2\gamma_A - \beta = \gamma_A + \gamma_{AB} - \gamma_B < 0$. For Ge onto Si(111) one obtains $4 < z_o < 5$ [56] in agreement with experimental data.

In fact (38) is only valid for the exponential behaviour of the long-range inter-layer potential we have chosen. More generally the equilibrium number of wetting layers depends upon the form of the long-range inter-layer potential. For instance if no such long range interactions is accounted for and only short range interactions are supposed to act as in first neighbours model, there is a cut-off for z=1 that means $\Phi(z=1) = \Phi_\infty$ but $\Phi(z>1) = 0$. Then it results only one equilibrium wetting layer z=1 (see appendix E). This is however an extreme prevision since to such short-range forces adds some long-range contribution.

In the case of long range contribution and of finite misfit it is seen that when $\Delta\mu = \mathcal{E}_o$ the number of equilibrium layers z^* given by (35) becomes infinite (see fig. 7). In the case of a vanishing misfit $m_o \to 0$ the number of layers z_{SK} given by (35) tends towards infinity too, even at $\Delta\mu=0$. Obviously latter layer-by-layer growth is quite normal but in the $m_o \neq 0$ case the elastic energy stored increases with the number of layers z so that the system has to relax either by plastic deformation or islanding.

* Let us first consider *relaxation by interfacial dislocation insertion*. The critical number of layers beyond which dislocations may appear can be obtained following a simple treatment of Matthews [80]. From a thermodynamical point of view the number of interfacial dislocations may pass from N to N+1 when the total elastic energy change due to the introduction of the (N+1)th dislocation is negative. It is easy to show (see end of appendix F with K=1 and b=a) that the critical number of layers z_{disl} beyond which dislocations may thermodynamically insert roughly is the solution of the following equation

$$z_{disl} \approx \frac{1}{4\pi} \frac{1+\ln z_{disl}}{m_o} \qquad (39)$$

Conversely this relation tells that for $z > z_{disl}$, the misfit m_o or the mean strain in the layers decreases roughly as the inverse of the number of deposited layers and thus reduces the elastic energy and therefore the equilibrium number of layers given by (35) for fixed $\Delta\mu$. Obviously owing to kinetic reasons dislocations may only enter for greater thicknesses.

* The case of *elastic relaxation by islanding* concerns the Stranski-Krastanov transition and will be treated in section 2.2.3.3. At this stage we will only notice that beyond some new critical number of layers we will call z_{SK}, 3D islanding may occur. Thus according to the relative values of z_{disl} (varying as $1/m_o$ see (39) and z_{SK} (varying as $\ln(1/m_o^2)$ see formula (a) table II) relaxation takes place, at thermodynamical equilibrium, either by islanding or dislocation entrance. Furthermore since activation energies for dislocation entrance and 3D islanding behave as m_o^{-2} and m_o^{-4} respectively [81] there is really also a kinetics competition between these two modes of relaxation.

As a partial conclusion at equilibrium, elasticity modifies the chemical potential of each layer and fixes the number of layers at $\Delta\mu=0$ (see (38)). However since elastic energy diverges with z, beyond some critical number of layers elastic relaxation by dislocation entrance or islanding occurs.

2.2.3.2. 3D growth on a bare substrate (z=0, Φ_∞>0): Volmer-Weber case

* In this case the relation (d) in table II gives the equilibrium volume V* of the island. This relation says:

(1) The value of the chemical potential $\Delta\mu$ selects the volume of the crystal. More precisely the greater $\Delta\mu$, the smaller the size of the equilibrium crystal. Typically this effect is the usual Gibbs Thomson behaviour for first order phase transitions.

(2) However here 3D crystals can only exist when V*(r)>0, that means when

$$\Delta\mu > \mathcal{E}_o R^{3D}(r). \tag{40}$$

If the elastic relaxation effect is neglected, $R^{3D}(r)=1$, the 3D crystal can only exist at $\Delta\mu > \mathcal{E}_o$ [87].

* Relation (e) in table II describing the equilibrium shape ratio of the 3D crystal is more interesting. It can be rewritten as a parametric system in r with h* and ℓ^* the equilibrium height and length of the 3D crystal [65,79,87]

$$\begin{cases} h^* = -\frac{4\gamma'}{3\,\mathcal{E}_o}\left[1-\frac{r_0}{r}\right]\left(\frac{dR^{3D}}{dr}\right)^{-1} \\ \ell^* = h^*/r \end{cases} \tag{41}$$

Where r_o is given by (34)(25). For $m_o=0$ the system gives the usual Wulf-Kaishew theorem [82-86]

$$h^*/\ell^* = r_0 = \frac{2\gamma_A - \beta}{2\gamma'_A} \tag{42}$$

whose principal meaning is that the aspect ratio r=r_0 is size independent (for a discussion see [65,87]).

In the presence of misfit, $\mathcal{E}_o \neq 0$, the system (41) only has a solution for r>r_0. This means that *epitaxially box shaped strained crystals must have greater aspect ratio than the strain-free crystal*. More precisely, (1) epitaxial strain acts against wetting (adhesion) so that globally it leads to a thickening of the equilibrium shape; (2) owing to strain this equilibrium shape becomes size dependent.

Obviously relation (41) can only be used practically when the relaxation factor for Volmer Weber growth has been calculated for the shape family r under study. For a box shaped crystal an analytical form of $R^{3D}(r)$ has been calculated [65,87]. In figure 7 we plot from [65] the relaxation factor of a box shaped crystal when deposit and substrate have the same elastic properties. It is seen that when r=0, a uniform coherent film is not relaxed ($R^{3D}(0)=1$). However as the deposit becomes a rectangular box (finite r) it relaxes from its borders so that $R^{3D}(r)<1$ and therefore it stresses the underlying substrate (for a more complete discussion about relaxation factors for various shapes see [88]). *Let us note that for $5<1/r<50$ $R^{3D}(r)$ can be roughly fitted by $R^{3D}(r) \approx 0.14r^{-1/2}$.*

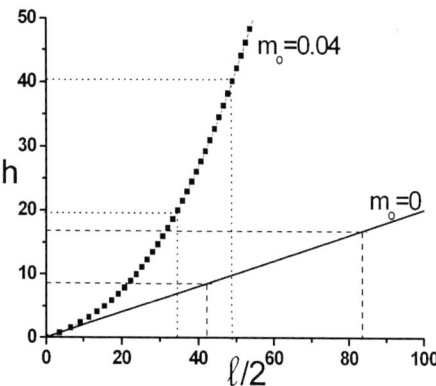

Figure 9: *Half equilibrium shape of a 3D box shaped crystal (discontinuous lines) h: height, $\ell/2$: half basis in atomic units. Continuous curves are the corner trajectories calculated for $r_o=0.1$, $m_o=0$ (straight line) or $m_o=4\%$ (squares). The epitaxial misfit leads to a thickening of the ES. Furthermore similarity is lost.*

Calculating then dR/dr from figure 7 we can plot from (41) $h_{eq} = h_{eq}(\ell_{eq})$ for a given wetting factor r_o and $\gamma_{//}/E$ value, a so-called elasto-capillary length where E is Young's modulus in an isotropic surface of surface energy γ. This length scales $10\gamma/Y \approx 10^{-8}$ cm that means with the size of an atom as mentioned by F.C.Frank [45]. We will use it in the following discussion. The result of the resolution of (41) is shown in figure 9 for $r_o=0.1$, for $m_o=0$ and $m_o=4\%$. Each curve figures the trajectory of the edge of the half equilibrium shape (ES) with size.

For $m_o=0$ the usual Wulf-Kaishew theorem still holds and the equilibrium aspect ratio $r=r_o$ is size independent (ES are obtained from the straight line on figure 8) or more generally they have similarity. For $m_o \neq 0$ the ES ratio increases with size so that similarity is no more preserved. A more complete discussion about wetting and relative stiffness can be found in [65,87].

Thus as a partial conclusion, in presence of elasticity 3D growth takes place when supersaturation overpasses the bulk elastic energy (see (40)) and the ES of the growing crystal is modified in such a manner the greater the misfit, the higher the equilibrium shape. For other polyhedral shapes the relaxation factor can be calculated by numerical methods [61-65]. Elastic effects on truncated pyramids [60,62,79,87], 2D cylinders [63] or 3D spheres [64] have been considered. Nevertheless the main effect is the same (excepted when the island shape has been fixed as in [64]): a thickening of the ES but furthermore the various facets extension changes with size, some facets decreasing, other increasing [65,87].

* Obviously this scenario cannot be valid whatever the size. Indeed as for 2D film (see section 3.2.3.1.) the growing crystal accumulates elastic energy, in spite of elastic relaxation which can become prohibitive so that plastic relaxation occurs. The treatment is similar as in section 3.2.3.1. One finds (see appendix F) that a first dislocation orthogonal pair may thermodynamically enter in the island interface as soon as a critical height h_C is reached. It is obtained as a solution of:

$$h_c \approx \frac{1}{4\pi} \frac{1+\ln h_c}{(m_o-2/\ell)R^{3D}(r)} \tag{43}$$

Each supplementary dislocation entrance abruptly drops the strain from m_o to $m'=m_o-Nb/\ell_N$ (see appendix F) where b is the Burgers vector modulus of the interfacial dislocation and ℓ_N the lateral size of the crystal in which enters the Nth dislocation. *Then since we have seen that the equilibrium shape is strain dependent, each dislocation entrance abruptly modifies the equilibrium shape [65,87,88,89] (see figure 10 and its caption). More precisely since the smaller the misfit, the flatter the crystal, the main effect of dislocation entrance thus is a back flattening of the equilibrium shape. Such a quick variation of the equilibrium shape at each dislocation entrance has been experimentally shown by F.Legoues et al. [90,91]. For shapes more complex than box shaped crystals there is a jerky modification of the various facets extension [65,88].*

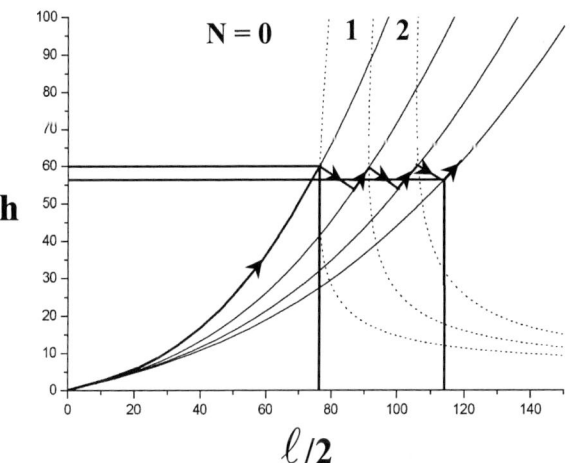

Figure 10: *Effect of dislocation entrance on the ES. The box shaped crystal accumulates strain energy when growing along the $h(\ell/2)$ curve with an arrow until it reaches some critical size (here for $r_o=0.1$ and $m_o=4\%$, $h_o=61$, $\ell_o=155$) where a first dislocation may thermodynamically enter (dotted curves represent the thermodynamic criterion for the first (N=0) dislocation entrance (eq. iii appendix F). If the dislocation effectively enters for this size, according to equation (ii) appendix F the misfit passes from m_o to m_o-1/ℓ_o. The ES of the growing crystal thus must follow a new ES trajectory re-calculated from (41) but with the new misfit .If it is assumed that the island changes its shape at constant number of atoms, the crystal abruptly flattens. The same thing occurs for the second (N=1), third (N=2), and fourth (N=3) dislocation entrance where in the case under study the misfit passes respectively from 3.4 % to 2.9% then 2.6%.*

2.2.3.3. 2D relayed by 3D growth ($\Phi_\infty<0$): Stranski-Krastanov case

* From the condition of existence of $z^*>0$ and $V^*>0$ obtained from equations (b) and (c) in table II Stranski Krastanov growth can only occur, in near equilibrium conditions, when the following relation is fulfilled:

$$\mathcal{E}_o R^{SK}(r,z) < \Delta\mu < \mathcal{E}_o \qquad (44)$$

that means in a finite domain of chemical potential $\Delta\mu$ where now $R^{SK}(r,z)<1$ describes the elastic relaxation of a 3D crystal sitting on z pseudomorphic layers covering the lattice mismatched substrate. In absence of elastic relaxation ($R^{SK}(r,z)=1$) according to (44) there is no more place for Stranski Krastanov mode in near equilibrium conditions. *Elastic relaxation thus is a prerequisite for the simultaneous existence of 2D layers and 3D crystals.*

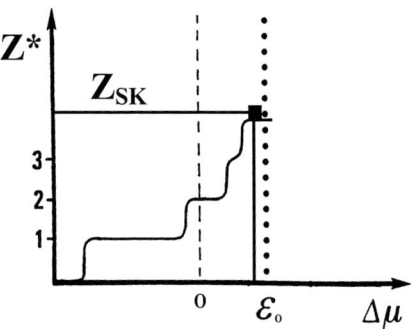

Figure 11: Number of equilibrium layers z^* versus the chemical potential $\Delta\mu$ in case of SK growth. For $m_o \neq 0$, 3D islands may appear as soon as $\Delta\mu > \mathcal{E}_o R$. corresponding to $z > z_{SK}$.

In figure 11 we schematically plot the number of 2D layers as a function of the chemical potential $\Delta\mu$. To each layer formation z^* corresponds a constant value of $\Delta\mu$ given by $\Delta\mu(z^*)= \mathcal{E}_o - e^{-z^*}\Phi_\infty(1-\theta)$. For $m_o=0$ the number of steps becomes infinite at saturation $\Delta\mu = \mathcal{E}_o = 0$ (fig.7b). For $m_o \neq 0$ 3D islands may appear as soon as $\Delta\mu > \mathcal{E}_o R^{SK}(r,z)$. Therefore in figure 11 for increasing $\Delta\mu$ there is a cut-off at $\Delta\mu = \mathcal{E}_o R^{SK}(r,z)$ where 3D crystals may appear on the z_{SK} underlying layers. Beyond $\Delta\mu = \mathcal{E}_o$ the representation does not make sense, exactly as when $m_o=0$ for $\Delta\mu > \mathcal{E}_o = 0$ (see fig. 7b). Let us remark that because of the limitation $\Delta\mu \leq \mathcal{E}_o$ the smallest volume a 3D crystal can reach is obtained by injecting $\Delta\mu = \mathcal{E}_o$ in the expression of V^* in table II. For a given aspect ratio r and for $z \to \infty$ this minimum volume reads:

$$V_{min}*(r) = \left[\frac{\frac{8}{3}\gamma_A'}{\mathcal{E}_o(1-R^{SK}(r,z))} \right]^3 r \qquad (45)$$

Here again we see that elastic relaxation $R^{SK}(r,z) < 1$ is a prerequisite for Stranski Krastanov mode. Indeed for $R^{SK}(r,z) = 1$, relation (45) says that the 3D crystal must have an infinite volume! However, because of the fact that the usual activation barrier for 3D nucleation is proportional to the one third of the total surface energy of the nucleus [6,92], nucleation of such large crystals should be difficult. But since SK mode exists the true growth mechanism must minimize this activation energy. We will come back to this point in section 3.2.3.5.

In fact, at equilibrium the chemical potential $\Delta\mu$ must be the same for the 2D layers and the 3D crystal. Combining thus z* and V* expressions (a) and (b) of table II, $\Delta\mu$-independent equilibrium values z* and V* can be easily obtained. *Thus for each value of z* there exists an aspect ratio r which minimises the crystal volume V*(r) which can co-exist on z* layers.*

At this stage we can summarise the conditions of Stranski-Krastanov transition onto a dislocation free 2D film:

$$\begin{cases} \mathcal{E}_o R^{SK}(r) < \Delta\mu < \mathcal{E}_o \\ z_o \leq z(r) < z_{disl} \\ V(r) > V^*_{min} \propto m_o^{-6} \end{cases} \quad (46)$$

* Concerning the equilibrium shape the main differences between Volmer-Weber and Stranski-Krastanov cases are:

(1) In the expression of the equilibrium shape ratio r* the factor $1 - r_o/r$ which appear in formula (41) in case of Volmer Weber (VW) growth has to be be replaced by $1 + r_o e^{-z}/r$ in case of Stranski-Krastanov growth. This originates in the now negative wetting energy Φ_∞ (necessary to build z pseudomorphous layers) which decreases with the film thickness z. Owing to this difference, positive height h^*_{VW} can only exist for r>r$_o$ in the VW case whereas crystal flatter than r=r$_o$ can exist in the SK case.

(2) The relaxation factors $R^{3D}(r)$ and $R^{SK}(r,z)$ appearing in table II are not the same.

As still mentioned relaxation factor $R^{SK}(r,z)$ for 3D crystals onto z pseudomorphic layers have not been calculated. Nevertheless if the number of underlying layers is great enough, it can be considered that the 3D crystal grows onto a homogeneously strained semi-infinite substrate of A. In this case it must be $R^{SK}(r,z) = R^{3D}(r)$ and thus the equilibrium shape can be obtained (for z→∞) from the following parametric equations

$$\begin{cases} h^* = -\dfrac{4\gamma'_A}{3\,\mathcal{E}_o}\left(\dfrac{dR^{3D}}{dr}\right)^{-1} \\ \ell^* = h^*/r \end{cases} \quad (47)$$

Thus since for box shaped crystals an analytical form of $R^{3D}(r)$ has been found [87] (see also 2.2.3.2.) it is possible to plot the equilibrium shape of the SK 3D crystal.

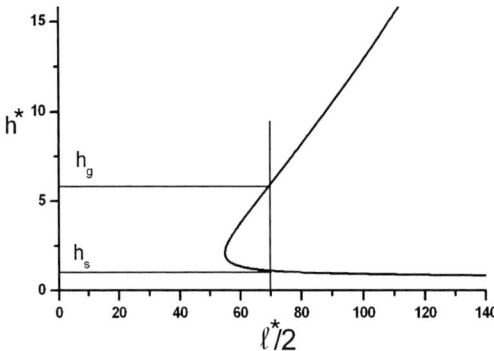

Figure 12a: *Corner trajectory of the ES $h = f(\ell/2)$ for Stranski Krastanov case ($r_o = 0$) calculated for $m_o = 4\%$.*

Figure 12b: *Free energy change (32) as a function of the aspect ratio of the growing crystal for $V < V^*_{min}(r)$ where there is no minimum (excepted in r=0) and for $V > V^*_{min}(r)$ where the minimum defines the equilibrium aspect ratio, the maximum a labile state to overpass.*

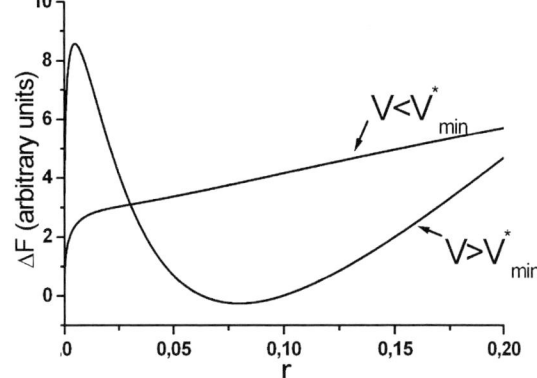

We plot on figure 12a such an ES, the graph $h^*(\ell^*)$ having now two branches which correspond to the extrema of $\Delta F(r)$ (32) and define two equilibrium aspect ratios r_g (great) and r_s (small) or two equilibrium heights h_g and h_s provided the volume V is greater than V^*_{min} (45). The lower branch (flat crystals with $r = r_{min}$) is the locus of the maximum of $\Delta F(r)$ (crest) (see figure12b) and thus describes a labile equilibrium whereas the upper branch (thick crystal with $r = r_g$) is the locus of the minimum of $\Delta F(r)$ (valley) and thus describes the stable equilibrium analogous to what we depicted for Volmer Weber growth (see section 3.2.3.2.). Obviously (see figure 12b) the transformation at constant volume from labile equilibrium (crest) to stable equilibrium (valley) is spontaneous [56,65].

From a theoretical point of view we have seen that since for SK growth z^* depends upon the 3D islands coverage $\theta = (\ell/L)^2$ (see z^* for SK growth in table II) the greater the volume of the 3D crystal the smaller the number of underlying layers at equilibrium. Most experiments agree with this description since they clearly show that Stranski-Krastanov transition can occur at a constant number of deposited atoms and that 3D growth occurs at the expense of the 2D layers. In other words *owing to the elastic relaxation of 3D crystals, some of the upper layers of a metastable 2D strained-film (thickness z') can transform into stable 3D islands leading to the Stranski-Krastanov situation where the remaining z* layers support these 3D* *islands.* In a paper we have shown how owing to strain relaxation sufficiently large 2D islands of simple height double their height and start the SK transition [56].

* Obviously, once more the stored elastic energy increases with the number of deposited atoms so that beyond some critical size dislocations may enter the system. If $z_{Disl}<z_{SK}$ the dislocations may enter in the film before Stranski-Krastanov transition takes place. Thus since dislocation entrance decreases the misfit, Stranski-Krastanov transition can no more occur even for further growth. Since on one hand the greater the wetting the greater z_{SK} whereas on the other hand the smaller the relative rigidity K (see appendix F) the smaller z_{Disl}, for weak wetting and sufficiently soft substrate there can also be $z_{Disl}>z_{SK}$. Obviously as for Volmer-Weber growth kinetics may modify this condition. When $z_{Disl}>z_{SK}$, naturally arises the question of the localisation of the dislocations. Are they at the bottom of the layers or at the bottom of the deposited crystals? The question remains open and the answer must depend, at equilibrium, upon the shape and the density of islands (in section 4. we will say some words on interacting crystals) as well as on the relative substrate to deposit rigidity since dislocations always go towards the softer material.

2.2.3.4. Comments on surface stress effects

Until now we have not considered surface stress effects in the just foregoing discussion. When such effects are included in the formulation of ΔF of (32) by means of the surface work during accommodation and relaxation, elastic relaxation and equilibrium shape calculations can no longer be explicitly solved (see for example [93]). Nevertheless surface stress has several predictable main effects

(1) For 2D growth, since the surface stress modifies the wetting factor from Φ_∞ to $\Phi_\infty^{m_o}$ (see (25) and (27)) it modifies the number of equilibrium layers. Surface stress also plays a role on the critical number of layers beyond which dislocations may thermodynamically appear. However since in this case the film relaxes by dislocation entrance the interfacial stress s_{AB}[††] also works [93]. It is thus easy to show (see appendix G) that for positive natural misfit m_o a positive $\Delta s^\infty = s_A + s_B - s_{AB}$ value (where s_A and s_B are the surface stresses of deposited crystal A and substrate B respectively) lowers the critical thickness h_c. On the contrary a negative Δs^∞ increases the critical thickness beyond which dislocations may appear. For Ge/Si(100) (m_o =-4%) Floro et al. [94] give Δs^∞ =2.3 Jm^{-2}. In this case m_o Δs^∞ <0 so that the critical thickness h_c is decreased by surface and interface stresses from 5.4 monolayers to 2.6 monolayers (see figure 13 where we plot the equilibrium strain versus the deposit height h for Δs^∞=0 and Δs^∞=±2.3 Jm^{-2}). Nevertheless such a reversible critical thickness dependence with misfit and surface stress is difficult to put in evidence since dislocation formation is an activated process so that the kinetic critical thickness beyond which dislocations effectively occur generally is greater than the predicted thermodynamical value.

(2) For 3D growth let us recall that taking into account surface stress work against surface deformation is equivalent to taking into account the first order development with strain of the surface energy (see comments just after (27)). Thus the gamma plot of the strain free equilibrium shape is modified by strain. Nevertheless since in most cases the surface energy

---

[††] In fact for semi coherent interface composed of a grid of dislocations one needs to distinguish two interfacial stresses. Indeed the usual interfacial stress is defined as the work done to deform both facing phases by the same amount, whereas in the presence of dislocations changing the in plane parameter of the substrate needs dislocation introduction (See [95])

change versus strain remains weak (of the order of $(s-\gamma)\varepsilon$ see (27)) surface stress is not really effective when elastic relaxation operates (see for example [96]). Nevertheless things can be different when new surface phases can be stabilized by external stress. Indeed in this case stress-induced changes of surface structures (first order transition) may lead to a discontinuous change of surface energy and surface stress and thus of the gamma plot. In this case new cusps could appear on the gamma plot and thus new stress-stabilized faces may appear on the equilibrium shape. We believe [89] that this happens for the well defined {105} facets of the so called "huts" appearing during the first stages of the growth of Ge/Si(100) [97,98].

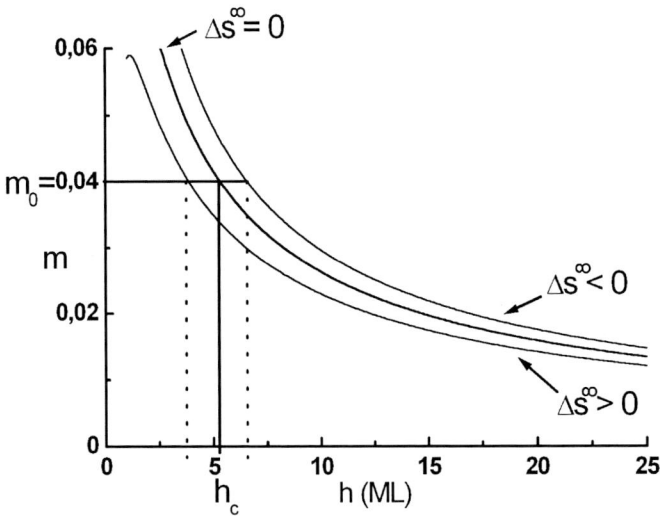

Figure 13: Residual misfit m' versus the film thickness h calculated for $m_o=-4\%$ and $\Delta s^\infty=0\pm2.3\ Jm^{-2}$ (see appendix G). The growing film remains pseudomorphous to its substrate up to $h= h_c$. Beyond this critical thickness h_c the misfit is partially accommodated by dislocation entrance and the residual elastic misfit decreases with the thickness $h>h_c$ of the film. For $\Delta s^\infty=0$ and $m_o=-4\%$, $h_c\approx 5.4$ ML. A positive surface stress change $\Delta s^\infty=2.3\ Jm^{-2}$ decreases h_c to 2.6 ML whereas negative surface stress change $\Delta s^\infty=-2.3 Jm^{-2}$ increase h_c to $h_c\approx 7.5$ ML. The case Ge/Si (100) corresponds to $m_o=-4\%$, $\Delta s^\infty=2.3\ Jm^{-2}$ [94] and thus $h_c\approx 2.6$ ML.

It was found in Molecular Beam Epitaxial growth that some foreign adsorption plays a role on the equilibrium state. In the case of 2D growth, such additives can modify the number of equilibrium layers z [99,100], whereas for 3D growth they can modify the equilibrium shape of the growing crystal [101]. Such additives are known to lower the surface energy [5] without bulk incorporation and thus were called surfactants by the semiconductor community. Nevertheless since surface energy γ is lowered by adsorption, and since Shuttleworth relation (10) connects surface energy to surface stress, consequently surface stress changes with adsorption. More precisely since there is $\partial\gamma/\partial\mu = -\Gamma$ where μ is the chemical potential of the adsorbed species and Γ the adsorption density (generally positive see section 1.4.) there is $\partial s/\partial\mu=-\Gamma-\partial\Gamma/\partial\varepsilon$. The adsorption may thus reduce or enhance the surface stress. Such surface stress changes with adsorption have been reported (for example see the important review paper by Ibach [102] its erratum [103] and [104]). Therefore surfactants not only play on the surface energy but also change the mechanical state of the crystal.

3. ELASTIC EFFECTS ON ATOMISTIC MECHANISMS

We recall some well-known facts for non-stressed crystals.

In the case of a flat (F) defect-free surface[‡‡] the rate of growth is determined by the frequency of formation of 2D or 3D nuclei. The nuclei formation requires overpassing an activation barrier. Supersaturation decreasing this barrier, a critical supersaturation has to be overcome for the growth to take place. We will see in section 3.1 how elasticity can influence the 2D and 3D nucleation process. For a stepped face (S) with an orientation along an inward cusped valley, the growth may occur by step flow mechanism as first depicted by Burton Cabrera and Frank (BCF) [105]. In their seminal paper the authors consider kinetics of growth of a vicinal surface as a balance between adatom deposition, adatom diffusion and adatom attachment to steps (for a review see [92,106]). If the sticking probabilities of adatoms from the upper terrace and lower terrace are equal, all the steps have the same velocity. Such growth mode is known as step flow and occurs at small supersaturation. In fact at higher supersaturation there is a transition from this step flow mode to 2D nucleation mechanism in between the steps. When no reevaporation takes place, or the diffusion length onto the terrace is much greater than the distance between steps, 2D nucleation on the terrace is more favourable. When the nucleus only contains one atom (very high supersaturation, see appendix H) this transition occurs when $(D/F)^{1/6} \approx L$ where L is the step to step distance, D the surface diffusion constant and F the impinging flux [106,107]. Such transition from step flow to 2D nucleation is easy to detect experimentally. Indeed in presence of 2D nucleation birth and spread of 2D islands give oscillations of the RHEED intensity at each completion of layer whereas in step flow mode there are no oscillations [108,109]. In the case of a flat (F) face but with screw dislocations 2D nucleation is shunted at low supersaturation [105] and steps with high Burgers modulus $b_\perp$ propagate laterally each anchored at the dislocation so that equidistant steps are winded as helices. This gives very flat growth pyramids around each non-cooperating screw dislocation. Their slope is proportional to $1/\Delta\mu$ [105].

Obviously elasticity may influence some of these elementary processes involved in step flow. In the framework of these lectures the various elastic effects on adatom detachment rate and surface diffusion will be discussed separately (section 3.2. and 3.3.). In each case we will not give the details of the calculations but only try to capture the essential physics for simple cubic material ($\gamma_A = \gamma'_A$).

3.1. Nucleation barrier

* On a flat perfect surface (F), crystal growth takes place after 2D or 3D nucleation. On average, clusters smaller than the so-called critical nucleus spontaneously disappears whereas clusters larger than the critical nucleus spontaneously grow. The critical nucleus size depends on supersaturation $\Delta\mu$ and its formation requires overpassing an activation barrier. In principle the activation barrier ΔF^* is obtained by injecting the equilibrium nucleus parameters h^*, ℓ^* or V^* r^* (table II) in the free energy change ΔF (32) due to the nucleus

---

[‡‡] An F face is a face having an inward cusp in its gamma-plot.

formation. However this cannot be done explicitly here in the epitaxial case. Furthermore we have to distinguish the Volmer-Weber (VW) and Stranski-Krastanov (SK) cases having very different behaviour.

* For the Volmer-Weber case ($\Phi_\infty>0$ or $r_o>0$) with table II and (32) the nucleation barrier can be factorised [87] as three terms of distinct physical meaning

$$\Delta F^*_{VW}/kT = (\Delta F^*_{hom}/kT) r_o \mathcal{F}_{ro,K}(\Delta\mu/\mathcal{E}_o) \tag{48}$$

The first and leading factor is the well known so called homogeneous nucleation barrier (without surface β→0 (($\Phi_\infty=0$, $r_o=1$) (see appendix H) which is reduced by the second one r_o, ($0<r_o\leq1$) (34) due to substrate wetting. The last term due to epitaxial strain at contrary opposes to the former one since $\mathcal{F}_{ro,K}<1$ is a function of $\Delta\mu$ which becomes very close to 1 for $\Delta\mu/\mathcal{E}_o \approx 10$. Homogeneous nucleation in vapour phase can be estimated to be effective at $\Delta F^*_{hom}/kT \approx 30$ (see appendix H) with a critical number of atoms $15<N^*<60$ and driven by supersaturation much higher then $\mathcal{E}_o$ (see (40)), $6<\Delta\mu/\mathcal{E}_o<10$. The function $\mathcal{F}_{ro,K}$ for these ratios is very close to unity so that *the activation barrier for classical nucleation is not influenced by misfit strain*!

* In contrast, elasticity is the driving force of the Stranski-Krastanov transition ($\Phi_\infty<0$) as underlined before. Let us demonstrate it by the following process. The free energy change of the transformation of z' 2D layers into a 3D island (volume V) sitting on z layers ΔF' is that one of (32) ΔF(z,V) minus ΔF(z',0) where the 3D crystals are absent. At constant volume of material A there is for $V/aL^2 = \theta^2 h/a <<1$ that means for a small fraction of the film surface covered by 3D islands :

$$\Delta F' = -\mathcal{E}_o V[1-R(r)] + |\Phi_\infty|\left[V-\left(\frac{V}{r}\right)^{2/3}\right]e^{-z/\zeta} + 4\gamma'_A V^{2/3} r^{1/3} \tag{49}$$

The activation energy for Stranski-Krastanov transition $\Delta F^{**}(r)$ can thus be obtained by injecting the equilibrium values V* and r* of the table I for $\Phi_\infty<0$ in the previous relation. For $\Phi_\infty e^{-z} \to 0$ (that makes sense at the SK transition) the barrier reads

$$\Delta F^{**}(r) = \frac{4}{3}\frac{(4\gamma'_A)^3}{(3\mathcal{E}_o)^2}\frac{r^*}{[1-R(r^*)]^2} \tag{50}$$

As usually ([92] and appendix H) the activation barrier $\Delta F'^*(r)$ is proportional to γ'^3_A but $\mathcal{E}_o(1-R(r^*))$ obviously plays the role of a driving force. On one hand $\Delta F'^*(r)$ is proportional to $\mathcal{E}_o^{-2}$ that means to m_o^{-4}, on the other hand inhomogeneous relaxation $R(r)\to 0$ pushes it [§§]. (The lowest value of $\Delta F'^*(r)$ is roughly reached for r*=0.05 that gives for

---

[§§] Let us note that in absence of elastic relaxation (R(r)=1 in (50)) the barrier becomes infinite. Thus clearly again it appears that the inhomogeneous relaxation of the 3D islands is the driving force for Stranski-Krastanov transition.

Cu(111) where $\gamma \approx 1300$ ergcm^{-2} [110] and $\mathcal{E}_o/m_o^2 = 2.310^{-12}$ ergcm^{-3} [17] so that $\Delta F^*/kT \approx 100$ for $m_0=1\%$, $\Delta F^*/kT \approx 30$ for $m_0=2\%$ but $\Delta F^*/kT \approx 2$ for $m_0=8\%$). *Thus generally Stranski Krastanov transition can only occur for a sufficiently high misfit $|m_o| > 2\%$.* Nevertheless some cases of SK growth are well known for misfit of the order of 1.10^{-2} where activation barrier calculated from the previous relation seems to be too high. Nevertheless ΔF^* could be lowered by other mechanisms. Furthermore our box shaped model is not the most flexible one. When considering truncated pyramids (see [65]) the γ'_A in (50) is reduced by some factor, vanishing when the summital facet disappears. This is however also the sign of non-stability of this face even in absence of stress. Furthermore since the true ES is that which minimises the activation barrier, the activation barrier could be lowered by other specific shapes. Last but not least SK transition could start on some point defects so that 2D or 3D nucleation activation barriers are lowered. This is however only possible at very high supersaturation on a F face [92] that means nuclei of several atoms. *In fact the problem of the real amount of the activation barrier remains widely an open question.*

3.2. Strain effects in irreversible condensation by growth simulation

Equilibrium thermodynamics describes statistically nucleation and growth at low supersaturation. Far from equilibrium or really irreversible growth studies are only possible by numerical resolution of kinetical systems [111-113] in the mean field approximation or by simulation catching more or less the collective nature. Latter studies started in their most simplest form by Monte Carlo technics in the 71th [114-116] bringing the now classical and eventually fascinating images of birth, spread and coalescence of islands on a growing compact crystal face, leading to the surface roughening divergence at some critical temperature.

Introduction of elastic strain started only in the 90th. Let's report about some studies of Ratsch and Zangwill [117-119] who took a very simple scheme. Atoms are randomly put on the nodes of a quadratic grid with a rate of F atoms per second per site. When accumulating, eventually at different levels, there is applied the prescription to avoid overhangs and holes in the so generated cubic 3D lattice (solid on solid or SOS model). By this simple scheme column clusters of various shapes and random diverging heights are generated (see [106]). However atoms can move away from the landing site, except to go back to the vapour phase, in the extreme case of complete condensation (in fact the vapour phase is reduced to a "directed beam" so that the lateral faces of the columns don't receive atoms, since no overhangs have to be created). Single atoms migrate to next neighbours sites at the highest rate say D per second (short range surface diffusion). Usually D/F>>1 so that surface diffusion is very active. Clusters are not allowed to migrate but they loose single atoms to neighbouring sites with a smaller rate $D\exp(-n\varphi/kT)$, depending on the number n=1,2,3,4 of lateral bonds of strength φ which have to be broken (φ stays for an activation energy). This so generated atoms migrate with rate D and fall by chance in traps of n=1,2,3,4 bonds φ where they reside thus longer thus stronger they are bonded. When all the atoms are bonded vertically by $E_s=\varphi$ it results normal crystal growth (homoepitaxy) of a flat face (F) that means a new layer starts when the other comes to completion. However single atoms settle also on higher levels thus more F/D is high and multilayer growth may occur. Very flat pyramids (up to 3 layers) may form. Rough kinetics estimations [106,107,113] confirm the mean nucleation density ℓ^{-2} and the mean coalescence size $\ell/a = (D/F)^{1/4}$.

Formulating epitaxial growth there is to choice **(i)** the bond energy $E_s \neq \varphi$, either $E_s < \varphi$ for VW growth (equivalent to $\Phi_\infty > 0$) or $E_s > \varphi$ for Frank-van der Merwe and SK growth (equivalent to $\Phi_\infty < 0$); **(ii)** strain energy has to be accounted too. The studies [117,118] considered SK growth $E_s >> \varphi$ and due to the model of first neighbours interactions, only one SK wetting layer A is grown on the substrate B (see also our analysis close to (38) and appendix E). The atoms in the second layer are vertically bonded with $E_s = \varphi$ so that lateral layer-by-layer growth should follow. Due to misfit however the SK layer is strained and strain weakening of lateral bond energies is taken as guiding principle by the authors:

$$\frac{1}{2} 4\varphi \to 2\varphi - \mathcal{E}_o \qquad (51)$$

However 3D box shaped clusters are elastically relaxed by $R(h/\ell) \propto (\ell/h)^{1/2}$ (a crude approximation of figure 7) so that the frequency prescription for the growth simulation is $D \exp[-n(\varphi - \mathcal{E}_o R(r)/2)]$ for detaching an atom on a summital layer of a cluster [119]. Atoms thus detach more frequently a cluster is flat, so that taller cluster are favoured during the evolution; this trend being thus stronger the misfit square is high. Simulated images [117,118] show slightly dispersed rectangular near quadratic based clusters with mostly complete layers. At increasing total coverage above the SK layer of Ft=1/4, ½ ¾ monolayers, single and double layered clusters appear progressively. At Ft=1, 2D islands of 2,3, and 4 layers are formed so that only some half of the SK layer is covered. This thickening of the clusters is clearly due to the strain relaxation $\mathcal{E}_o R$ even if somehow exaggerated by the type of simulation.

The authors observed that thickening starts only at "some critical misfit" of 3%. This surprising result may not be general and needs our comments. Consider a distribution of box shaped crystals above their SK wetting layers so that the energy to spent is in average, that to create lateral faces plus the elastic energy written in the former mentioned approximation ($R^{3D}(r) \approx 0.14/\sqrt{r}$ see 2.2.3.2.)

$$\Delta F = 4\gamma_A h\ell + 0.14\, \mathcal{E}_o V \sqrt{\ell/h} \equiv 4\gamma_A V/\ell + 0.14\, \mathcal{E}_o \sqrt{V\ell^3} \qquad (52)$$

At constant volume $V = h\ell^2$ for spontaneous thickening there must be $\partial \Delta F/\partial h|_V < 0$. This is realised when for this volume $\ell/a > [19\gamma_A/\mathcal{E}_o a]^{2/3} (h/a)^{1/3}$. Doubling the height of a 2D island occurs when its size exceeds $\ell/a|_1 > [19\gamma_A/\mathcal{E}_o a]^{2/3}$. Clearly it has to be smaller than the coalescence size $\ell/a|_c = (D/F)^{1/4}$ (if not the single height islands annihilate mutually) the authors simulations took $D/F=10^6$ so that $\ell/a|_c \approx 30$ atomic units. There is according to Frank's rule [45] $\gamma/E \approx 10^{-9}$ cm for clean crystal faces, but 2D edges being thermally roughened we take $\gamma_1/E \approx 1/3.10^{-9}$ cm so that with a=3Å there is a critical size $\ell/a|_c \approx 170$ (for $m_o=1\%$), 40(3%), 26 (4%), 15 (6%). Thus doubling can only start for misfit greater than roughly 3% which corresponds to what the authors "observed". Conversely it can be foreseen that *"SK roughening" can be avoided when $\ell_c < \ell_1$ that means by decreasing the coalescence* *size or increasing the density of critical nuclei.* This can be done by increasing the reduced flux F/D >$60\,m_o^{16/3}$ or by increasing the nuclei density by other means. Lets add that these simulations have been done on a flat F face. When done on a vicinal with terraces width L, the coalescence size has an upper limit for 2D islands <L.

3.3 Growth instability induced by strain on vicinal faces

3.3.1 Growth of a vicinal face without strain

Growth instabilities are very frequent in bulk growth when diffusion-convexion of matter-heat are involved. In vapour growth, even outside any strain considerations, surface diffusion coupled with interfacial kinetics leads to instabilities we have to mention first. In the pioneer work of BCF [105] a vicinal face receives (or looses) adatoms from the terraces where they migrate towards (or away) the steps making them to advance or to recede. In any case, steps exchange (from their kinks) their atoms with the two adjacent terraces from the top side (+) and from the low side (-) with frequencies D^+ and D^- leading to a very quiet step flow when $D^+=D^-\leq D$. Disymmetric step kinetics may result from adsorption of impurities in the kinks [105] $D^+\neq D^-\leq D$. But there exists also some intrinsic effects for clean steps with $D^+<D^-\leq D$ due to some activation barrier near the upper ledge, the so-called Schwoebel [120]- Erhlich [121] barrier[***].

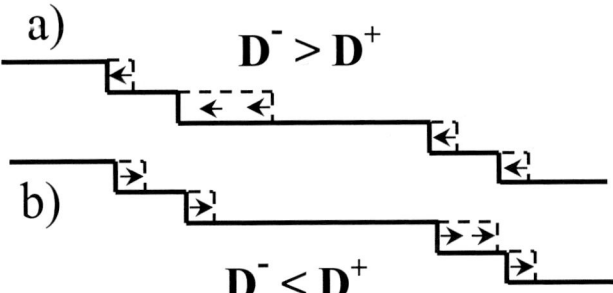

Figure 14: a/ Adatom sticking is easier onto an upper step ($D^->D^+$). Then since the number of atoms that reach a terrace is proportional to the terrace area, a terrace larger than its neighbours becomes smaller, thus all terraces reach the same size, each step reaches the same velocity and step flow mechanism occurs.

b/ Adatom sticking is easier onto a lower step ($D^-<D^+$), a terrace larger than its neighbours becomes larger and larger and step bunching occurs. The opposite scenarios are valid for evaporation. This concerns a non-strained solid.

In BCF's theory the lateral step velocity is an increasing function of the size of the adjacent upper and lower terraces: $v = D^+ f(L^+) + D^- f(L^-)$ [†††]. Suppose a given step (figure 14) in a

---

[***] When atoms approach or leave an upper ledge second attractive neighbour bonds have to be cut, but those from the lower ledge not.
[†††] In fact v depends upon the normalised distances $L/2x_s$, where x_s is the mean diffusion distance before desorption of an atom.

regular train $L^+=L^-=L$, during growth the mean speed $\bar{v}>0$ fluctuates, say $\Delta L>0$ so that the lower terrace becomes smaller, the upper one wider. The velocity of this step is

$$v = \bar{v} + \Delta v = D^+ f(L+\Delta L) + D^- f(L-\Delta L) \approx (D^+ + D^-)f(L) + (D^+ - D^-)\frac{df}{dL}\bigg|_L \Delta L \qquad (53)$$

The symmetric case $D^+=D^-$ doesn't change the velocity from the mean value of the train, $D^+<D^-$ slows down the step, $D^+>D^-$ boosts it and may lead to instability. For evaporation, $\bar{v}<0$, the opposite happens.

P.Bennema and G.H.Gilmer [122] showed that the governing differential equation is similar to that of a chain of masses connected by springs, having for growth (evaporation) an exponential damping regime for $D^+<D^-$ ($D^+>D^-$) and an exponentially increasing perturbation for $D^+>D^-$ ($D^+<D^-$) leading to step bunching (see also [106]). In summary ordinary (usual) Schwoebel-Ehrlich effect renders, fortunately, normal crystal growth stable.

The case of strained layers of vicinal nature has been studied by Duport et al. in 1994 [123,124] and Tersoff in 1995 [125]. Because of the height h discontinuities, the border of these strained layers bear elastic monopoles in excess (see (16)) with the elastic dipoles existing without strain (see 2.). There is first to study the thermodynamic stability of such a train, secondly its stability versus growth or evaporation.

3.3.2. Tersoff's step bunching driven by step-step interaction due to bulk strain

(i) We have seen (1.3.2.) that a step on a vicinal surface can be described as a line of dipoles at which adds a line of monopoles. The interaction energy per unit length of a pair of identical, L-apart, steps (h=a) can thus be separated in four terms. Each of these terms and its physical origin are described in table III when surface stress is neglected ($F_3=0$ in (14))

mo-mo		$: \alpha_1 \ln(L/2\pi a)$	$\alpha_1 = \dfrac{Ea^2 m_o^2}{\pi(1-v^2)}$
di-di		$: \alpha_2 (a/L)^2$	$\alpha_2 = \dfrac{2A^2(1-v)^2}{\pi E a^4}$
mo-di		$: \alpha_3 (a/L)$	$\alpha_3 = -\dfrac{Am_o}{\pi a}(1+v)$
di-mo		$:-\alpha_3 (a/L)$	

<u>Table III:</u> Sketch of the various interactions in between elastic monopoles (mo) and dipoles (di), with the analytical expressions of the interaction energies in column 3. Here $m_o<0$ and $A>0$

In the given analytical expressions of the four terms (see third and fourth column of table III) let us recall that A is a dipole moment as in I32. Here $A>0$ describes a dilatation centre. Identical monopoles are said long range attractive (lnL) whereas the identical dipoles, whatever their sign, are said short-range repulsive (L^{-2}). The medium range monopoles-dipoles interaction depends on the sign of Am_o, but for the pair of steps the total interaction is zero. In appendix J we derive these interactions.

Consider now a step m at position x_m in between two others steps (of same sign) at x_{m-1} and x_{m+1}, with $x_{m+1}-x_{m-1}=2\overline{L}$ (x_m is the deviation of step m from its middle position taken as origin see figure 15). The step-step interaction per unit length versus the deviation x_m from table III thus reads

$$U^{s-s}(x_m) = U_1(x_m) + U_2(x_m) = \alpha_1 \ln\left[\frac{L-x_m^2}{(2\pi a)^2}\right] + \alpha_2\left[\left(\frac{a}{L-x_m}\right)^2 + \left(\frac{a}{L+x_m}\right)^2\right] \quad (54)$$

and is drawn as a full line on figure 15. The step m at $x_m=0$ thus is mechanically unstable since by some fluctuation it is attracted either on one side or the other side and finally trapped in the left or right minimum due to the repulsive potential. For figure 15 we took usual values of the various ingredients $E=5.10^{11}$ erg cm^{-3}, $\nu=1/3$, $a=2.10^{-8}$ cm. For the dipole moment $A=0.7eV=1.10^{-12}$ erg consistent with E and a (see appendix I). With a misfit $m_o=2.10^{-2}$ one thus obtains $\alpha_1=2.10^{-8}$ erg cm^{-1}, $\alpha_2=1.10^{-5}$ erg cm^{-1} and $\alpha_2/\alpha_3=1/3.10^3$. From figure 15 where $L/a=100$ the pairing distance is seen to be $(L-x_m)/a=23$ atomic units. A good approximation is $(2\alpha_2/\alpha_1)^{1/2} \approx 2|A|/(Ea^3|m_o|)=25$. From latter result one understands that epitaxial strain m_o whatever its sign compresses the pair. Adding to α_2 a $\alpha_2(T)$ term[‡‡‡], a temperature increase may further separate the pair.

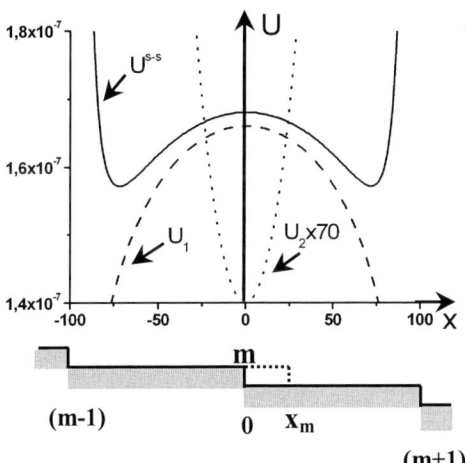

Figure 15: *Interaction energy of a step of a strained solid U^{s-s} (54) versus its the position in between two others, U_1 attracting part due to elastic monopoles, U_2 repulsive part due to elastic dipoles (units: erg cm^{-1}).*

[‡‡‡] The elastic interactions as all elastic properties are slightly decreasing with temperature. An independent dipole-dipole repulsion $\alpha_2(T)(a/L)^2$ exists for $T\neq 0$ due to kink formation [105] so that the steps meander and due to their mutual confinement there results [126] an exponentially increase of $\alpha_2(T)$ with T. Obviously these "dipoles" don't couple with elastic monopoles or dipoles.

(ii) How does this mechanical instability lead to step bunching? This is a matter of cooperative kinetics of transport of atoms by surface diffusion in between steps. Tersoff [125] considers the adatoms coming from exchange with the kinks on the steps and eventually from an incoming flux Fsec^{-1} per site. No atom leaves the crystal, complete condensation is thus supposed so that the time dependant adatom density θ is governed by the simplified BCF [106] diffusion equation. The boundary conditions at the steps are simplified too: **(1)** The transfert of atoms from or to the kinks and nearest adatoms positions $x_m \pm a$ has no kinetical barrier (at least that of the usual surface diffusion coefficient D). There is no interfacial kinetics, no Schwoebel or other retardation effects. **(2)** The local adatom density at the steps is that of thermodynamical equilibrium

$$\theta(x_m + a) = \theta(x_m - a) = \exp\left[-\Delta E\big|_{x_m,\overline{L}} / kT\right] \tag{55}$$

where ΔE is the bond energy of a kink atom minus that one of the adjacent adatom at ±a (in fact the adatom creation energy from a kink atom). This quantity is however modified by the elastic field where the kink is located at x_m inside the train of steps $\overline{L}$ as quoted in (55) by the indices x_m and $\overline{L}$. A kink atom at $x_m \neq 0$ is submitted to a net attractive force $af(x_m) = -a\partial U^{s-s}/\partial x\big|_{x_m} > 0$ towards the nearest of its neighbouring step. When gaining an atom the kink goes ahead by $\Delta x_m = a$ so that its energy changes by $\Delta W \approx af(x_m)\Delta x_m = a^2 f(x_m)$ so that §§§

$$\Delta E\big|_{x_m,\overline{L}} \approx \Delta E\big|_{0,\overline{L}} - a^2 \frac{\partial U^{s-s}}{\partial x}\bigg|_{x_m} \tag{56}$$

where now $\Delta E\big|_{0,\overline{L}}$ is the adatom creation energy from a kink when this force vanishes. From (55) and (56) there is

$$\theta(x_m \pm a) = \theta(\pm a)\exp\left[-a^2 \frac{\partial U^{s-s}/\partial x\big|_{x_m}}{kT}\right] \tag{57}$$

which means that these elastic interactions due to epitaxial strain boost the equilibrium adatom density at the step when the step deviates from its middle position. This is a simple way to formulate how a mechanical effect is transformed in a chemical one.

For testing stability let us consider at the time t=0 a regular train of steps of mean distance $\overline{L}$. Let us assume a small displacement of every N steps by $u_m(o) = \Delta \cos(2\pi m/N)$, at a small time t>0 the following time evolution (58) is found to be valid [125]

$$u_m(t) \approx \Delta e^{rt} \cos\left[\frac{2\pi}{N}(m + Ft)\right] \tag{58}$$

---

§§§ In fact this is only true when each step site is a kink site. When the kink density is 0<ε<1, the local displacement is reduced to $\Delta x_m = a\varepsilon$

The exponent is given in (59) for widely spaced steps ($L/a > \alpha_2/\alpha_1$) and since positive, leads to an amplification of the bunching rate r we factorise:

$$r = \frac{Ea^3 m_o^2}{1-v^2}\left(a^2 D_M/2kT\right)\left[\left(\frac{2\pi a}{N\overline{L}}\right)^3 \pi\left(1-\frac{1}{N}\right)\right] \tag{59}$$

First there is the "driving energy" of epitaxial strain per atom, then the "kinetic resistance" due to the material transport. Appears the mass diffusion coefficient

$$D_M = D\exp\left(-\Delta E\big|_{x_m,\overline{L}}/kT\right) = D_o \exp\left(-\left(\Delta E\big|_{x_m,\overline{L}} + E^*\right)/kT\right) \tag{60}$$

composed of the adatom creation energy $\Delta E\big|_{x_m,\overline{L}}$ usually greater than the activation barrier E^* for surface diffusion of atoms. $\Delta E\big|_{x_m,\overline{L}}$ is able to freeze this rate process when temperature is not high enough, e.g. when $T < 2/3 T_{melting}$. The last term in (59) shows that bunching starts for N=2, that puts in evidence the cooperative nature of the process. N=2 is the most efficient mode, the amplification slows down rapidly for modes N>2. This initial bunching rate is flux independent. At F=0 (but zero evaporation is prescribed) numerical resolution [125] showed, starting with a random train, the development of the bunches with time. The bunch size (mean number of steps in a bunch) varies monotonously as $<n> \propto t^{1/4}$. Atomically flat zones can be obtained separated by bundles of many steps. Curiously it was observed (but not systematically studied) that when flux is put on, maintaining the same other conditions (1) bunching progresses less rapidly $<n> \propto t^{1/6}$ (2) bunching saturates at some small value $<n>_{sat}=3$ for F=25. Such simulations should be reactivated in parallel with in situ experiments similar to those of Métois et al. [127] under well controlled flux F<0, F=0 or F>0.

3.3.3. Duport's strain driven surface diffusion instability

The Grenoble group [123,124] predicted first in 1994 an other instability we are now able to qualify more precisely. The phenomenon concerns strained vicinal faces under biaxial misfit m_o with an incoming flux F>0 but without re evaporation (complete condensation) as in 3.3.2.

The authors considered the elastic interaction of an adsorbed atom on a terrace in-between two consecutive L apart steps (see figure 16) described as a dipole A^{ad} located in-between two lines of identical monopoles m on the right and m-1 on the left. Taking the middle point as origin of x_{ad}, the interaction energy per atom reads in absence of surface stress and thus with $F_3=0$ in (14) (see table III)

$$U_L^{ad-s}(x_{ad}) = U_3(x_{ad}) + U_2(x_{ad}) = \alpha'_3\left[\frac{a}{L/2+x_{ad}} - \frac{a}{L/2-x_{ad}}\right] + \alpha'_2\left[\left(\frac{a}{L/2+x_{ad}}\right)^2 + \left(\frac{a}{L/2-x_{ad}}\right)^2\right] \tag{61}$$

where $\alpha'_3 = -\dfrac{A^{ad} m_o}{\pi a}(1+v)$ [123,124]**** for the dipole-monopoles interaction and $\alpha'_2 = 2A^{ad} A/(\pi E a^3)$ for the dipole-dipoles interaction (see table III). In principle $A^{ad} \neq A$ but are of same sign so this interaction again is repulsive. Mostly $|A^{ad}|<|A|$.

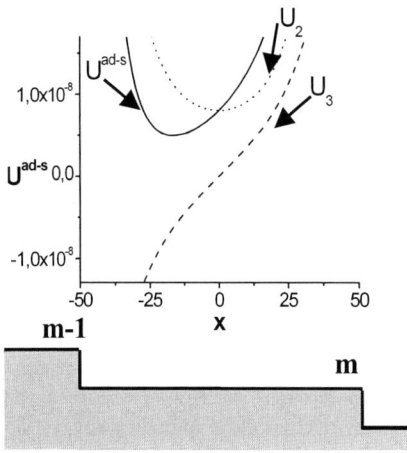

Figure 16: Interaction energy U^{ad-s} of (61) erg cm^{-1} of an adatom (dipole) and two steps of same sign of a strained body. U_2 repulsive part due to the two dipole–lines with the ad-dipole interaction, U_3 attractive part due to the two monopole-lines and the ad-dipole

In figure 16 we draw (61) as a full line taking the same numerical values as in figure 15 of E, v, a, m_o, L/a and A, furthermore we put $A^{ad}=A$ (making therefore an underevaluation). Thus $\alpha'_3 = 8.4.10^{-15}$, $\alpha'_2 = 2.10^{-13}$ erg per adatom. For $A^{ad}m_o > 0$ the adatom is attracted to the lower terrace of the steps the short range repulsion creating a single well located near the lower terrace of the step, (specifically here $x_{ad}/(L/2) = -20$ atomic units). When $A^{ad}m_o$ changes sign, the well switches to the symmetrical position. The stationary adatom density at the position x along the terrace is given by the resolution of

$$F - \dfrac{\partial J_m(x)}{\partial x} = 0 \qquad (62)$$

where $J_m(x)$ is the net current proportional to the gradient of adatom density $\theta(x)$ to which is added a drift term due to the elastic force $f_{ad} = -\partial U^{ad-s}/\partial x$ acting on the adatoms:

$$J_m(x) = -D\dfrac{\partial \theta}{\partial x} + \dfrac{D}{kT}\theta(x) f_{ad} \qquad (63)$$

where D/kT is the so-called Einstein mobility of the adatoms when submitted to the force $f_{ad}(x)$.

The Einstein-Focker-Planck equation (63) thus gives a purely kinetics effect as can be easily seen since when this force derives from the equilibrium distribution $\theta(x) = \exp[-U^{ad-s}/kT]$

**** We put here minus sign since our misfit convention is of opposite sign of Duport's $\delta a/a = -m_o$.

there will be no current on the terraces. Putting (63) in (62) and integrating along a terrace gives

$$Fx + D\left[\theta(x) + \int \frac{\theta(x)}{kT} \frac{\partial U^{ad-s}}{\partial x} dx\right] + C = 0 \tag{64}$$

The constant C is obtained from boundary conditions at the step. Obviously they have to be non-equilibrium conditions: for the left step (m-1), $\frac{D^-}{a}\theta(L/2) = C - FL/2$; for the right step $\frac{D^+}{a}\theta(L/2) = C + FL/2$. The condition for a step m to collect from its front and its back terrace leads again to a differential equation very similar to (59) with a stability exponent r

$$r = -\left\{\left(\frac{D}{D^+}\right)^2 - \left(\frac{D}{D^-}\right)^2 + \frac{2(1+\nu)}{\pi}\frac{A^{ad}m_o}{kT}\frac{L}{a}\right\}\left(\frac{2\pi a}{NL}\right)^2 F \tag{65}$$

Physically however the result is very different. **(i)** The oscillation of the train suffers damping or amplification only when a flux F≠0 exists (clearly only an incoming flux since evaporation was precluded). **(ii)** The shortest mode, even N=1, is the most efficient one. **(iii)** For $m_o=0$, an asymmetric adatom integration ($D^+<D^-<D$) is stabilising (similar to Schwoebel effect). For $A^{ad}m_o>0$ (for self-adsorption $A^{ad}>0$ it means $m_o>0$) stability is increased even for symmetric integration ($D^+=D^-$). **(iv)** For the opposite case $A^{ad}m_o<0$ (that means $m_o<0$ for self adsorption) and with still $D^+<D^-<D$, the epitaxial strain drives so much the adatoms downward the step that they overcome easily the Schwoebel barrier. However this only happens provided the mean step distance L/a overpasses a critical value depending on the height of the barrier.††††

From our discussion here and the one in III32 it is clear that Duport's and Tersoff's instabilities are not of the same nature. *The Tersoff's one is misfit square dependent, exists at zero flux and slows down for increasing flux. The Duport's one is misfit sign dependent, does not exist for $m_o>0$ and when exists ($m_o<0$) is boosted by increasing flux.* This gives a contradictory feeling about both effects. Furthermore in Tersoff's theory the adatoms have not been supposed subjected to the elastic field of the steps and thus are not dragged towards the steps. In Duport's theory the steps are supposed do not interact by their elastic field as of course they should. The situation is however not so bad since Tersoff's and Duport's instabilities *in fact occur in different temperature ranges*. Indeed in Tersoff's formula (59) appears the mass diffusion coefficient $D_M = D_o e^{-E^*/kT} e^{-\Delta E/kT}$, which is very temperature dependent through $\Delta E/kT$, and thus only works at very high temperature. On the contrary, in Duport's formula (65) appears the surface diffusion coefficient $D = D_o e^{-E^*/kT}$. Thus since $E^*/\Delta E \approx 1/5$ for stable faces the Duport instability works at low temperature where kinks can't produce adatoms by their own (Adatoms in this case are only provided by the incoming flux).

---

†††† Thus wafers with very small miscuts and the use of very weak flux would be helpful for avoiding this instability.

3.3.4. Miscellaneous kinetics effects

* In the previous section we have seen that strain can modify surface diffusion by way of the supplementary elastic forces along the steps acting on adatoms. Nevertheless strain can also have an effect on the diffusion coefficient itself as shown by Schroeder and Wolf [128] who calculated activation barriers for diffusion on strained high symmetry plane surfaces (without steps) of simple cubic, fcc and bcc crystals. For this purpose they described pair-wise interactions by means of an anisotropic Lennard-Jones potential with a strain-modified distance in between atoms; then they placed an adatom on a binding site and moved it by small steps. The activation energy for surface diffusion is calculated by a conventional minimal energy path saddle point. *The main result is that generally for tensile stress ($\sigma>0$) the diffusion barrier is increased whereas compressive stress ($\sigma<0$) decreases the barrier.* (see figure 17). The diffusion barrier change is mainly due to a change of the saddle point energy whereas the minima are shifted only very little. As said by the authors this behaviour can be naïvely understood on the basis of limiting cases. Indeed in the limit of large compressive stress ($\sigma<0$) the surface becomes continuous and thus there is no longer a diffusion barrier. On the contrary, within the limit of large tensile stress ($\sigma>0$) the surface consists of isolated atoms and diffusion becomes equivalent to breaking a pair of atoms and building a new pair[‡‡‡‡]. In the same paper the authors have also studied theoretically diffusion on top of a stressed island. In this case, since the finite size island can elastically relax by its free edge the strain along the top surface of the island becomes inhomogeneous and thus diffusion may vary from the centre of the island towards its edges. *For compressive strain the diffusion is faster near the island centre whereas for tensile stress it is faster towards the edges.* Thus it should be easier to nucleate on top of a tensile strained island than on top of a compressive strained island[§§§§].

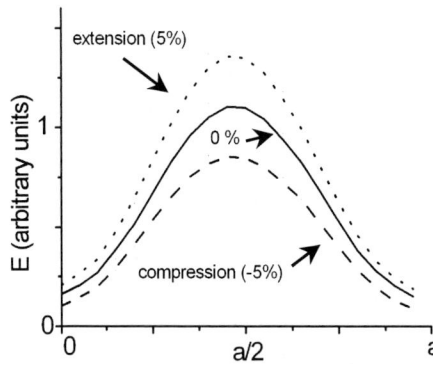

Figure 17: Diffusion barrier path for an adatom on an homogeneously strained cubic lattice (parameter a) calculated by [128]. Upper curve: tensile strain (5%), middle curve $m_o=0$, lower curve: compressive strain (-5%). In fact for a homogeneous strained (001) surface of a simple cubic crystal the diffusion barrier depends linearly upon strain which can be easily shown by a simple first order development of pair-wise potential in respect to strain [128].

These two effects are second order effects in comparison to elastic interaction in between adatoms and steps as described in the previous section. Furthermore the *simple surface diffusion change cannot modify the growth mechanism*. More precisely since adatom density essentially depends upon the ratio D/F a strain-induced increase (decrease) of the surface

---

[‡‡‡‡] In other words in this latter case the diffusion barrier becomes equal to the pair-binding energy

[§§§§] More precisely for an inhomogeneous strained surface the saddle point energy and the binding energy may vary as well. The saddle point energy change leads to an inhomogeneous nucleation whereas the binding energy change leads to a drift term in the diffusion current. Nevertheless for cubic crystals the binding energy change remains weak [128].

diffusion constant (D) is exactly equivalent to an appropriate increase (decrease) of flux (F) and thus may only weakly shift the transition between step flow and 2D nucleation or change the nucleation density. At the same since the taller a crystal, the more relaxed its top face, this kind of elasticity-induced Schwoebel barrier thus can only help the first stages of the thickening of tensile islands.

* We avoided in these lectures to treat alloy formation and especially the effect of strain. Let's fisrt mention some facts.
(i) Deposition of A pure on a B pure substrate forming either 3D crystals (VW) or thin epitaxial layers (F-vdM). In the absence of epitaxial misfit and defects the interface in both cases moves, in a planar way in the first case, rebuilding the substrate in the vicinity of the 3D crystals for the second case. One speaks about compositional strain due to the different atomic radius of A and B.
(ii) Deposition of an alloy A_xB_{1-x} on a substrate B. Interface diffusion is common for planar Ge/Si or Si/Ge systems and starts at temperature higher than 650°C but is difficult to follow. Some other systems are more accessible and much more brilliant for demonstration. It is the case of epitaxial deposit $BaTiO_3$ film deposited on MgO buffered sapphire substrate: $BaTiO_3(100)//MgO(100)Al_2O_3(11\bar{2}0)$ [129] where the interface $BaTiO_3$/MgO is very abrupt whereas $MgAl_3O_4$ layers appear at the interface MgO/Al_2O_3 (see figure 18). This alloy layer thickness depends on the subsequent deposition time of $BaTiO_3$ deposited at 1100°C, the MgO layer having been deposited on Al_2O_3 at 650°C only.

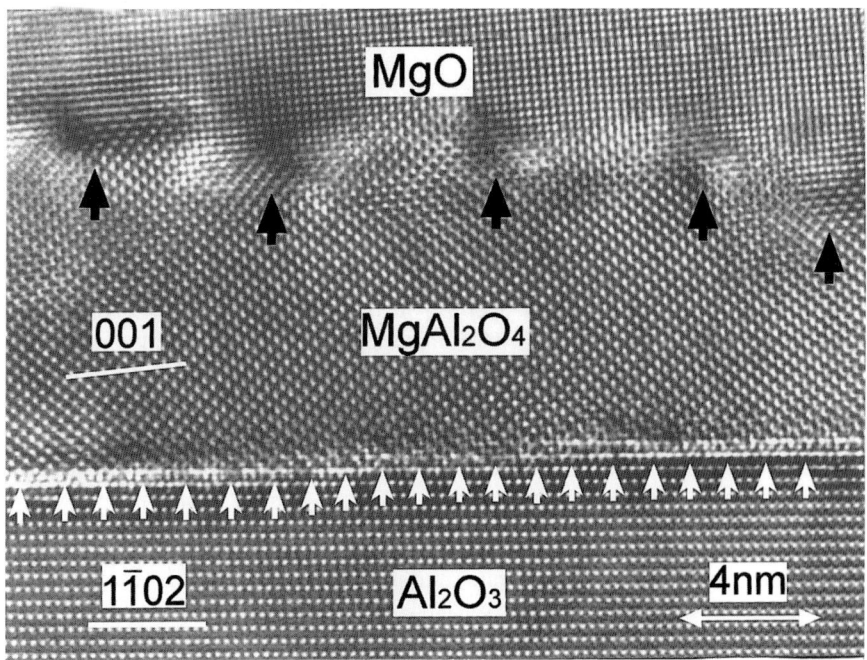

Figure 18: $BaTiO_3(100)//MgO(100)Al_2O_3(11\bar{2}0)$: *Lattice image of the $MgAl_2O_4$ spinel reaction layer between the MgO buffer layer and the sapphire substrate (courtesy of C.H. Lei et al [129])*

(iii) Interesting to study is the surface of a strained alloy where compositional and morphological instabilities occur both [130-141]. Undulations become totally unstable for all wavelengths. Islands nucleate at different composition than the alloy layer, stress induced nucleation rate is drastically increased. Experiments are far behind theory.

Last but no least lets mention very important practical but trivial effects: misfit changes that occur at the end of a growth process when temperature comes back to normal temperature and the different dilatation coefficient are not adjusted.

4. A FEW REMARKS ABOUT INTERACTING CRYSTALS

Up to now we have only described isolated epitaxial crystals. Generally one has to do with a collection of crystals so that when their mean distance $\overline{L}$ approaches their mean size $\overline{\ell}$, these crystals may interact. We will distinguish two types of interactions. Even when the crystals are far one from another, $\overline{\ell} \ll \overline{L}$, they may exchange atoms by surface diffusion on the substrate provided the surface diffusivity is high enough. At smaller distances $\overline{\ell} \approx \overline{L}$, there is furthermore to consider elastic interaction of the crystals via the substrate.

* When the equilibrium shape of a 3D crystal (Volmer Weber case) is realised putting equation (d) in equation (e) of table II it follows the generalised Gibbs Thomson equation [87]:

$$\Delta\mu = 4\gamma'/\ell_{eq} + \mathcal{E}_o \left[R(r_{eq}) + r_{eq} \frac{\partial R}{\partial r}\bigg|_{r_{eq}} \right] \tag{66}$$

Relation (66) says that an equilibrium crystal of finite size ℓ_{eq} has an excess chemical potential $\Delta\mu$ with respect to a bulk and non-strained crystal. The first term of (58) corresponds to a hyperbolical decrease of $\Delta\mu$ with size ℓ_{eq} as in the classical Gibbs-Thomson equation. The second term represents the contribution to chemical potential of the strained but relaxed crystal at its equilibrium shape ratio r_{eq}. For usual elasto-capillar length γ/Y (see section 3.2.3.2.) this term contributes to less than 5% for misfits as high as $m_o=5.10^{-2}$. Thus the smaller the equilibrium shape ℓ_{eq}, the higher the chemical overpotential. So at thermodynamic equilibrium, the usual Ostwald ripening still holds: small crystals loose molecules in favour of the bigger ones[*****]. However kinetics when limited by surface diffusion may oppose to this thermodynamic tendency towards Ostwald ripening.

For coherent epitaxies a continuous layer is fully strained at its natural misfit m_o (see section 3.2.3.1.) Thus a collection of islands initially relaxed, when coalescing, have to strain back to m_o at layer completion. This was first mentioned by Cabrera [141] who said that this proceeds by the overlapping of the substrate strain fields when the island borders come closer. This means that, when close enough, two islands communicate by the substrate. This must affect the equilibrium shape of each crystal since the greater the coverage, the greater the

---

[*****] This may be different when the various crystals have different shapes. In this case, owing to its specific shape, a great crystal should have a greater chemical potential than a small one. In this case there could be a size selection.

elastic energy. In [87] we have studied the equilibrium shape ratio of interacting box shaped crystals (Volmer Weber growth) and shown that interacting crystals have a shape ratio r which deviates from that of isolated crystals. More precisely, the near equilibrium growing crystals prefer to thicken rather than to come closer to the borders and finally to coalesce. Obviously the harder the substrate, the smaller the deviation from the equilibrium shape of isolated crystals. Nevertheless these elastic interactions should also affect Ostwald ripening since now the mean elastic energy per atom that contributes to the total energy becomes coverage dependant. Floro et al. [142] tried to introduce such a contribution to the mean energy per atom by writing the usual chemical potential as $\Delta\mu = 4\gamma'_A [1/\ell_{eq} + p(\theta)]$. The mean field term $p(\theta)$ they introduced was obtained by finite elements calculations and found to be proportional to $\exp(\theta^2) - 1$ where θ is the surface fraction covered by the islands. In this case the authors found that Ostwald ripening is enhanced by elastic repulsion as soon as elastic interaction in between deposited islands plays a role, that means for high coverage close to coalescence. The results thus obtained are compatible with their experiments [142].

- Such elastic interactions have also been proposed to be the driving force for self-organised growth [143,144]. Indeed, if in an array of islands one of the island deviates in size or shape, its neighbouring islands feel the change. It therefore installs a driving force for material transport restoring a uniform size and shape distribution, if temperature is high enough.

In the case of multilayers films, the repeated deposition of layers enhances the self-organisation, so 3D islands may organise progressively in a uniform and regular pattern [150]. The theoretical description of such organisation in multilayers has been given by Xie et al. [143] then Tersoff [145], and called vertically self organised growth.

ACKNOWLEDGEMENTS

We thank A. Ranguis who has drawn some of the figures and helped us for editorial work.

APPENDIX A

In the case of a (001) biaxially strained ($\varepsilon_1 = \varepsilon_2 = m_0$) layer ($\sigma_3 = 0$ at the free surface) of a cubic material, the relation (3) can be written:

$$\begin{pmatrix} \sigma_1 \\ \sigma_2 \\ 0 \\ 0 \\ 0 \\ 0 \end{pmatrix} = \begin{pmatrix} C_{11} & C_{12} & C_{12} & 0 & 0 & 0 \\ C_{12} & C_{11} & C_{12} & 0 & 0 & 0 \\ C_{12} & C_{12} & C_{11} & 0 & 0 & 0 \\ 0 & 0 & 0 & C_{44} & 0 & 0 \\ 0 & 0 & 0 & 0 & C_{44} & 0 \\ 0 & 0 & 0 & 0 & 0 & C_{44} \end{pmatrix} \begin{pmatrix} m_o \\ m_o \\ \varepsilon_3 \\ 0 \\ 0 \\ 0 \end{pmatrix}$$

where we use the C_{ij} matrix for a cubic material (x_1, x_2, x_3 being the fourfold axis). The compliance matrix has the same form. For cubic material S_{ij} and C_{ij} are connected via the following relations [15]:

$$S_{11} = \frac{C_{11} + C_{12}}{(C_{11} + 2C_{12})(C_{11} - C_{12})}; \quad S_{12} = \frac{-C_{12}}{(C_{11} + 2C_{12})(C_{11} - C_{12})}; \quad S_{44} = 1/C_{44}$$

Notice that C_{ij} and S_{ij} coefficients can be inverted.
The previous relations between σ and ε thus give

$$\begin{cases} \sigma_1 = \sigma_2 = (C_{11} + C_{12})m_o + C_{12}\varepsilon_3 \\ 2C_{12}m_o + C_{11}\varepsilon_3 = 0 \end{cases}$$

and thus

$$\begin{cases} \sigma_1 = \sigma_2 = (C_{11} + C_{12} - 2\frac{C_{12}^2}{C_{11}})m_o \\ \varepsilon_3 = -2\frac{C_{12}}{C_{11}}m_o \end{cases}$$

The elastic energy density (5) thus reads
$$w_{el} = E_{cub}^{2D} m_0^2$$

where $E_{(001)}^{2D} = C_{11} + C_{12} - 2\frac{C_{12}^2}{C_{11}}$ and $v_{(001)}^{2D} = -2\frac{C_{12}}{C_{11}}$ define the two dimensional Young's modulus and Poisson's ratio for this orientation [19,20].

In the case of (111) strained layer one obtains $Y = 6C_{44}(C_{11} + 2C_{12})/(C_{11} + 2C_{12} + 4C_{44})$.
For less simple cubic orientations see [146].

APPENDIX B

The mathematical definition of the principal value (vp) is:

$$vp\left[\int_{-\infty}^{\infty} \frac{f(x')}{x'-x} dx'\right] = \lim_{\delta \to 0}\left[\int_{-\infty}^{x-\delta} \frac{f(x')}{x'-x} dx' + \int_{x+\delta}^{\infty} \frac{f(x')}{x'-x} dx'\right]$$ which can be written under the Hilbert

form with t=x-x' $vp\left[\int_{-\infty}^{\infty} \frac{f(x')}{x'-x} dx'\right] = \lim_{\delta \to 0}\left[\int_{\delta}^{\infty} \frac{f(x+t) - f(x-t)}{t} dt\right]$. Thus in the case under

study there is

$$vp\left[\int_{-\infty}^{\infty} \frac{\sin(\omega x')}{x'-x} dx'\right] = \lim_{\delta \to 0}\left[\int_{\delta}^{\infty} \frac{\sin\omega(x+t) - \sin\omega(x-t)}{t} dt\right] = \lim_{\delta \to 0}\left[2\cos\omega x \int_{\delta}^{\infty} \frac{\sin\omega t}{t} dt\right]$$

Thus for $\omega > 0$ there is $vp\left[\int_{-\infty}^{\infty} \frac{\sin\omega x'}{x'-x} dx'\right] = \pi \cos\omega x$

APPENDIX C

By writing the strain tensor in terms of displacement:

$$\frac{1}{2}\int_V \sigma_{ik}\varepsilon_{ik}dV = \frac{1}{2}\int_V \sigma_{ik}\frac{1}{2}\left[\frac{\partial u_i}{\partial x_k}+\frac{\partial u_k}{\partial x_i}\right]dV$$

Integration by parts transforms the second member in a surface integral:

$$\frac{1}{2}\int u_i\sigma_{ik}n_k dS - \frac{1}{2}\int \frac{\partial \sigma_{ik}}{\partial x_k}u_i dV$$

The last integral is zero since in the bulk of the solid the bulk density of force components (1), $f_i = \partial\sigma_{ik}/\partial x_k$ have to be zero when no body forces as gravity or others are acting. In the first integral n_k are the components of the unit vector normal to the surface so that $\sigma_{ik}n_k = f_i$ are surface force density components creating the displacement $w_{el} = \frac{1}{2}\int_S f_i u_i dS$.

APPENDIX D

For semi-conductor films, the situation may appear simple since it seems that a film of roughly 1 to 3 monolayers can already be considered as an elastic continuum where bulk constants are roughly valid [147,148]. This may come from the short-range potential describing semi conductor bondings. For other materials the situation is more complex since now the bulk elastic properties must be size dependent. However a simple model of size dependence of the biaxial modulus of thin film has been published [71]. In this paper Streitz et al. have calculated the thickness-dependent biaxial modulus Y(h) of thin metal films (thickness h) as the second derivative of the total energy ΔU per unit volume with respect to strain. They show that for Co, Ni, Ag and Au with (001) or (111) orientations, size dependent biaxial moduli obtained from atomistic simulations are perfectly fitted by a simple analytical model where $\Delta U = \Delta U_B + \Delta U_S$ with ΔU_B the volume strain density energy of an infinite material (characterised by the usual bulk biaxial modulus) and ΔU_S the work done against surface and interface stress. More precisely using their relation (18) the biaxial modulus of a thin supported film reads in the framework of the assumptions herein (linear elasticity, strain independent surface stress) $Y(h) = Y_\infty[1-\varepsilon_{//}(2\eta-3)]$ with $\varepsilon_{//} = -\frac{1-v_A}{E_A}\frac{s_A+s_{AB}}{h_A}$ and where $\eta = -\varepsilon_{zz}/\varepsilon_{xx}$ is a function of the Poisson ratio. Thus the biaxial modulus Y(h) of a thin film scales with the reciprocal of the film thickness and reaches the usual bulk value Y_∞ for increasing thickness. Thus in the framework of linear elasticity it seems that it is formally equivalent to use size-dependent bulk elastic constants or to properly consider surface stress.

APPENDIX E

The proof is: the short range behaviour reads $\Phi(z) = \Phi_\infty \prod_0^z (z-1)$, $\prod$ the Heaviside function and from (32) with $\theta=0$, $V=0$

$$\frac{\partial \Delta F}{\partial z} = -(\Delta\mu - \mathcal{E}_0)L^2 a + L^2 \frac{\partial \Phi(z)}{\partial z} = 0$$

so that now in table II for $\Phi_\infty < 0$ there is $\delta(z-1) = \dfrac{\Delta\mu - \mathcal{E}_0}{-|\Phi_\infty|}$. Since $\Delta\mu < \mathcal{E}_o$ the ratio is positive so that the solution is $z^* = 1$ and also for $\Delta\mu = 0$, $z_o = 1$.

APPENDIX F

From a thermodynamical point of view the number of interfacial dislocations may pass from N to N+1 when the total energy change due to the introduction of the (N+1)th dislocation is negative. The elastic energy stored by the system of 2xN orthogonal dislocations exist can be roughly written [80] for isotropic solids

$$E_N = 2\frac{m_o - m'}{a} b^2 \frac{E_{AB}}{2\pi}(1+\ln\lambda)S_{AB} + Ym'^2 VR \tag{i}$$

where the first term is the energy of a double array of perpendicular non interacting dislocations with $1/E_{AB}=1/E_A+1/E_B$ the reciprocal "interfacial modulus" ($E_A=Y_A/(1-\nu_A)$ and $E_B=Y_B/(1-\nu_B)$ are the elastic modulus of A and B respectively), b the Burgers vector component in the interface, $m_o - m'$ the part of the misfit accommodated by the (N+1)th dislocation pair and λ a cut off. When h<d where d is the equidistance in between dislocations there is $\lambda=h$, if not there is $\lambda=d/2$. When there are N interfacial dislocations the released elastic misfit is obtained from Vernier considerations. It reads

$$m' = m_0 - Nb/\sqrt{S_{AB}}\Big|_N \tag{ii}$$

where $S_{AB}\big|_N$ is the interfacial area for a crystal having N interfacial dislocations. The thermodynamical criterion for the (N+1)th dislocation entrance is thus obtained from $E_{N+1} - E_N < 0$ that means when

$$\frac{h}{a} = \frac{V}{S_{AB}\big|_N} > \frac{b}{a}\frac{1}{2\pi}\frac{K}{1+K}\frac{1+\ln\lambda}{\left(m_o - \dfrac{N+1/2}{\sqrt{S_{AB}\big|_N}/b}\right)R} \tag{iii}$$

Where $K=E_B/E_A$ is the relative rigidity, and R the relaxation factor. For a thin pseudomorphous infinite film there is $R=1$, $S_{AB} \to \infty$ and $\lambda=h/a$ so that the previous relation reads

$$\frac{h}{a} > \frac{b}{a}\frac{1}{2\pi}\frac{K}{1+K}\frac{1+\ln(h/a)}{m_o} \tag{iv}$$

which is nothing other than the usual Matthews relation giving the critical height beyond which dislocations may appear in a pseudomorphous film [80]. Let us note that more precise expressions of h have been given in literature (for a review see [149]) but above relations roughly give the good order for reasonable misfits (a few %).

APPENDIX G

The elastic energy of a film in presence of surface stress can be written

$$E_{film}(m') = \mathcal{E}_0 m'^2 h_A + 2m' \Delta s^\infty (1 - \exp(-h_A/a\zeta))$$

The first term is the bulk elastic energy stored by the film of height h_A. The second term is the work against surface and interface stresses corrected for long range exponential inter layers forces [69,70]. In this latter term $\Delta s^\infty = s_A^\infty + s_{AB}^\infty - s_B^\infty$ is the surface stress change due to the film. The value of m' which minimises $E_{film}+E_{disl}$ is: $m' = m'_o - \dfrac{\Delta s^\infty (1 - \exp(-h_A/\zeta a))}{\mathcal{E}_0} \dfrac{1}{h_A}$

With $m'_o = \dfrac{1}{h_A} \dfrac{b}{2\pi} \dfrac{K}{1+K} \left[1 + \ln\left(\dfrac{h_A}{b}\right) \right]$ where $K = E_B/E_A$ is the relative rigidity of the substrate with respect to deposit and where we put for the cut off distance $\lambda = h_A$ in the expression of E_{disl}.

The film thickness h_c beyond which the first dislocation entrance becomes energetically favourable is then obtained by setting $m' = m_o$ in the previous relation. Thus there is:

$$h = \dfrac{1}{|m_o|} \dfrac{b}{2\pi} \dfrac{K}{1+K} \left[1 + \ln\left(\dfrac{h}{b}\right) \right] - \dfrac{1}{m_o} \dfrac{\Delta s^\infty (1 - \exp(-h/\zeta a))}{\mathcal{E}_0}$$

The critical thickness is therefore amended (compare with last equation in appendix F) by the surface stress effect.

APPENDIX H

As known [6,92] homogeneous *classical nucleation* barrier amounts to 1/3 of the total surface energy of the nucleus :

$$\dfrac{\Delta F^*}{kT} = \dfrac{1}{3} \gamma_A \dfrac{6(\ell/a)^2}{kT} = \dfrac{2\gamma_A a^2}{kT}(N^*)^{2/3} \quad (i)$$

with N^* the number of molecules in the critical nucleus.

From $\dfrac{\partial \Delta F}{\partial N^*} = 0$ there is

$$N^* = \left(\dfrac{4\gamma_A}{\Delta\mu}\right)^3 \text{ and } \dfrac{\Delta F^*}{kT} = \dfrac{a^2}{2kT} \dfrac{\gamma_A^3}{\Delta\mu^2} \quad (ii)$$

The nucleation rate reads $dN^*/dt \approx v \exp[-\Delta F^*/kT]$ with $v=10^{13}$ sec^{-1} an attempt frequency. Thus a nucleation rate $dN^*/dt \approx 1$ which is quite reasonable gives

$$\Delta F^*/kT \approx 30 \quad (iii)$$

Since $2 < (2\gamma_A a^2)/kT < 5$ is a quite usual surface energy at evaporating temperature the operative critical nucleus contains

$$15 < N^* < 60 \quad (iv)$$

molecules. From (ii) it results the supersaturation range $3kT/a^2 < \Delta\mu < 5kT/a^2$ or compared to $\mathcal{E}_0$, $\Delta\mu/\mathcal{E}_0 = \Delta\mu/(Yam_o^2)$, $\mathcal{E}_0 = 10^{12}$ erg cm^{-3}, $m_o = 2.10^{-2}$ gives at $T \approx 10^3$K

$$6 < \dfrac{\Delta\mu}{\mathcal{E}_0} < 10 \quad (v)$$

Non-classical nucleation is that one where the nucleus size N* is so small (N*=1,2,3…) that the macroscopic concepts of surface energy no more holds. Then additive bond energies are convenient to define each cluster (see Walton's simple theory in [92, 106] or papers as [113].

APPENDIX I

Exact calculations of the dipole moment A_{ij} of adsorbed atoms on a substrate can be done precisely when the interactions in-between two ions (i) and (j), $x^{i,j}$ apart, are well represented by a pair potential $\Phi(x^{i,j})$. Indeed, using the concept of point forces the components of the force of an atom (i) acting at $x^{i,j}$ is: $f_\alpha^{i,j}\delta(\vec{x}-\vec{x}^{i,j})$ with $f_\alpha^{i,j}=-\dfrac{\partial\Phi(x^{i,j})}{\partial x^{i,j}}\dfrac{x_\alpha^{i,j}}{|\vec{x}^{i,j}|}$. Far from the point of application of the force $x >> x^{i,j}$ this force reads up to the first order $f_\alpha^{i,j}\delta(\vec{x}-\vec{x}^{i,j})=f_\alpha^{i,j}\left[\delta(\vec{x})+\sum_\beta x_\beta^{i,j}\dfrac{\partial}{\partial x_\beta^{i,j}}\delta(\vec{x})\right]$. For all i-j bonds the total distribution of forces thus reads $F_\alpha^i(\vec{x}-\vec{x}^i)=\sum_j f_\alpha^{i,j}\left[\delta(\vec{x})+\sum_\beta x_\beta^{i,j}\dfrac{\partial}{\partial x_\beta^{i,j}}\delta(\vec{x})\right]$. At mechanical equilibrium the first term of the development vanishes and there is $F_\alpha^i(\vec{x}-\vec{x}^i)=A_{\alpha\beta}\dfrac{\partial}{\partial x_\beta^{i,j}}\delta(\vec{x})$ with the moment of the elastic dipole $A_{\alpha\beta}=-\sum_j\dfrac{\partial\Phi(x^{i,j})}{\partial x^{i,j}}\dfrac{x_\alpha^{i,j}x_\beta^{i,j}}{|\vec{x}^{i,j}|}$

In [26] appendix IV one can find for an ion self adsorbed, but non relaxed on the (001) face of the NaCl-structure type : A_{ii}=0.10 e^2/a±2%, i=1,2 with a the shortest equilibrium distance between opposite ions, e^2=1.5 10^{-7} eVcm per ion so that for a= 2 10^{-8} cm A_{ii}=0.75 eV (i=1,2). More generally, comparing with the cohesion energy of this structure type $W_{coh.}=\dfrac{1}{2}M\dfrac{e^2}{a}\left(1-\dfrac{1}{m}\right)$ with M=1.7486±0.05 % the Madelung number per ion, m the Born repulsion exponent (8<m<12). Therefore $A_{ii}/W_{coh}=\dfrac{0.20}{M}\dfrac{m}{m-1}$, i=1,2 or the narrow estimation valid for all the alkali-halide series, 0.12< $A_{ii}/W_{coh.}$ <0.13 (i=1,2). Notice the peculiarity of the (001) faces $A_{33}/W_{coh.}<10^{-3}$.

More crude estimations have been done with Lennard-Jones (6-12) interactions. [25] gives A=4.5 eV for Xe on (111)Au, [23] gives A=0.23 eV for Ar/(111)Ar. Finally Duport et al. (see [124] appendix A2) scaled A_{11}/W_{coh} =0.17 and $A_{22}/W_{coh.}$ =0.07 for the self adsorbed atom on the border of a hypothetic compact two-dimensional crystal.

APPENDIX J

Using (12), the expression (17) of the elastic interaction of 2 elastic defects (1) and (2) (located at the surface z=0) reads:

$$U = \frac{1}{2}\sum_{\alpha}\sum_{\beta}\iint F_{\alpha}^{(1)}(\vec{x})D_{\alpha\beta}(\vec{x},\vec{x}')F_{\beta}^{(2)}(\vec{x}')dVdV'$$

which when forces have only x components reads

$$U = \frac{1}{2}\sum_{x}\sum_{x}\iint F_{x}^{(1)}(\vec{x})D_{xx}(\vec{x},\vec{x}')F_{x}^{(2)}(\vec{x}')dVdV \qquad (i)$$

with $D_{xx}(\vec{x},\vec{x}') = D_{xx}(x,x') = \frac{1-v^2}{\pi E}\left[\frac{1}{r}+\frac{v}{1-v}\frac{(x-x')^2}{r^3}\right]$ [13]

and $r = \left[(x-x')^2 + (y-y')^2 + (2a)^2\right]^{1/2}$ the in-plane distance in between the two defects. A cut off distance 2a is introduced to avoid local divergences.

For our purpose the step on a stressed body is described by an elastic monopole whose x component (perpendicular to the step and directed towards the lowest terrace) reads $F_{x}^{mo}(\vec{x}) = F\delta(\vec{x})$ with $F = \frac{Ea}{1-v^2}m_o a$ whereas the elastic dipole is described by $F_{x}^{di}(\vec{x}) = A\frac{\partial}{\partial x}\delta(\vec{x})$ where $\delta(\vec{x})$ is the Dirac function. Thus injecting these expressions in (i) one obtains easily by using substitution properties of the Dirac function:

$$U^{mo-mo} = \frac{1}{2}F^2 D_{xx}(x^{(1)},x^{(2)}); U^{mo-di} = \frac{1}{2}FA\frac{\partial}{\partial x^{(1)}}D_{xx}(x,x')\bigg|_{x^{(1)},x^{(2)}}; U^{di-di} = \frac{1}{2}A^2\frac{\partial^2}{\partial x^{(1)}\partial x^{(2)}}D_{xx}(x,x')\bigg|_{x^{(1)},x^{(2)}} \qquad (ii)$$

The interaction in between a lines of defects and a single defect located at a distance x=L of the line is thus obtained by integrating (ii) along the line y with x=L. Then the interaction in between two lines of defects is obtained by multiplying the previous result by the number of defects in the second line. The expressions of table III are the first order development of these expressions for y→∞.

REFERENCES

1. L. Royer, Bull. Soc. Fr. Min. Crist., 51 (1928) 7.
2. S. Jain and W. Hayes, Semic. Sci. Techn., 6, (1991) 547.
3. D.Sander, R. Skomski, A.Enders, C.Schmidthals, D. Reuter and J.Kirschner, J. Phys. D, Appl. Phys. 31 (1998) 663.
4. G.Wulf, Z.Krist. 34 (1901) 449.
5. C. Herring in Gomer and Smith Eds. Structure and Properties of Solid Surface Univ. Chicago, Press 1953.
6. J.W. Gibbs in The collected works of J.W. Gibbs, p.314 (Longmans, Green and co, New York, 1928).
7. R. Shuttleworth, Proc. Roy. Soc. London 163, 644 (1950).
8. A.F. Andreev, Y.A. Kosevitch, JETP 54, (1981) 761.
9. P. Nozières, D.E. Wolf, Z. Phys B 70 (1988) 399, (1988) 507.
10. P. Nozières, D.E. Wolf, Z. Phys B 70 (1988) 507.

11. E.Bauer, Z.Krist 110 (1958) 372.
12. E.Bauer, Z.Krist 110 (1958) 395.
13. L.D. Landau and E.M. Lifshitz, Theory of elasticity (Oxford Pergamon) 1970.
14. A. Love "A treatise on the mathematical theory of elasticity" (New York, Dover publication 1927).
15. J.F. Nye, Physical properties of crystals (Oxford, University press) 1985.
16. A. Angot, "Compléments de mathématiques" (Ed. revue d'optique, Paris) 4° edition, 1961 page 272-282.
17. H.B. Huntington, "the elastic constants of crystals" in Solid state Physics vol. 7 (Academic Press, New York) 1958, 213.
18. R.F. Hearmon, Rev. of Modern physics, 18 (1946) 409.
19. D. Sanders, Rep. Prog. Phys. 62 (1999) 809.
20. W.A. Brandtley, J. Appl. Phys. 44, (1973) 534.
21. R. Kern, in I. Sunagawa (Ed.) The equilibrium form of a crystal in Morphology of crystals Vol A, Terra Tokyo (1987) 79.
22. R.D. Mindlin, Physics, 7, (1936) 195.
23. A. Maradudin, R. Wallis, Surf. Sci. 91 (1980) 423.
24. V.I.Marchenko, A.Y.Parshin, JETP 52 (1980) 129.
25. K.Lau, W.Kohn, Surf. Sci. 65 (1977) 607.
26. R.Kern, M.Krohn, Phys. Stat. Sol. 116 (1989) 23.
27. P. Nozières, Lectures at the Beg Rohu Summer School, Solid far from equilibrium (Cambridge University Press, Ed. C. Godrèche) 1993.
28. J.M. Rickman, D.J. Srolovitz, Surf. Sci. 284 (1993) 211.
29. W. Kappus, Z. Phys B 29 (1978) 239.
30. W. Kappus, Z. Phys B 38 (1980) 263.
31. P. Peyla, A. Vallat, C. Misbah, J. Cryst. Growth 201/202 (1999) 97
32. P. Peyla, A. Vallat, C. Misbah, H. Müller-Krumbhaar, Phys. Rev. Lett. 82 (1999) 787.
33. R.J. Asaro, W.A. Tiller, Metall. Trans. 3 (1972) 1789.
34. M.A. Grienfeld, Dokl. Akad. Nauk. SSSR 283 (1985) 1139.
35. M.A. Grienfeld, J. of Intelligent Material Systems and Structures, 4 (1993) 76.
36. H. Gao, J. Mech. Phys. Solids, 39 (1991) 443.
37. W.H. Yang, D.J. Srolovitz, J. Mech. Phys. Solids, 42 (1994) 1551.
38. L.B. Freund, Int. J. Solids Structures 32 (1995) 911.
39. D.J. Srolovitz Acta Metall. 37 (1989) 621.
40. B.J. Spencer, P.W. Voorhees, S.H. Davis, J. App. Phys. 73 (1993) 4955.
41. B.J. Spencer, D.I. Meiron, Acta Metall. 42 (1994) 3629.
42. M. Thiel, A. Willibald, P. Evers, A. Levchenlo, P. Leiderer, S. Balibar, Europhys Lett. 20 (1992) 707.
43. R.H.Torii, S. Balibar, J. of Low Temperature Physics, ½ (1992) 391.
44. V.Marchenko, JETP 54 (1981) 605.
45. F.C. Frank, A.N. Stroh, Proc. Phys. Soc. B65 (1952) 811.
46. J. Grilhe, Acta Metall.Mater., 37 (1993) 909.
47. C.H. Wu, J.Hsu, C.H.Chen, Acta. Mater 11 (1998) 3755.
48. K. Kassner, C. Misbah, Europhys. Lett. 28 (1994) 245.
49. O. Alerhand, D. Vanderbilt, R. Meade, J. Joannopoulos, Phys. Rev. Lett. 61 (1988) 1973.
50. F. Hen, W. Pachard, M. Webb, Phys. Rev. Lett, 61 (1988) 21
51. B. Croset, C. de Beauvais, Phys. Rev. B, 61 (2000) 3039

52. E. Williams, R. Phaneuf, N. Bartelt, Mat. Res. Soc. Symp. Proc. 238 (1992) 219
53. E. Williams, Surf. Sci. 299/300 (1994) 502
54. E. Bauer, Appl. Surf. Sci. 11/12 (1982) 479.
55. M. Dupré, Théorie mécanique de la chaleur, Paris, (1969)
56. P. Müller, R. Kern, Appl. Surf. Sci. 102 (1996) 6
57. I. Stranski, L. Krastanov, Sitz. Ber. Akad. Wiss Wien (1938) 145
58. R. Kern, P. Müller, J.Cryst. Growth 145 (1995) 193.
59. J. Tersoff, R. Tromp, Phys. Rev. Lett. 70 (1993) 2782.
60. C. Duport, C. Priester, J. Villain in Morphological organisation in epitaxial growth and removal. Ed. Z. Zhang, M. Lagally, World Scientific (1998).
61. S. Christiansen, M. Albrecht, H. Strunk, P. Hansson, E. Bauser Appl. Phys. Lett 66 (1995) 574.
62. E. Pehlke, N. Moll, A. Kley, M. Scheffler Appl. Phys.A 65 (1997) 525.
63. L. Freund, H. Johnson, R. Kukta, MRS bull. 399 (1996) 259.
64. D. Wong, M. Thouless J. Mater. Sci. 32 (1997) 1835.
65. P. Müller, R. Kern, Surf. Sci. 457 (2000) 229.
66. R. Kern, JJ. Métois, G. Lelay in Current Topic in Material Science vol. 3, Kaldis (Ed.) North Holland, Amsterdam, 1979, p. 196.
67. D. Nenow and A. Trayanov, J. Cryst. Growth 79, (1986) 801.
68. J.G. Dash, Cont. Phys. 30, (1989) 801.
69. P. Müller, R. Kern, Appl. Surf. Sci. 102 (1996) 6.
70. P. Müller, O. Thomas, Surf. Sci. Lett., 465 (2000) L 764.
71. F.H. Streitz, R.C. Cammarata and K. Sieradzki, Phys. Rev. B 49, 10699 (1994).
72. C. Duport, "Elasticité et croissance cristalline" PhD University J. Fourier, Grenoble 1996.
73. S.M. Hu, J. Appl. Phys. 50, 4661 (1979).
74. R. Kern, P. Müller, Surf. Sci. 392 (1997) 103.
75. R. Kern, JJ. Métois, G. Lelay in Current Topic in Material Science vol. 3, Kaldis (Ed.) North Holland, Amsterdam, 1979, p. 131-419.
76. A. Thomy, X. Duval, Surf. Sci. 299/300415 (1994) 797.
77. J.W. Schultze, D. Dickerman, Surf. Sci. 54 (1976) 489.
78. G. Gerth, V. Abelman, Cryst. Res. Techn. 24 (1989) 35.
79. P. Müller, R. Kern, A. Ranguis, G. Zerwetz, Eur. Phys. Lett. 26 (1994) 461.
80. J.W. Matthews in Dislocations in Solids, F. Nabarro (Ed.), Vol. 2 (1989) 461. and in Epitaxial Growth vol. A and B, Materials Science Series, (Acad. Press) (1975).
81. J. Tersoff, F.K. LeGoues, Phys. Rev. Lett. 72, 3570 (1994).
82. R. Kaishew, Bull. Acad. Sc. Bulg. (Ser. Phys.) 2 (1951) 191.
83. R. Kaishew, Arbeitstatung Festköper Physik, Dresden (1952) 81.
84. R. Kaishew, Comm. Bulg. Acad. Sci. 1 (1950) 10.
85. R. Kaishew, Fortscht. Miner. 38 (1960).
86. W. Winterbottom, Acat. Metall. 15 (1968) 303.
87. P. Müller, R. Kern, J. Cryst. Growth 193, (1998) 257.
88. P. Müller, R. Kern, Appl. Surf. Sci. 162/163 (2000) 133.
89. P. Müller, R. Kern, Microsc. Microanal. Microstruct. 8 (1997) 229.
90. M. Hammar, F. Legoues, J. Tersoff, M.C. Reuter, R. Tromp, Surf. Sci. 349 (1996) 129.
91. F. Legoues, M. Hammar, M. Reuter, R. Tromp, Surf. Sci. 349 (1996) 249.
92. I.V. Markov, Crystal growth for beginners, World Scientific, Singapore (1995).
93. P. Müller, R. Kern, Appl. Surf. Sci. 164 (2000) 68.

94. J. Floro, E. Chason, R. Twesten, R. Hwang, L. Freund, Phys. Rev. Lett. 79 (1997) 3946.
95. J. W.Cahn, Acta Metall. 28 (1980) 1333.
96. N. Moll, M. Scheffler, E. Pehlke, Phys.Rev. B 58 (1998) 4566.
97. Y. Mo, D. Savage, B. Schwartzentruber, M. Lagally, Phys. Rev. Lett. 65 (1990) 1020.
98. C. Aumann, Y. Mo, M. Lagally, Appl. Phys. Lett. 59 (1991) 1061.
99. D. Steigerwald, I. Jacob,W. Egelhoff, Surf. Sci. 202 (1988) 472.
100. M. Copel, M.C. Reuter, E. Kaxiras, R. Tromp, Phys. Rev. Lett. 63 (1989) 632.
101. D. Eaglesham, F. Unterwald, D. Jacobson, Phys. Rev. Lett 70 (1993) 966.
102. H. Ibach, Surf. Sci. Reports, 29 (1997) 193.
103. H. Ibach, Surf. Sci. Reports, 35 (1999) 71.
104. D. Sander and H. Ibach in Landolt-Börnstein New Series vol. III/42, Physics of covered solid surfaces, H. Bonzel (Ed.).
105. W.K. Burton, N. Cabrera, F.C. Frank Phil. Trans. Roy. Soc. 243 (1951) 299.
106. J. Villain, A.Pimpinelli, "Physique de la croissance cristalline" Aléa Saclay Eyrolles (Ed. C. Godréche) (1995).
107. S. Stoyanov, D. Kashiev, in "Current topics in material Science" vol. 7 Ed. E. Kaldis (North Holland Publishing) 1981, 69.
108. J.J. Harris, B.A. Joyce, Surf. Sci. 103 (1981) L90.
109. Y. Horio, A. Ichimaya, Surf. Sci. 298 (1993) 261.
110. A.R. Miedena, Z. Metallkde, 69 (1978) 287.
111. G. Zinsmeister, Thin Solid Films, 2 (1968) 497.
112. G. Zinsmeister, Thin Solid Films, 7 (1971) 51.
113. J. Venables, Phil. Mag. 27 (1973) 697.
114. F. Binsberger, in Crystal growth, ICGG III Conference, North Holland Pub.(1971) 44.
115. G. Gilmer, P. Bennema, ICGG III Conference, North Holland Pub. (1971) 147.
116. H.J. Leamy, G.H. Gilmer, J. Cryst. Growth 24/25 (1974) 499.
117. C. Ratsch, P. Smilauer, D. Vvedensky, A. Zangwill, J. de Phys. I, 6 (1996) 575.
118. C. Ratsch, A. Zangwill, P. Smilauer, Surf. Sci 314 (1994) L942.
119. C. Ratsch, A. Zangwill, Appl. Phys. Lett. 63 (1993) 2348.
120. R.L. Schwoebel, E.J. Shipsey, J. Appl. Phys. 37 (1966) 3682.
121. G. Ehrlich, Surf. Sci. 63 (1977) 422.
122. P. Bennema and G.H. Gilmer in Crystal Growth Ed. P. Hartman, North Holland, 1973, p. 263-327.
123. C. Duport, P. Nozières, J. Villain, Phys. Rev. Lett. 74 (1995) 134.
124. C. Duport, P. Paoliti, J. Villain, J. de Phys. I, 5 (1995) 1317.
125. J. Tersoff, Phys. Rev. Lett. 75 (1995) 2730.
126. E. Gruber et M. Mullins, J. Phys. Chem. Solids 28 (1967) 875.
127. J.J. Métois, S. Stoyanov, Surf. Sci. 440 (1999) 407.
128. M. Schroeder, D. Wolf, Surf. Sci. 375 (1997) 129.
129. C. Lei, C. Jia, J. Lisoni, M. Siegert, J. Schubert, C. Buchal, K. Urban, J. Cryst. Growth, in press(2000).
130. F. Glas, Phys. Rev. B, 55 (1997) 11277.
131. F. Glas, Appl. Surf. Sci. 123/124 (1998) 298.
132. F. Ross, J. Tersoff, R. Tromp, Phys. Rev. Lett. 80 (1998) 984.
133. F. Glas, J. Appl . Phys. 62 (1987) 3201.
134. F. Glas, J. Appl. Phys. 70 (1991) 3556.
135. I. Patova, V. Malyshkin, V. Schukin, J.Appl. Phys. 74 (1993) 7198.
136. J.E. Guyer, P.M. Voorhes, Phys. Rev. Lett. 74 (1995) 4031.

137. B.J. Spencer, D.I. Meiron, Acta Met. Mater 42 n°11 (1994) 3629.
138. J.E. Guyer, P.M. Voorhes, J. Cryst. Growth 187 (1998) 149.
139. C. Priester, J. Vac. Sci. Technol. B 16 (1998) 2421.
140. J. Tersoff, Phys. Rev. Lett. 81 (1998) 3183.
141. N. Cabrera, Mem. Sci. Rev. Metal. Fr., 62 (1965) 205.
142. J. Floro, M. Sinclair, E. Chason, L. Freund, R. Twesten, R. Hwang, G. Lucadamo, Phys. Rev. Lett. 84 (2000) 701.
143. Q. Xie, A. Madhukar, P. Chen, N. Kobayashi, Phys. Rev. Lett. 75 (1995) 2542.
144. V.A. Shukin, N.N. Ledentsov, P.S. Kop'ev, D. Dimberg, Phys. Rev. Lett., 75 (1995) 2968.
145. J. Tersoff, C. Teichert, M. Lagally, Phys. Rev. Lett. 76 (1996) 1675.
146. L. De Caro, L. Tapfer, Phys. Rev. B 48 (1993) 298.
147. O. Brandt, K. Ploog, R. Bierwolf, M. Hokenstein, Phys. Rev. Lett. 68 (1992) 1339.
148. J.C. Woicik, J.G. Pellegrino, S.H. Southworth, P.S. Shaw, B.A. Karlin, C.E. Bourlin, K.E. Miyano, Phys. Rev. B 52 (1995) 52.
149. D.J. Duntan, J. of Mat. Sci. : Mat. in Electronics, 8 (1997) 33.
150. J.Moison, F.Houzay, F.Barthe, L.Leprince, E.André, O.Vatel, Appl. Phys. Lett. 64 (1994) 19.

Introduction to the atomic structure of surfaces: the theoretical point of view

M.C. Desjonquères[a] and D. Spanjaard[b]

[a]DSM/DRECAM/SPCSI, CEA Saclay
F-91 191 Gif sur Yvette, France

[b]Laboratoire de Physique des Solides
Université Paris Sud, F-91 405 Orsay, France

An introduction to surface atomic structure and vibrations is presented. After a brief recall of two-dimensional Bravais lattices, relaxation and reconstruction phenomena are discussed and interpreted qualitatively, both for clean and adsorbate covered surfaces. Then the principle of surface diffraction, which is the phenomenon at the basis of many experimental techniques for the determination of surface structure, is explained. Finally surface vibrations and their consequences on mean square displacements of surface atoms are studied.

1. INTRODUCTION

In Physics and Chemistry a surface is defined as the frontier between two three-dimensional homogeneous phases with different physical or chemical properties. Strictly speaking, this is always an interface (solid-vapour, solid-atmosphere, liquid-gas...) and the domain of surface physics is thus very large. In the following we will limit ourselves to the surfaces of crystalline solids in contact with ultra high vacuum or with a controlled atmosphere.

Then a surface corresponds to a discontinuity in the periodicity of the crystal in the direction perpendicular to it. This discontinuity may lead to atomic rearrangements. This phenomenon will be the subject of Sect.2 in which we will describe these rearrangements and show qualitatively how they are related to electronic properties. Indeed, atomic structure and electronic structure have a mutual influence and must be determined self-consistently. Then we will explain the theoretical principles of surface diffraction which is at the basis of many techniques currently used to study the atomic structure of surfaces. We will end this section by a short discussion on the influence of temperature: appearance of surface defects and vibrations. The perturbation of vibrational properties due to the surface will be discussed in more details in Sect.3. For a more advanced study the reader is referred to the books quoted in [1]

2. ATOMIC STRUCTURE OF SURFACES

2.1. The five two-dimensional Bravais lattices

The first idea that one can have of the surface of a crystal is the following: let us consider two adjacent crystallographic planes P and P' and imagine that we are able to break all the bonds between these two planes and separate them to obtain two semi-infinite crystals. If we assume that in each of them the relative positions of the atoms are unchanged, the atomic structure of the surface will be described by a lattice with a two-dimensional (2D) translational symmetry which is completely determined by the crystallographic orientation of P and P', i.e., by their Miller indices. Let us recall that each site in the 2D lattice is obtained from the site taken as the origin by a translation vector:

$$\mathbf{T}_{//} = m\mathbf{a} + n\mathbf{b} \qquad (1)$$

where m and n are integers. The vectors **a** and **b**, called elementary translations, define the unit cell of the 2D lattice.

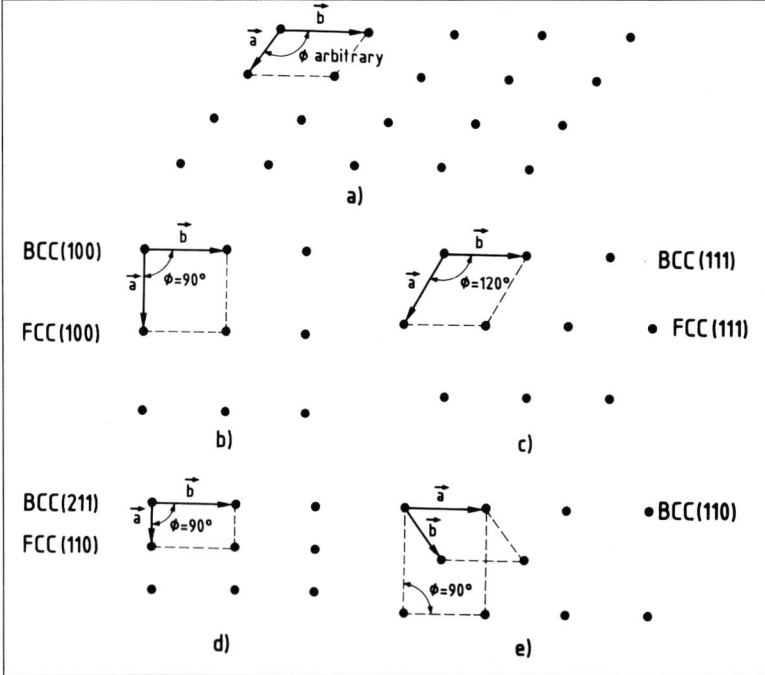

Figure 1. The five two-dimensional Bravais lattices. Examples of lattices found in low index crystallographic planes of cubic crystals are indicated (FCC: Face Centered Cubic crystal, BCC: Body Centered Cubic Crystal).

It is possible to show that there are only five possible 2D lattices: these five Bravais lattices are drawn in Fig.1 with their unit cell. The atomic structure of the surface is then completely determined by the set of atoms (or molecules) and their coordinates, called the basis, attached to each lattice site. Depending on this basis, the symmetry of the surface may be reduced or not with respect to that of the Bravais lattice. In the following, save for a few exceptions that we will point out, we will limit ourselves to systems with a single atom per unit cell. The symmetry of the surface will then be that of the Bravais lattice.

2.2. Relaxation and reconstruction of clean surfaces

Infortunately, even though the picture of the surface described in the preceding section is a useful starting point to understand the structure of monocrystalline surfaces, most often it is not sufficient. Actually, since some bonds have been broken in order to create the two surfaces, the sum of forces acting on atoms no longer vanishes. Consequently, at least in the vicinity of the surface, the atoms undergo atomic displacements to reach a new equilibrium position. Two cases are possible:

- the 2D translational symmetry is not modified by the atomic displacements. This phenomenon is known as *relaxation*: each atomic plane is rigidly displaced with respect to its bulk position. This relaxation is said to be *normal* (Fig.2) or *parallel* depending on the direction of the displacement (normal or parallel to the surface). Of course the displacement can have both normal and parallel components.

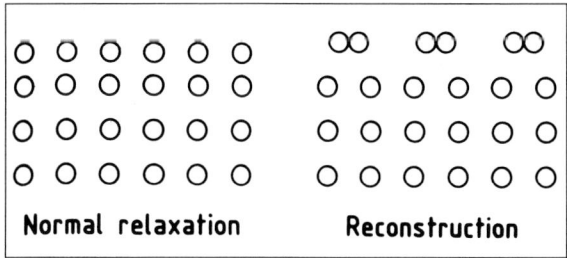

Figure 2. Examples of a section normal to the surface of a crystal showing normal relaxation and reconstruction.

- the 2D translational symmetry is modified by atomic displacements. Consequently the 2D lattice has a new unit cell: the phenomenon is then called a *reconstruction* (Fig. 2) and it is said that the surface presents a *surstructure*. It is then necessary to introduce new notations to define this surstructure. These notations give the size, the nature and the orientation of the new unit cell with respect to that, defined by the elementary translations **a** and **b**, of the inner layers.

When the elementary translation vectors of the unit cell of the surstructure are $\mathbf{a}' = n\mathbf{a}$, $\mathbf{b}' = m\mathbf{b}$ (n, m positive integers), the surstructure is denoted as $(n \times m)$. If this surstructure has 2 atoms per unit cell, one at the origin, the other at the center, it is denoted as $c(n \times m)$ (Fig. 3).

When the new unit cell is rotated by an angle θ relative to that of the underlying lattice with $|\mathbf{a}'| = \sqrt{n}|\mathbf{a}|$ and $|\mathbf{b}'| = \sqrt{m}|\mathbf{b}|$, the structure is denoted as $(\sqrt{n} \times \sqrt{m})R\theta$ (Fig.3).

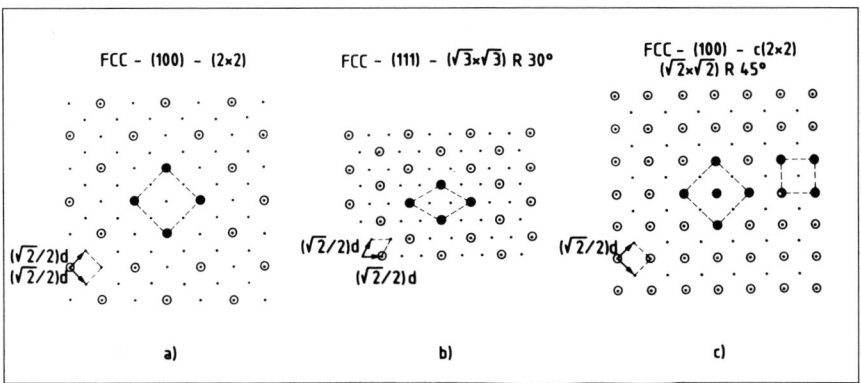

Figure 3. Schematic diagrams of surface surstructures (circles), the structure of the underlayers is given by points. The unit cells of both lattices are indicated, d is the bulk parameter.

Let us note that the same surstructure may sometimes be indicated with different notations (see Fig.3c) and that the notations specifying the surstructure do not give the atomic positions with respect to the underlying solid: it only gives the translational periodicity. It should be noticed that the same notations are also used to describe the structure of ordered overlayers of adsorbates on the surface.

Even though the physical origin of surface relaxation and reconstruction is quite clear, the atomic displacements are rather difficult to predict. Indeed, for this purpose, a precise description of the forces acting on atoms is needed, and this can only be achieved by total energy calculations based on the study of electronic structure. We will now illustrate this point qualitatively.

Let us start with the case of *normal relaxation*. According to the most simple energetic model, the total energy is written as the sum of pair interactions, depending only on the interatomic distance, with a single minimum. If the range of forces is limited to first nearest neighbours, the relaxation must vanish since, at equilibrium, each neighbour exerts a zero force (Fig.4a). This is not the case when the range of forces is extended to second nearest neighbours. The interactions between first nearest neighbours are necessarily repulsive to compensate the attraction due to second nearest neighbours (Fig.4b). Since the ratio of the numbers of each type of neighbours is usually different at the surface and in the bulk, a normal relaxation is expected. However, save for some peculiar cases in which the relaxation vanishes ((100) surface of a simple cubic lattice, for instance), this type of model usually leads to an expansion of the first interlayer spacing [2].

Let us take, as an example, the case of the (110) surface of a simple cubic lattice. We

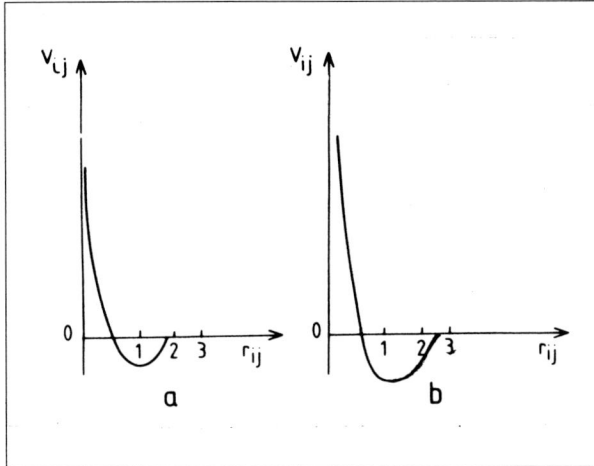

Figure 4. Two types of central force pair interactions.

call $v(r)$ the central force pair potential and a the lattice parameter. In the bulk there are 6 first nearest neighbours at the distance a and 12 second nearest neighbours at the distance $a\sqrt{2}$, the energy per atom is then written as:

$$E = 3v(a) + 6v(a\sqrt{2}) \tag{2}$$

The bulk equilibrium condition is thus:

$$3v'(a) + 6\sqrt{2}v'(a\sqrt{2}) = 0 \tag{3}$$

or:

$$v'(a\sqrt{2}) = -\frac{v'(a)}{2\sqrt{2}} \tag{4}$$

with $v'(a) < 0$ and $v'(a\sqrt{2}) > 0$.

Let us now consider the (110) surface and assume that it has undergone a normal relaxation, limited to the first interlayer spacing, that we will denote as h, the bulk interlayer spacing being $a\sqrt{2}/2$. It is easily seen that the only pair interactions that are modified are 2 first nearest neighbour and 4 second nearest neighbour interactions between the first (surface) and the second layer, and 1 second nearest neighbour interaction between the first and third layer. The total energy of the semi-infinite crystal as a function of h is then written as:

$$E_{tot}(h) = 2v(\sqrt{h^2 + \frac{a^2}{2}}) + 4v(\sqrt{h^2 + \frac{3a^2}{2}}) + v(\frac{a\sqrt{2}}{2} + h) + C^{ste} \tag{5}$$

The force acting on the first layer when $h = a\sqrt{2}/2$ is:

$$F = -\frac{dE_{tot}}{dh}(h = a\sqrt{2}/2) \qquad (6)$$

and taking (4) into account, we obtain:

$$F = -\sqrt{2}v'(a)/4 > 0 \qquad (7)$$

i.e., an expansion of the first interlayer spacing. Unfortunately, in many metals for instance, this is in contradiction with experiments which, except for a few cases (such as Pd(100) and some closed packed surfaces of simple metals), show always a contraction [3] which increases when the atomic density of the surface layer decreases (Fig.5) [4]. Thus the pair interaction model is too simple to describe this phenomenon and it is necessary to rely on electronic structure calculations.

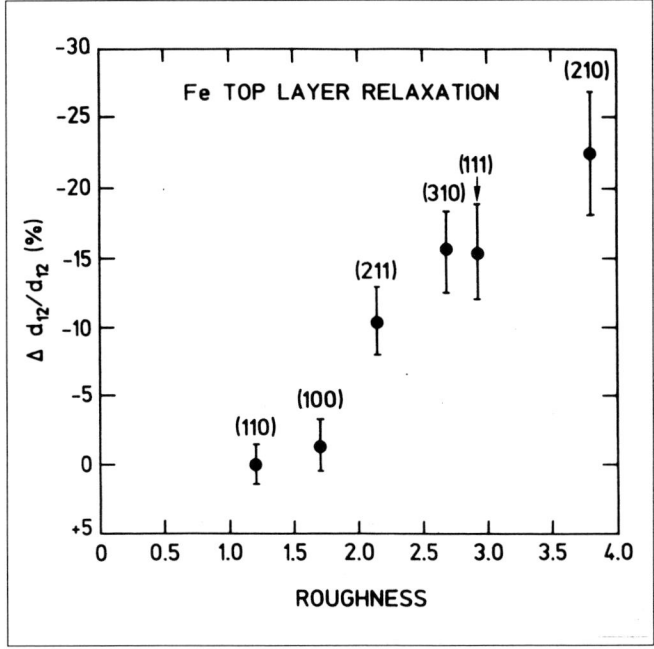

Figure 5. Relaxation of the first interlayer spacing for various orientations of Fe surfaces. The horizontal scale (roughness) is inversely proportional to the atomic density of the surface layer [4].

The first idea which has been put forward to explain relaxation is the formation of a dipole layer at the surface. Two effects can give rise to a dipole layer. On the one hand,

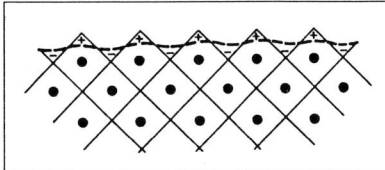

Figure 6. Formation of a dipole layer at the surface due to the smoothing of the modulation of the charge density.

there is a smoothing of the modulation of the electron density at the surface (Fig.6) which tends to produce a contraction. On the other hand, since the potential which maintains the electrons inside the crystal is finite, the electron density extends slightly into vacuum [5] (pressure effect of s electrons) dragging the ion cores outwards. Thus the two effects are opposite and, since the first one is certainly rather weak on a closed packed surface but stronger on an open surface, they may explain the experimental results.

More recently, two interpretations have been proposed applying to simple metals and transition metals, respectively:

- for simple metals, a good approximation (called Effective Medium Theory) [6] is to consider that each atom is embedded in a jellium, the density $\bar{n}$ of which is an average of the tails of the electronic densities due to its neighbours. The cohesive energy is then equal to the embedding energy of the atoms into this jellium, corrected by an electrostatic term due to the absence of positive jellium in the volume occupied by each atom. The cohesive energy goes through a minimum when $\bar{n} = n_0$ which fixes the equilibrium interatomic distance. Let us consider now the atoms belonging to an unrelaxed surface. These atoms have lost some neighbours and are embedded in an average electronic density smaller than n_0. Consequently, they tend to come closer to the second layer in order to recover the optimum density n_0. This explains the tendency to a contraction of the first interlayer spacing.

- for transition metals and in the tight-binding approximation, the energy of an atom is written as the sum of a repulsive pair interaction, thus proportional to the coordination number Z, and of an attractive term due to the broadening of the valence d level into a partially filled band. This last term, roughly proportional to the mean broadening, varies as $\sqrt{Z}$. It is easily seen that due to the lowered coordination of surface atoms the attractive force decreases less than the attractive one and a contraction occurs [7].

The case of *surface reconstruction* is more complicated since surfaces with the same crystalline structure may behave differently. For instance:

- W(100) is reconstructed at low temperature and exhibits a $c(2 \times 2)$ structure corresponding to the formation of zig-zag chains on the surface (Fig.7). On the opposite, the (100) surface of Ta, which has also the BCC structure in the bulk, does not reconstruct. This different behaviour has been explained in the following way [8]. In the absence of reconstruction a band of states localized at the surface exists giving rise to a sharp central peak in the local density of states of surface atoms (Fig.7). The formation of zig-zag chains lowers and broadens this peak. In W the d band is approximately half filled and, consequently, this peak is partially occupied and the gain in band energy is large enough to overcome the variation of repulsive energy. On the opposite in Ta, which has one electron less than W, the central peak is almost empty, thus Ta(100) does not reconstruct since the gain in band energy is too small.

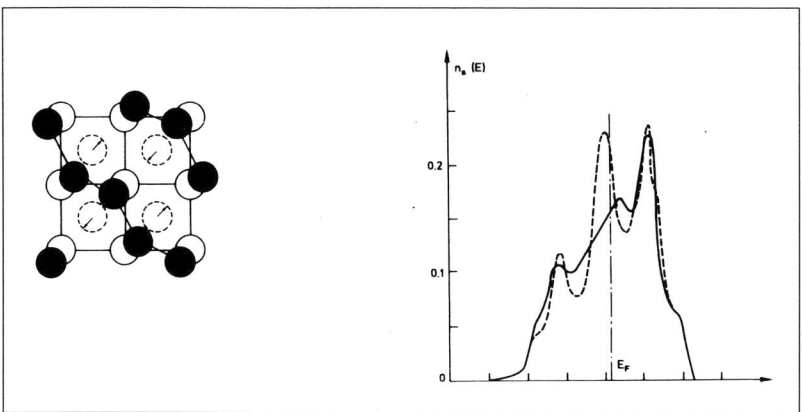

Figure 7. Reconstruction of W(100) at low temperature (left hand side) and surface local density of states (right hand side): unreconstructed surface (dashed line), reconstructed surface (full line).

- The three noble metals (Cu, Ag, Au) have all the FCC structure and a filled d band. Nevertheless, Au(110) exhibits a missing row reconstruction (Fig.8) whereas Cu(110) and Ag(110) do not reconstruct. This has been explained in the framework of the tight-binding approximation, by different laws of variation with distance of the dd hopping integrals [9].

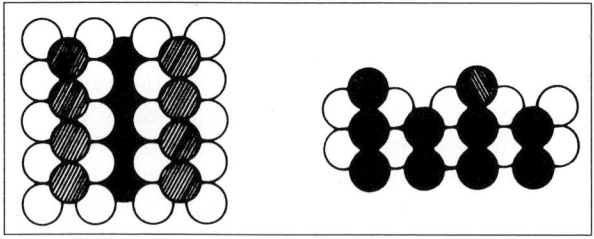

Figure 8. Missing row reconstruction of (110) FCC surfaces.

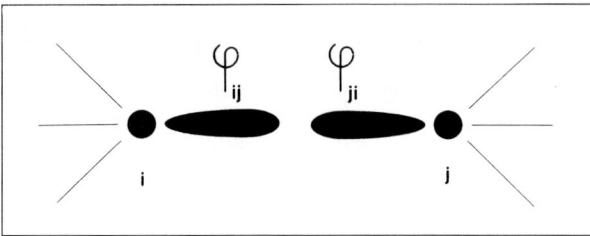

Figure 9. The molecular model.

The surfaces of semiconductors often show very complex reconstructions due to the covalent and highly directional character of their bonds. The mechanism of reconstruction is here mainly due to the fact that the dangling bonds try, by means of a distorsion, to saturate themselves at least partially. Indeed it is rather easily seen that a dangling bond costs a lot of energy.

Let us consider the simple case of homopolar semiconductors Si and Ge. In the free state, their atomic configuration is ns^2np^2 and, in the bulk, their band structure can be described in the tight-binding approximation by using s and p atomic orbitals, with respective levels E_s and E_p, as a basis set. Both elements crystallize in the diamond cubic structure and have directional bonds. Thus the use of sp^3 hybrids on each atom, i.e., of the four linear combinations of s and p orbitals pointing towards the four apices of the tetrahedron formed by its nearest neighbours, is more appropriate for qualitative reasonings. Indeed in a pair of nearest neighbour atoms, the hopping integral β ($\beta < 0$) between the hybrids pointing towards each other (Fig.9) is by far the largest and as a first approximation, it is possible to neglect the others. Moreover $|\beta| >> |E_s - E_p|$, thus E_s and E_p can be replaced by their average value $E_0 = (E_s + 3E_p)/4$. This very rough model is called *molecular model* since, in these conditions, each pair of hybrids pointing towards each other does not interact with the other pairs [1]. As a result the electronic structure is simply given by a filled bonding level at $E_0 + \beta$ and an empty antibonding level at $E_0 - \beta$. These two levels give rise to the valence band and conduction band, respectively, when all interactions are taken into account. At the surface some sp^3 hybrids are unpaired (dangling bonds) and they remain a solution of the Schrödinger equation at the energy E_0. Then a dangling bond costs the energy $|\beta|$ which is of the order of 1eV. This is the reason why the system distorts to try to saturate at least some of these dangling bonds. A typical example is the (2×1) reconstruction of Si(100) in which one of the 2 dangling bonds per surface atom is suppressed by the formation of dimers (Fig. 10).

As a conclusion, the study of surface reconstruction is even more complex than that of normal relaxation. Finally one must note that the atomic displacements in relaxation or reconstruction are not limited to the surface layer but decrease exponentially, often with oscillations, on the underlying planes as we will see below.

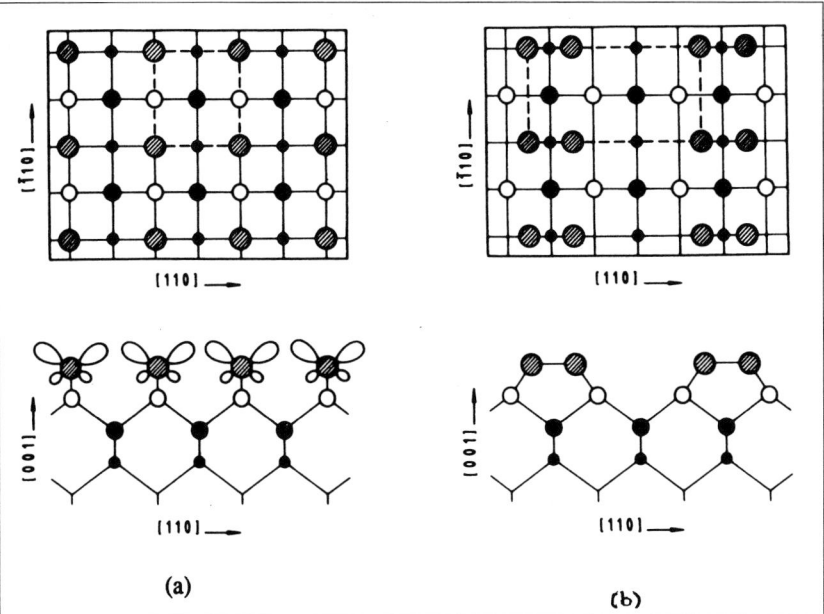

Figure 10. The (100) surface of Si: (a) the ideal surface in which each surface atom has two dangling bonds, (b) the dimerized (2×1) surface.

2.3. Relaxation and reconstruction in the presence of adsorbates

The presence of chemisorbed atoms has an influence on surface relaxation and may completely modify surface reconstruction. Indeed the chemisorption bond involves a large variation of energy which can reach several eV. It is thus expected that the corresponding forces may lead to surface restructuring.

Let us first consider the case of *normal relaxation*. Due to chemisorption the number of bonds of the surface atoms increases and, using the same arguments as those developed for clean surfaces, the surface atoms tend to move outwards. As a consequence the first interlayer spacing which is generally contracted with respect to the bulk interplanar distance for the clean surface tends to recover this last distance and may even overcome it. Thus the adsorption of half a monolayer of hydrogen on Cu(110) cancels the contraction which was about 4% for the clean surface [10]. An expansion (about 6%) of the first interlayer spacing of Ni(110) is observed when a $c(2 \times 2)$ sulphur overlayer is adsorbed [11].

The effect of chemisorbed species on surface reconstruction is much more difficult to predict. In order to understand the possible restructuration, the reconstruction energy of the clean surface must be compared with the variation of the adsorbate chemisorption energy on the ideal unreconstructed surface.

Let us illustrate this point on the case of Cu(110). This surface does not reconstruct when it is clean. Let us consider first the case of the (2 × 1) reconstruction (Fig.8) where one closed packed row of atoms out of two is missing (or added!). Calculations show that the energy needed to trigger such a reconstruction is very small (about 0.02eV per surface atom) [12]; a very small variation of chemisorption energy may thus be sufficient to induce it. This is actually found for small coverages ($\theta < 0.35$) of K on Cu(110), the gain in adsorption energy being due to a larger effective coordinence of the adsorbate on the reconstructed surface. For larger values of θ the reconstruction is destabilized by K-K electrostatic repulsions [12](Fig.11).

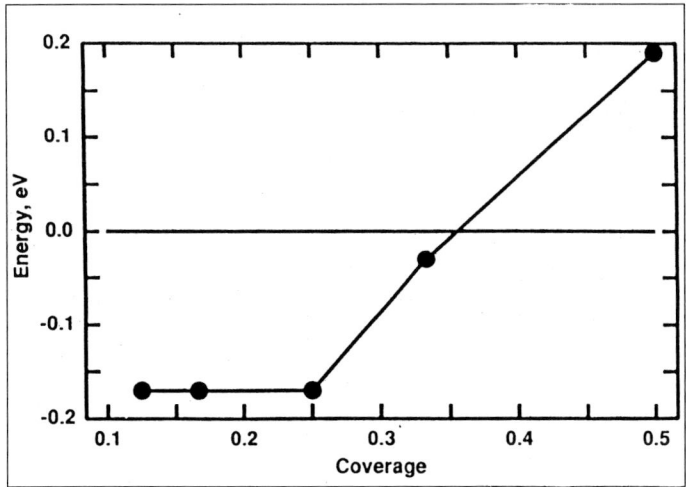

Figure 11. Adsorption of K on Cu(110): the difference between the adsorption energies (≥ 0) on the unreconstructed surface and on the (2 × 1) reconstructed surface.

Consider now the case of the (1 × 2) reconstruction where one non closed packed atomic row out of two is removed from the surface. The energy needed to produce this reconstruction is much higher than in the previous case since a larger number of bonds is broken and it reaches 0.3eV per surface atom. Nevertheless this reconstruction becomes possible in the presence of half a monolayer of oxygen (Fig.12). This reconstruction has been interpreted [12] as due to a strong interaction between the oxygen p states and the metal d states. This interaction leads to the existence of a *surface molecule* with the antibonding state empty on the unreconstructed surface, but partially filled on the ideal one.

Let us note that the presence of chemisorbed species may also modify existing reconstructions. For instance, the presence of hydrogen on W(001) leads to a surface which, even though keeping a $c(2 \times 2)$ symmetry, is no longer characterized by zig-zag chains but by a dimerisation of surface atoms. Finally, chemisorption may also suppress existing reconstructions, the substrate surface recovering its ideal structure. This is observed when H is adsorbed on the (100) and (111) surfaces of Si due to the saturation of the dangling bonds. This occurs on metals as well, for example when adsorbing O on the (110) surfaces of Ir and Pt which, when they are clean, exhibit a (2 × 1) missing row reconstruction.

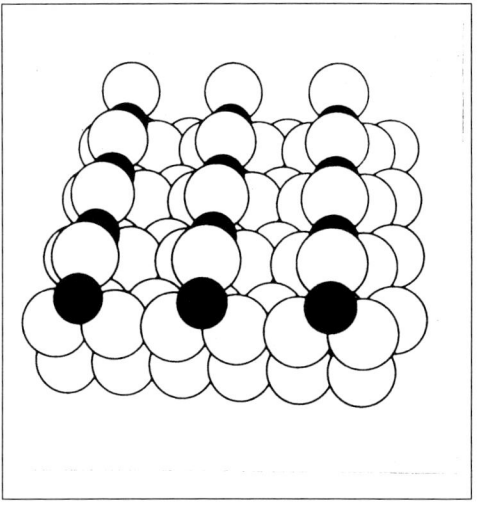

Figure 12. A view of Cu(110)(1 × 2)-O, the oxygen atoms are drawn as small full circles.

2.4. 2D Reciprocal lattice and Brillouin zone

The 2D unit cell (defined by the elementary translations **a, b**) being known, it is useful to determine the 2D reciprocal lattice. Indeed, as in the 3D case, this lattice can be observed directly in diffraction experiments. The 2D reciprocal lattice is given by the set of vectors:

$$\mathbf{G}_{//} = h\mathbf{A} + k\mathbf{B} \tag{8}$$

h, k being integers, and:

$$\begin{aligned} \mathbf{A}.\mathbf{a} &= 2\pi & \mathbf{A}.\mathbf{b} &= 0. \\ \mathbf{B}.\mathbf{a} &= 0. & \mathbf{B}.\mathbf{b} &= 2\pi \end{aligned} \tag{9}$$

The example of the (110) surface of a BCC crystal is shown in Fig.13. The construction of the 2D first Brillouin zone is quite similar to that of the 3D case. We draw the lines connecting a given point to all nearby points in the reciprocal lattice. At the midpoint and normal to these lines we draw new lines. The first Brillouin zone is the smallest polygon enclosed in this way. The example of the first Brillouin zone of the (110) surface of a BCC crystal is given in Fig.14.

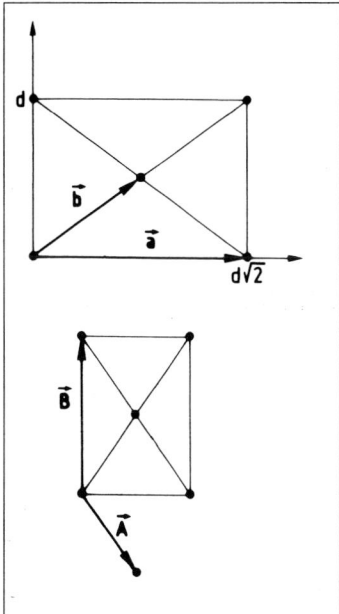
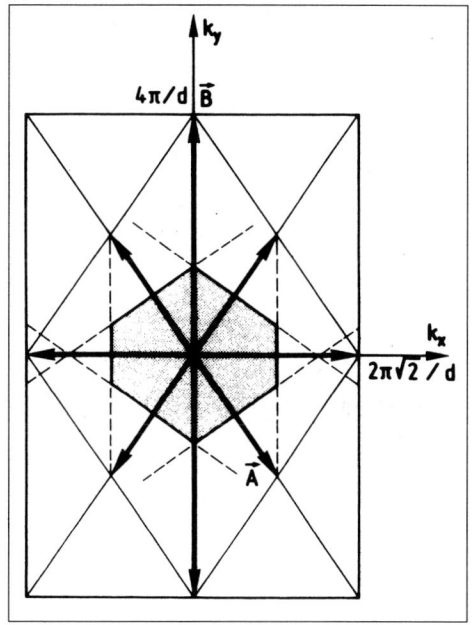

Figure 13. The elementary translation vectors **a,b** (**A,B**) of the direct (reciprocal) lattice of the (110) surface of a BCC crystal.

Figure 14. Construction of the 2D Brillouin zone of the (110) surface of a BCC crystal.

2.5. Diffraction theory
2.5.1. Principles

Let us consider a clean unreconstructed surface, with a single atom per unit cell for simplicity, and an incident particle of mass M and energy E (electron, atom,...) described by a plane wave:

$$\Psi_i = \exp(i\mathbf{k}.\mathbf{r}) \qquad E = \hbar^2 k^2 / 2M \tag{10}$$

In vacuum, far from the surface, the diffracted beam is the superposition of plane waves of the type:

$$\Psi_d = \exp(i\mathbf{k}'.\mathbf{r}) \tag{11}$$

Let us assume that each particle is diffracted only once (single diffusion hypothesis) by a surface atom and consider two waves with wave vector $\mathbf{k}'$ diffracted by two surface atoms related by the translation $\mathbf{T}_{//}$. These two waves interfere constructively when (see Fig.15)

$$k(AH' - A'H) = (\mathbf{k}'_{//} - \mathbf{k}_{//}).\mathbf{T}_{//} = 2n\pi \qquad n = 0, \pm 1, \pm 2... \tag{12}$$

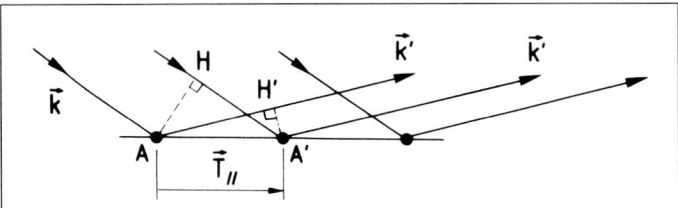

Figure 15. Principle of surface diffraction. The path length difference between the two beams diffracted by A and A' is (AH' -A'H).

Taking equations (8) and (9) into account leads to:

$$\mathbf{k}'_{//} = \mathbf{k}_{//} + \mathbf{G}_{//} \tag{13}$$

The k'_z component is obtained from the conservation of energy:

$$k^2 = k'^2_z + (\mathbf{k}_{//} + \mathbf{G}_{//})^2$$
$$k'_z = -(k^2 - (\mathbf{k}_{//} + \mathbf{G}_{//})^2)^{1/2} sgn(k_z) \tag{14}$$

Each diffracted beam can be denoted by the corresponding 2D reciprocal lattice vector $\mathbf{G}_{//}$. The total wave field is thus formed by the superposition of the incident wave, a finite number of propagating diffracted waves (for which k'_z is real, i.e., such as $(\mathbf{k}_{//} + \mathbf{G}_{//})^2 < k^2$) and evanescent diffracted waves:

$$\Psi_{tot} = \exp(i\mathbf{k}.\mathbf{r}) + \sum_{\mathbf{G}_{//}} a_{\mathbf{G}_{//}} \exp(i(\mathbf{k}_{//} + \mathbf{G}_{//}).\mathbf{r}_{//}) \exp(ik'_z z) \tag{15}$$

This result can be generalized to lattices with several atoms per unit cell or when multiple diffusion is taken into account since this is a direct consequence of the Bloch theorem [13]. The total wave field must satisfy the Schrödinger equation of the particle in interaction with the surface:

$$((-\hbar^2/2M)\Delta + V(\mathbf{r}_{//}, z))\Psi_{tot} = E\Psi_{tot} \tag{16}$$

where $V(\mathbf{r}_{//}, z))$ is the potential energy which is invariant for a translation $\mathbf{T}_{//}$. According to Bloch theorem we have:

$$\Psi_{tot} = \exp(i\mathbf{k}_{//}.\mathbf{r}_{//})v_{\mathbf{k}_{//}}(\mathbf{r}_{//}, z) \tag{17}$$

with $v_{\mathbf{k}_{//}}(\mathbf{r}_{//} + \mathbf{T}_{//}, z) = v_{\mathbf{k}_{//}}(\mathbf{r}_{//}, z)$.
As a consequence $\Psi_s = \Psi_{tot} - \Psi_i$ is also a 2D Bloch wave:

$$\Psi_s = \exp(i\mathbf{k}_{//}.\mathbf{r}_{//})u_{\mathbf{k}_{//}}(\mathbf{r}_{//}, z) \tag{18}$$

and $u_{\mathbf{k}_{//}}(\mathbf{r}_{//}, z)$ can be decomposed into a Fourier series since it has the periodicity of the 2D lattice:

$$\Psi_s = \sum_{\mathbf{G}_{//}} \alpha_{\mathbf{G}_{//}}(z) \exp(i(\mathbf{k}_{//} + \mathbf{G}_{//}).\mathbf{r}_{//}) \qquad (19)$$

We want to know the functions $\alpha_{\mathbf{G}_{//}}(z)$ far in vacuum where the potential vanishes. Substituting $\Psi_i + \Psi_s$ for the total wave field into the Schrödinger equation we obtain:

$$\alpha_{\mathbf{G}_{//}}(z) = a_{\mathbf{G}_{//}} \exp(ik'_z z) \qquad (20)$$

where k'_z is given by (14). Thus we recover equation (15).

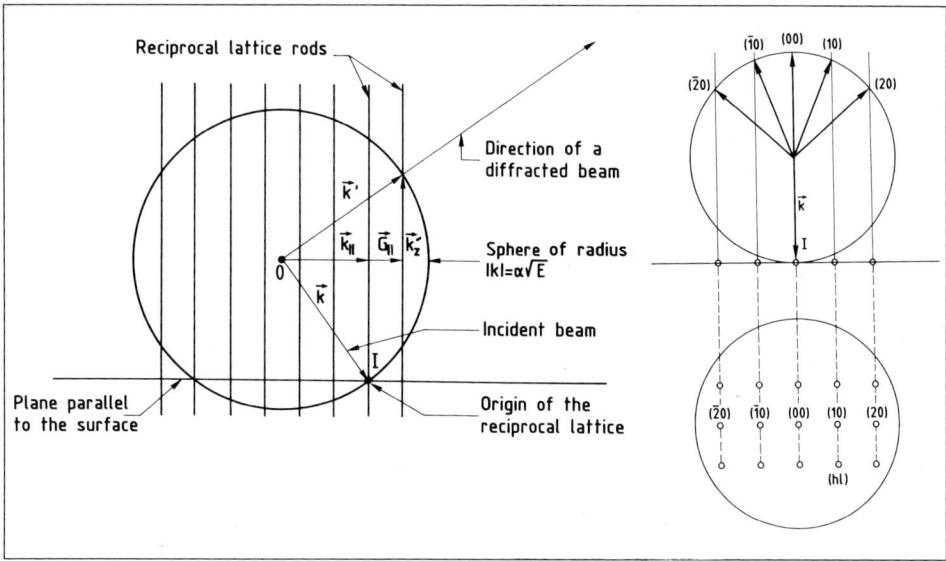

Figure 16. The Ewald construction for surface diffraction (left hand side) and the labelling of spots at normal incidence (right hand side).

Far from the crystal the total wave field gives rise to pin-point diffraction spots in the limit of an infinite 2D lattice. The diffraction pattern is obtained from the Ewald construction (Fig.16). Let I be a point of the 2D reciprocal lattice chosen as the origin and O the point in the surface reciprocal space such that **OI** = **k**. All the vectors connecting the point O to any point on the sphere of center O and radius OI satisfy the energy conservation rule (14). The points of intersection of this sphere with the lines perpendicular to the surface plane and passing through a surface reciprocal lattice point, called the Bragg or diffraction rods, determine the possible **k**′ which satisfy both (13) and

(14). At normal incidence the obtained pattern is just an image of the surface reciprocal lattice (Fig.16).

When the 2D lattice has a single atom per unit cell, the distance between diffracted spots and the symmetry of the pattern give the atomic structure of the surface layer but no information on its position relative to the underlayers. When atoms are added in the unit cell (without changing it), the diffraction pattern is unchanged but the intensity of spots is modified. Consequently the knowledge of the diffraction pattern gives information neither on the number of atoms in the unit cell nor on their relative position. In order to derive this information, an analysis of the intensity of the diffracted spots should be carried out, i.e., the coefficients $a_{\mathbf{G}_{//}}$ must be computed [13].

2.5.2. Diffraction by a surstructure

When the surface presents a surstructure there are two 2D lattices: the lattice of the surstructure, S, and the lattice of the underlayers, B. Let us consider the lattice with the largest unit cell (usually S). If we limit ourselves to commensurate structures two situations can occur [13,14]:

- *All* points of this lattice can be simultaneously put into coincidence with points of the finer lattice by a rigid translation parallel to the surface. The lattices S and B are said to be *simply related*.

- If S and B are not simply related, there is a third lattice L such that, by a rigid translation, *all* points of this lattice can be simultaneously put into coincidence with points in *both* S and B lattices. The lattices S and B are said to be *rationally related*.

In both cases the translation symmetry of the system is given by the coincidence lattice: S in the first case, L in the second. The diffraction pattern at normal incidence gives the reciprocal lattice of the coincidence lattice. Let us illustrate this with two examples:

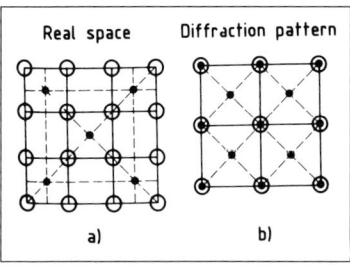

Figure 17. Square lattice with a c(2 × 2) surstructure: (a) real lattice, underlayer (open circles), surstructure (full circles); (b) diffraction pattern (full circles) compared with that of the underlayer (open circles).

- Consider a surstructure $c(2 \times 2)$ on a square lattice B (Fig.17). It is easily seen that S and B are simply related, S is the coincidence lattice. The diffraction pattern, compared to that of B alone, shows additional spots with half-integer indices (Fig.17).

- For a surstructure $c(2 \times 3)$ on a square lattice, S and B are rationally related. The coincidence lattice L has a (2×3) unit cell and the diffraction pattern is a rectangular lattice (Fig.18a,b).

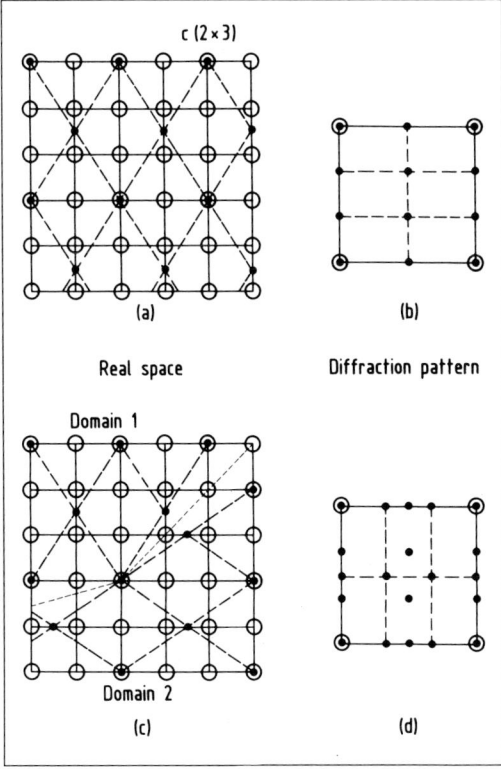

Figure 18. Same caption as Fig.17 but for a $c(2 \times 3)$ surstructure: (a), (b) for a single domain; (c), (d) for the two possible types of domains.

Finally, the orientation of the surstructure may vary by a symmetry operation of B from one region of the surface to another (domains). In the example given in Fig.17 this question does not arise since both S and B lattices have a fourfold symmetry. This is not the case in the second example since S has only twofold symmetry. Thus, two domains (at 90°) are possible (Fig.18c) and, if the incident beam interacts with several domains, we must superimpose the diffraction pattern of Fig.18b with the same figure but after a 90° rotation. The resulting pattern is given in Fig.18d.

2.6. Influence of temperature

Up to now we have only considered perfect surfaces, i.e., with full 2D translational periodicity. However, at finite temperature, due to entropic effects, such a surface does not exist. Thus, on true surfaces, defects are always present (Fig.19).

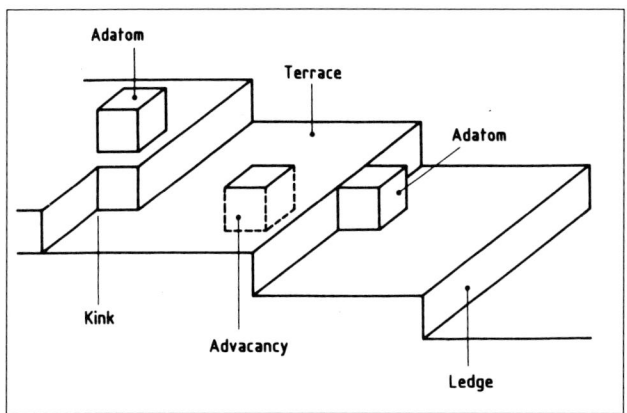

Figure 19. Schematic picture of a real surface.

There are point defects such as adatoms (of the same chemical species as the substrate or impurities), kinks, advacancies and linear defects like steps between two low index terraces. Steps are very important. They are present in the formation of vicinal surfaces (high index surfaces) which are slightly misoriented (misorientation $\leq 10°$) with respect to a low index surface. These surfaces present a periodic array of low index terraces and steps (Fig.20).

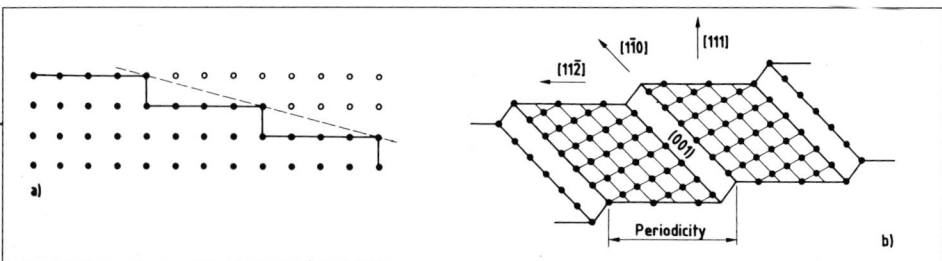

Figure 20. Vicinal surface: section (left hand side); structure of the (557) or [6(111) × (001)] FCC surface (right hand side).

The number of defects is of course a function of temperature. Indeed the formation of a defect produces an increase of the internal energy ($\Delta E > 0$) and of the entropy ($\Delta S > 0$). As a consequence the variation of free energy $\Delta F = \Delta E - T\Delta S$ associated with the creation of a defect decreases with temperature and this may completely modify the morphology of the surface. Consider for instance the step formation energy, it is possible that at some temperature T_R, ΔS vanishes. Then if T_R is below the melting point, steps proliferate on the surface at $T > T_R$ and the surface becomes rough at the atomic scale. This transition is called the roughening transition [15,16].

Finally, up to now, we have assumed that the atoms have fixed positions in the lattice. However, even at zero temperature, they present thermal vibrations which are modified by the surface. This is the subject of the next section.

3. Vibrations at surfaces

Let us now study the influence of the surface on the variational properties of the crystal. After a brief recall of the principles of the calculation of phonon spectra in bulk crystals, we first show that the presence of the surface leads to the existence of a new type of vibration eigenmodes: the surface modes. Then we explain that interesting information on surface relaxation and reconstruction can be drawn from the equations of motion. Finally we define the local spectral densities of modes which are convenient to calculate the mean square displacements of surface atoms. These last quantities are indeed important since they determine the Debye-Waller factors which describe, for instance, the influence of temperature on the intensity of diffracted beams.

3.1. Vibrations of bulk crystals

The vibrations of crystals are most often studied in the *harmonic approximation* which we will first summarize. Let us consider a set of N atoms i, the equilibrium position of each atom i, of mass M_i, being defined by the vector $\mathbf{R}_i$. Let us assume that at time t these atoms are slightly displaced at $\mathbf{R'}_i = \mathbf{R}_i + \mathbf{u}_i(t)$ and that the potential energy of the system is completely determined by $\mathbf{R'}_i$. The development of the potential energy into a Taylor series up to second order with respect to displacements $\mathbf{u}_i(t)$ gives:

$$V(\mathbf{R'}_1, ..., \mathbf{R'}_N) = V(\mathbf{R}_1, ..., \mathbf{R}_N) + \frac{1}{2} \sum_{i\alpha, j\beta} C_{i\alpha,j\beta} u_{i\alpha}(t) u_{j\beta}(t) \tag{21}$$

The quantities:

$$C_{i\alpha,j\beta} = \frac{\partial^2 V}{\partial R'_{i\alpha} \partial R'_{j\beta}}(\mathbf{R}_1, ..., \mathbf{R}_N) \tag{22}$$

are called *force constants* and the greek indices label the three components of the vectors. Note that there is no first order term since the development is carried out around the equilibrium positions. The solutions of the equations of motion are written:

$$\mathbf{u}_i(t) = \mathbf{u}_i(0) \exp(i\omega t) \tag{23}$$

The displacements $\mathbf{u}_i(0)$ will be denoted in the following by their components $u_{i\alpha}$. They are determined by the following ($3N \times 3N$) linear homogeneous system:

$$M_i \omega^2 u_{i\alpha} = \sum_{j\beta} C_{i\alpha,j\beta} u_{j\beta} \tag{24}$$

This system can be written as an eigenvalue equation:

$$(D - \omega^2 I)U = 0 \tag{25}$$

I is the unit matrix and D is the dynamical matrix with matrix elements:

$$D_{i\alpha,j\beta} = C_{i\alpha,j\beta}/(M_i M_j)^{1/2} \tag{26}$$

and

$$U_{i\alpha} = (M_i)^{1/2} u_{i\alpha} \tag{27}$$

The eigenfrequencies are the solution of the equation:

$$Det|D - \omega^2 I| = 0 \tag{28}$$

and the eigenvectors give the polarization of the modes.

In a bulk crystal with a single atom per unit cell, there is a full 3D periodicity, thus the Bloch theorem applies:

$$u_{i\alpha} = u_\alpha \exp(i\mathbf{k}.\mathbf{R}_i) \tag{29}$$

The wave vector $\mathbf{k}$ is real, since the displacement cannot diverge in any direction, and belongs to the first 3D Brillouin zone; u_α is the displacement of the atom located at the origin of coordinates. The solutions are thus plane waves propagating in the lattice with the wave vector $\mathbf{k}$. Using (29) the dynamical matrix D is reduced to a (3×3) matrix $D(\mathbf{k})$ which can be easily diagonalized to obtain the dispersion curves $\omega(\mathbf{k})$ from which it is possible to compute the density of modes $n(\omega)$ ($n(\omega)d\omega$ gives the number of eigenmodes with a frequency between ω and $\omega + d\omega$). Examples are given in Fig.21 [17].

Such a calculation needs the knowledge of the force constants. They are most often computed from central force pair potentials with a finite range, sometimes corrected by potentials giving rise to angular forces. Recently more elaborate empirical potentials have been introduced or even force constants have been derived from *ab-initio* methods.

3.2. Surface modes

In the presence of a surface, generally, there is still a 2D periodicity determined by the translations $\mathbf{T}_{//}$. The Bloch theorem gives:

$$u_{i\alpha} = u_{p\alpha} \exp(i\mathbf{k}_{//}.\mathbf{T}_{//}) \tag{30}$$

$u_{p\alpha}$ is the displacement of an origin atom in the layer p ($p = 0$: surface plane), $u_{i\alpha}$ is that of an atom i deduced from the origin atom by the translation $\mathbf{T}_{//}$. The knowledge of

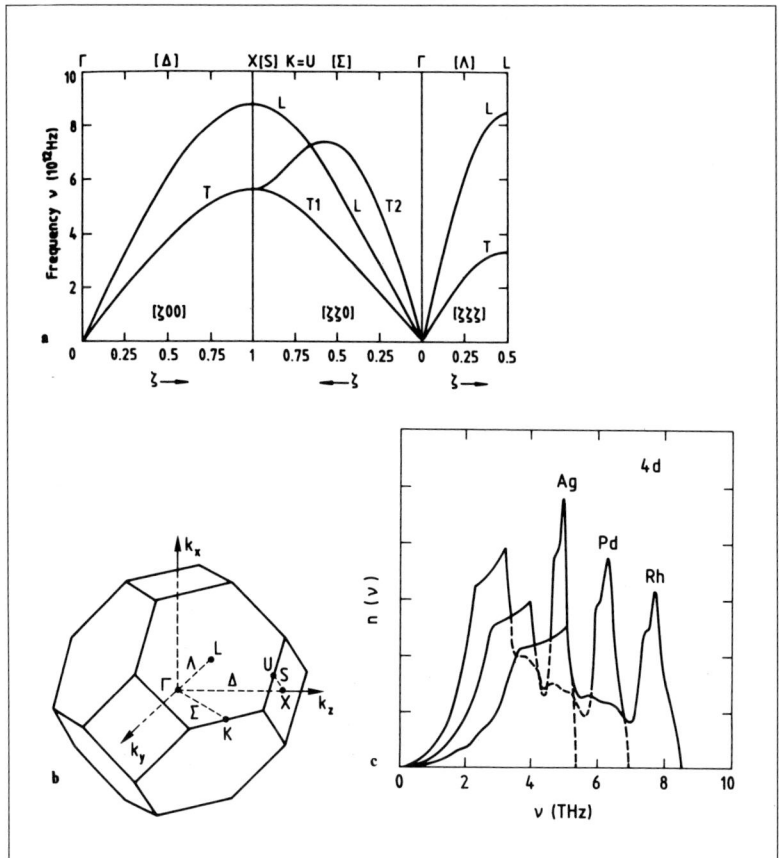

Figure 21. Phonon dispersion curves of a FCC crystal (Rh, upper curves) in the high symmetry directions of the Brillouin zone shown below, L and T label respectively the longitudinal and transverse modes; density of modes of 3 FCC transition metals (lower curves).

all $u_{p\alpha}$ and $\mathbf{k}_{//}$ determines then all displacements when the 2D lattice has a single atom per unit cell.

For each $\mathbf{k}_{//}$, the problem is formally identical to that of semi-infinite linear chain of atoms p with three degrees of freedom. Note that, far inside the crystal, the bulk modes with finite displacements ($\mathbf{k}$ real) must be recovered when p tends to infinity. However modes with a complex perpendicular component $k_\perp$ cannot be excluded if the sign of the imaginary part is such that the wave is damped in the crystal. By using the analogy with the linear chain, we will now show that such surface localized modes exist.

3.2.1. Vibrations of a semi-infinite linear chain

Let us consider a semi-infinite linear chain of atoms of mass M separated by a distance a and interacting with a force constant β between nearest neighbours. Assume that the first atom has a mass $M_0 \neq M$ (Fig.22).

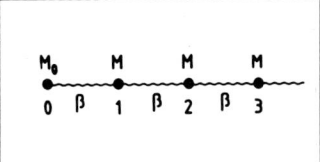

Figure 22. Semi-infinite linear chain.

Let us call u_0 and u_n the displacements of the first and $(n+1)$th atoms. The equations of motion are:

$$-\omega^2 M_0 u_0 = \beta(u_1 - u_0) \tag{31}$$
$$-\omega^2 M u_n = \beta(u_{n+1} + u_{n-1} - 2u_n) \qquad n \geq 1 \tag{32}$$

For an infinite chain all equations are identical to (32) and substituting $u \exp(ikna)$ for u_n, the dispersion relation is readily obtained:

$$\omega = \left(\frac{4\beta}{M}\right)^{1/2} \left|\sin \frac{ka}{2}\right|. \tag{33}$$

Thus the frequency spectrum of the semi-infinite chain has a continuous part extending from $\omega = 0$ to $\omega_{max} = (4\beta/M)^{1/2}$. Let us look now for solutions which are exponentially damped when n tends to infinity, i.e.:

$$u_n = u_0 \exp(-kna) \tag{34}$$

with $Re(k) > 0$. Substituting (34) for u_n into (32) we get:

$$M\omega^2 = 2\beta(1 - \cosh(ka)) \tag{35}$$

and substituting now (35) and (34) into (31) gives:

$$-2\beta \frac{M_0}{M}(1 - \cosh(ka)) = \beta(\exp(-ka) - 1) \tag{36}$$

This last equation is satisfied when:
- either $\exp(ka) = 1$, this solution has no interest since it corresponds to a rigid translation of the chain

- or:

$$\exp(ka) = 1 - \frac{M}{M_0} \tag{37}$$

Consequently when $M < M_0$, $\exp(ka)$ is real, positive and smaller than 1 thus $Re(k) < 0$, which contradicts our assumptions. On the contrary when $M > M_0$, $\exp(ka)$ is real and negative and we can write:

$$\exp(ka) = \exp(k_0 a + i\pi) \tag{38}$$

with:

$$\exp(k_0 a) = \frac{M}{M_0} - 1 \tag{39}$$

$k_0 = Re(k)$ is thus positive when $M > 2M_0$. When this condition is satisfied the solution is:

$$u_n = u_0(-1)^n \exp(-k_0 n a) \tag{40}$$

with an eigenfrequency:

$$\omega_s = (\frac{2\beta}{M})^{1/2}(1 + \cosh(k_0 a)) \tag{41}$$

above the continuous spectrum ($\omega_s > \omega_{max}$).

As a conclusion, when $M_0 < M/2$, a discrete eigenmode exists above the frequency band of the infinite chain. This mode is localized at the "surface" since the corresponding displacements decrease exponentially (with oscillations) when the distance to the surface increases.

3.2.2. Vibrations of a semi-infinite crystal

When (30) is taken into account, the size of the dynamical matrix D is reduced but remains infinite for a semi-infinite crystal. In practice, even though exact methods exist to solve the equations of motion of the semi-infinite system [18], most often a system of N_p atomic layers is used, N_p being chosen large enough to avoid interactions between the two surfaces ($N_p \approx 20$). The reduced matrix $D(\mathbf{k}_{//})$, of size $3N_p$, can be easily diagonalized on a computer. The frequency spectrum $\omega(\mathbf{k}_{//})$ for each $\mathbf{k}_{//}$ can be described in the following way (Fig.23):

- It exhibits the bulk frequency bands allowed for this value of $\mathbf{k}_{//}$ which may be separated by gaps. Indeed far from the surface, the bulk solutions remain valid.
- In the gaps discrete eigenfrequencies are sometimes present. They correspond to solutions of the type:

$$u_{p\alpha} = u_\alpha \exp(-qpd_\perp) \qquad Re(q) > 0 \tag{42}$$

where $d_\perp$ is the interlayer spacing. These solutions are surface modes since they are damped inside the crystal [19].

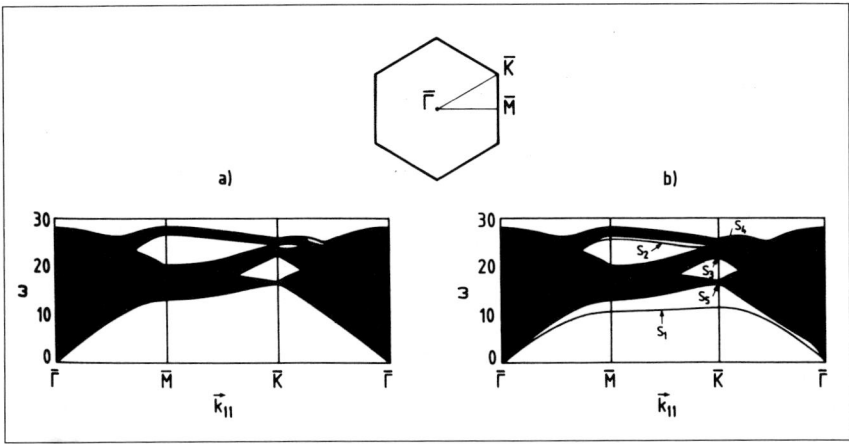

Figure 23. Eigenfrequency spectrum of a FCC(111) slab with 21 layers calculated with a Lennard-Jones potential in the high symmetry directions of the 2D Brillouin zone. The continuous part (shaded) corresponds to bulk modes and there are various surface modes labelled by S [19].

Let us remark that the existence of these surface modes was proved more than a century ago by Lord Rayleigh in the case of continuous media. We recover these Rayleigh modes when the wave length tends to infinity, i.e., $\mathbf{k}_{//} \to 0$ [20].

Note also that the force constants are often assumed unperturbed near the surface. The validity of this assumption can be checked by measuring the dispersion curves $\omega_s(\mathbf{k}_{//})$ of surface modes by means of Electron Energy Loss Spectroscopy (EELS) or Inelastic Diffusion of Atoms (IABS) (He for instance). The result of an EELS experiment on Ni(100) is given in Fig.24 [21]. The calculations show that the force constants between the surface and the first underlayer must be increased by about 25% to reproduce the experimental results. This is consistent with a small contraction of the first interlayer spacing. This type of experiments provides a good means to check the modification of interatomic forces near the surface.

3.3. Connection between phonons and relaxation and reconstruction phenomena

From the study of surface vibrations it is sometimes possible to get information on the possible reconstructions. Indeed, as shown above, the force constants may be different at the surface and in the bulk. If the surface force constants are taken as free parameters it may happen that for some values of these parameters the frequency of a surface phonon vanishes and becomes imaginary for a non vanishing value of $\mathbf{k}_{//}$ (*soft phonon*). The surface lattice becomes unstable and a reconstruction with wave vector $\mathbf{k}_{//}$ occurs, the direction of displacement being given by the corresponding solution of the equations of motion [22]. This type of mechanism has been suggested, for instance, to explain the

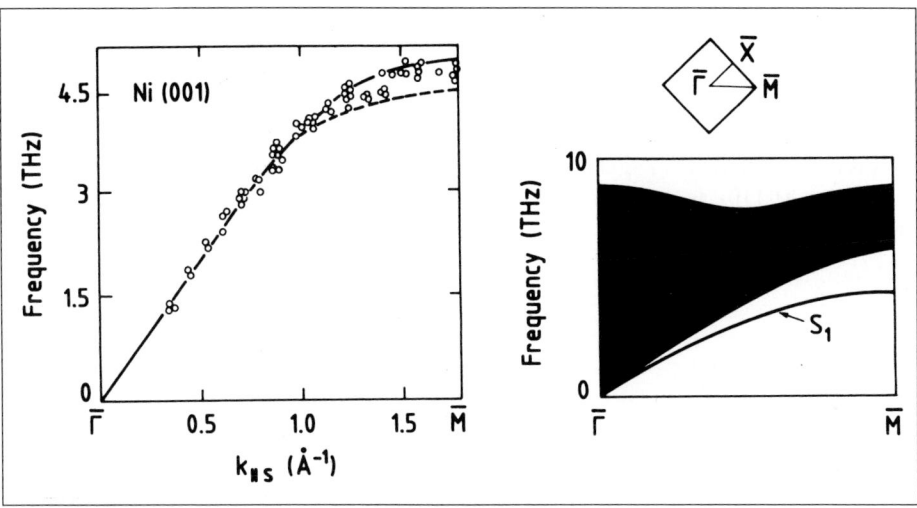

Figure 24. Dispersion curve (left hand side) of the surface phonon S1 (right hand side) of Ni(100) along $\bar{\Gamma}\bar{M}$ measured by electron energy loss spectroscopy (circles) and calculated without (dashed line) and with (full line) a change (25%) of the force constants between the first two layers [21].

reconstruction of W(100) [23].

Moreover, it has been shown that it is possible to derive the asymptotic behaviour of the static atomic displacements in a relaxation or a reconstruction from the bulk equations of motion [24]. Let us consider a semi-infinite crystal, with a given orientation, which is a perfect termination of a bulk solid, i.e., which is neither relaxed nor reconstructed. We have shown above (Sect 2.2) that the atoms of this semi-infinite crystal, located at $\mathbf{R_i}$, are no longer at equilibrium. When the equilibrium can be reached by small displacements the potential energy can be developed around the unrelaxed configuration (NR):

$$V = V_{NR} + \sum_{i\alpha} \Big(\frac{\partial V}{\partial R_{i\alpha}}\Big)_{NR} \delta R_{i\alpha} + \frac{1}{2} \sum_{i\alpha,j\beta} \Big(\frac{\partial^2 V}{\partial R_{i\alpha} \partial R_{j\beta}}\Big)_{NR} \delta R_{i\alpha} \delta R_{j\beta} \qquad (43)$$

At equilibrium V is at a minimum, thus:

$$-\Big(\frac{\partial V}{\partial R_{i\alpha}}\Big)_{NR} = \sum_{j\beta} \Big(\frac{\partial^2 V}{\partial R_{i\alpha} \partial R_{j\beta}}\Big)_{NR} \delta R_{j\beta} \qquad (44)$$

When the range of forces is finite and if atom i is far enough from the surface, (44) simplifies into:

$$0 = \sum_{j\beta} C_{i\alpha,j\beta} \delta R_{j\beta} \qquad (45)$$

which is nothing but the bulk equation of motion for $\omega = 0$. Save for $\mathbf{k} = 0$, the solutions correspond to complex wave vectors. For the relaxation of a clean surface the complex wave vectors must be of the form ($\mathbf{k}_{//} = 0, k_\perp$ complex) since all atoms in the same layer must have the same displacement. In the case of a surface reconstruction, one should have ($\mathbf{k}_{//} \neq 0, k_\perp$ complex). Among these solutions only those decreasing inside the crystal are retained. We show in Fig.25 the asymptotic behaviour of the normal relaxation of Al(110) as a function of the distance from the surface, derived from this method [24].

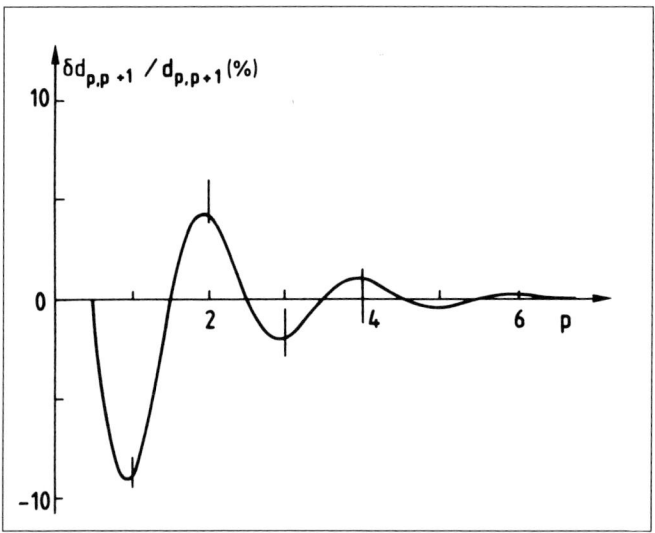

Figure 25. Asymptotic behaviour of the normal relaxation $\delta d_{p,p+1}/d_{p,p+1}$ ($d_{p,p+1}$: distance between the planes p and $p+1$) as a function of the distance to the surface for Al(110).

3.4. Mean square displacements

Let us first consider a single oscillator with frequency ω. It is known that the average internal energy of an oscillator is twice its kinetic energy:

$$< E > = M\omega^2 < u^2 > \tag{46}$$

This average energy at temperature T is given by:

$$< E > = Z^{-1} \sum_n E_n \exp(-E_n/k_B T) \tag{47}$$

where $E_n = (n + \frac{1}{2})\hbar\omega$ and Z is the partition function:

$$Z = \sum_n \exp(-E_n/k_B T) \tag{48}$$

$$= \frac{\exp(-\hbar\omega/2k_B T)}{1 - \exp(-\hbar\omega/k_B T)} = (2\sinh\frac{\hbar\omega}{2k_B T})^{-1} \tag{49}$$

From (47) it is easily shown that:

$$<E> = -\frac{\partial \ln Z}{\partial (1/k_B T)} = \frac{\hbar\omega}{2}\coth\frac{\hbar\omega}{2k_B T} \tag{50}$$

$$<u^2> = \frac{\hbar}{2M\omega}\coth\frac{\hbar\omega}{2k_B T} \tag{51}$$

and, taking into account that the eigenmodes can be considered as a set of independent oscillators, it is shown that, when all atoms are identical, the mean square displacement of atom i in the direction α is given by [25].

$$<u_{i\alpha}^2> = \frac{\hbar}{2M}\int_0^\infty \frac{1}{\omega}\coth\frac{\hbar\omega}{2k_B T} n_{i\alpha}(\omega)d\omega \tag{52}$$

where $n_{i\alpha}(\omega)$ is the local spectral density of atom i in the direction α defined as:

$$n_{i\alpha}(\omega) = \sum_n |e_{i\alpha}(\omega_n)|^2 \delta(\omega - \omega_n) \tag{53}$$

$e_{i\alpha}(\omega_n)$ are the components of the normalized eigenvector of the dynamical matrix corresponding to the eigenfrequency ω_n. Note that in a bulk crystal $n_{i\alpha}(\omega)$ is the same at all sites. Moreover when the crystal has a cubic symmetry $n_x(\omega) = n_y(\omega) = n_z(\omega) = n(\omega)$. This is not the case when the crystal has defects and, for instance in the case of the surface, the local spectral densities are modified, in particular by the existence of surface modes. Fig.26 shows an example of these surface local densities for Pd(111) [17]. It is seen that in the direction perpendicular to the surface the spectral density $n_\perp(\omega)$ exhibits a very sharp peak due to the surface mode denoted S_1 (Fig.23) whereas for displacements parallel to the surface $n_{//}(\omega) = (n_x(\omega) + n_y(\omega))/2$ is only slightly perturbed. Consequently the S_1 mode is mainly polarized perpendicular to the surface. Note that The center of gravity of $n_\perp(\omega)$ is located at a much lower frequency than that of $n_{//}(\omega)$ or $n(\omega)$ (Fig.26).

Let us now examine the consequences on the mean square displacements of surface atoms:

- If $T \to 0$:

$$<u_{i\alpha}^2> = \frac{\hbar}{2M}\int_0^\infty \frac{1}{\omega} n_{i\alpha}(\omega)d\omega \tag{54}$$

- If $T \to \infty$

$$<u_{i\alpha}^2> = \frac{k_B T}{M}\int_0^\infty \frac{1}{\omega^2} n_{i\alpha}(\omega)d\omega \tag{55}$$

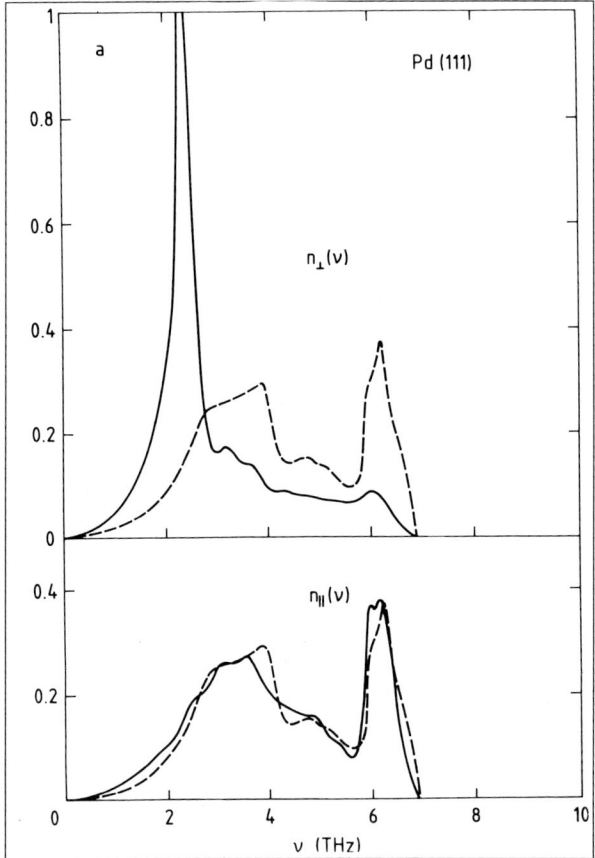

Figure 26. Spectral densities of modes of a surface atom for Pd(111): in the direction perpendicular to the surface $n_\perp(\omega)$, and parallel to the surface $n_{//}(\omega) = (n_x(\omega)+n_y(\omega))/2$. The bulk density of modes $n(\omega)$ is given (dashed lines) for comparison [17].

In both cases the integrals are mainly determined by the low frequency region in which $n_\perp(\omega)$ exhibits a sharp peak which is present neither in $n_{//}(\omega)$ nor in $n(\omega)$. The following inequalities are expected concerning mean square displacements of surface atoms $<u_{s\perp}^2>$ and $<u_{s//}^2>$:

$$<u_{s\perp}^2> \; > \; <u_{s//}^2> \; \geq \; <u_v^2> \tag{56}$$

where $<u_v^2>$ is the mean square displacement of bulk atoms.

Calculations show (Fig.27) that, for a given metal, $<u_{s\perp}^2>$ is rather insensitive to the orientation of the surface while this does not hold for $<u_{s//}^2>$ [17,26]. Note also that in agreement with (55) the mean square displacements increase linearly in the limit of high temperatures.

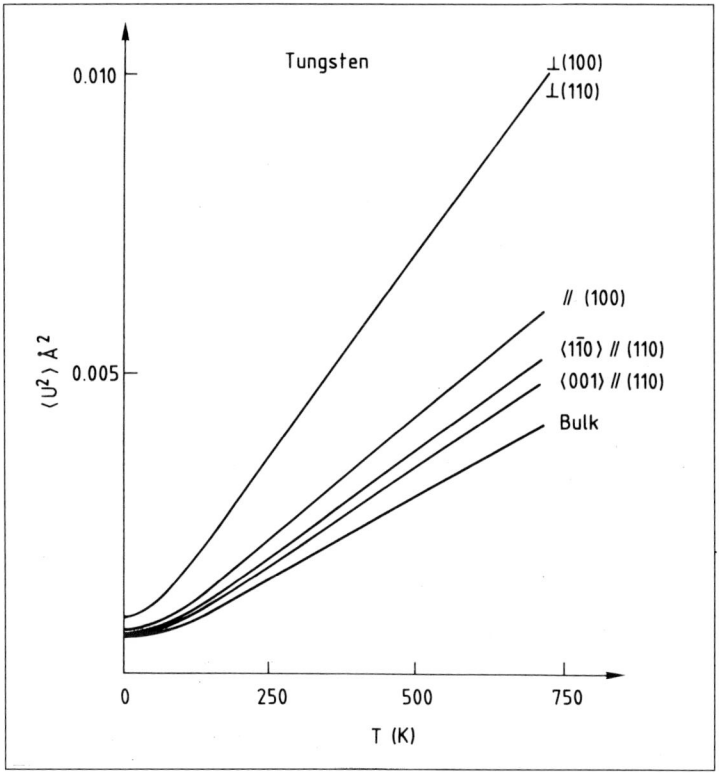

Figure 27. Variation of mean square displacements perpendicular and parallel to the (110) and (100) surfaces of W as a function of temperature [26].

Let us end this section by introducing the concept of surface Debye temperature. In the Debye model, justified at high temperature, the density of modes is written:

$$n(\omega) = 3\omega^2/\omega_D^3 \qquad 0 < \omega < \omega_D \qquad (57)$$
$$n(\omega) = 0 \qquad \omega > \omega_D. \qquad (58)$$

The bulk Debye temperature T_D^b is related to the Debye frequency ω_D by $k_B T_D^b = \hbar \omega_D$. By using (55) a simple relationship can be established between $<u_v^2>$ and T_D^b:

$$<u_v^2> = \frac{3\hbar^2 T}{M k_B (T_D^b)^2} \qquad (59)$$

Using the same model at the surface, the surface Debye temperature T_D^s is defined by the relation:

$$\frac{<u_s^2>}{<u_v^2>} = (\frac{T_D^b}{T_D^s})^2 \tag{60}$$

where the mean square displacements are here an average over all directions. Notice that it is also possible to define Debye temperatures for motions perpendicular or parallel to the surface.

3.5. Vibrations in systems with several atoms per unit cell: acoustical and optical phonons

3.5.1. Introduction

Consider a bulk crystal. The relationship:

$$\mathbf{u_i} = \mathbf{u_j} \exp(i\mathbf{k}.\mathbf{R_{ij}}) \tag{61}$$

only applies when the positions of the two atoms i and j are deduced from each other by a vector $\mathbf{R_{ij}}$ which is a period of the real lattice (Bloch theorem). When the crystal has p atoms per unit cell, p independent displacements $\mathbf{u_i}$ must be determined. The corresponding dispersion curves have $3p$ branches: three branches are said to be *acoustical* (i.e., those for which the frequency vanishes when $\mathbf{k} = 0$, i.e., with displacements corresponding to rigid translations of the lattice parallel to the three coordinate axes) and $3(p-1)$ branches said to be *optical*. Between these two types of branches a gap may exist for the bulk crystal in which it will be possible to find localized phonons when the crystal is limited by a surface. In order to illustrate this point we will first study the vibrations of a diatomic linear chain [27].

3.5.2. The diatomic linear chain
a) The infinite chain

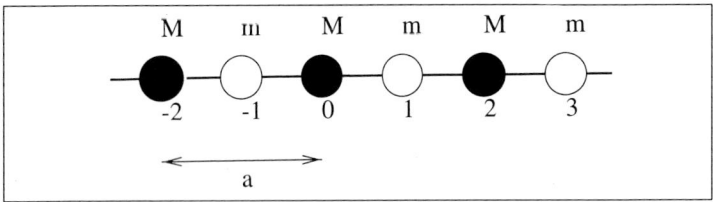

Figure 28. The diatomic linear chain.

Consider an infinite diatomic linear chain alternatively made of atoms of masses m and M ($M > m$), separated by the distance $a/2$ and interacting with a force constant β

(Fig.28). Let us determine first the vibration spectrum of this chain. The equations of motion are:

$$-M\omega^2 u_{2j} = \beta(u_{2j+1} + u_{2j-1} - 2u_{2j}) \tag{62}$$

$$-m\omega^2 u_{2j+1} = \beta(u_{2j} + u_{2j+2} - 2u_{2j+1}) \tag{63}$$

Setting:

$$v_{2j} = \sqrt{M} u_{2j} \qquad v_{2j+1} = \sqrt{m} u_{2j+1} \tag{64}$$

and:

$$\sqrt{\beta/M} = \Omega_0 \qquad \sqrt{\beta/m} = \omega_0 \tag{65}$$

and using the Bloch theorem, i.e.:

$$v_p = \exp(ika) v_{p-2} \tag{66}$$

the system (62) and (63) can be easily transformed into

$$(\omega^2 - 2\omega_0^2) v_{2j+1} + \omega_0 \Omega_0 (1 + \exp(ika)) v_{2j} = 0 \tag{67}$$

$$\omega_0 \Omega_0 (1 + \exp(-ika)) v_{2j+1} + (\omega^2 - 2\Omega_0^2) v_{2j} = 0 \tag{68}$$

The solution of this system leads to:

$$\omega^2 = \omega_0^2 + \Omega_0^2 \pm \sqrt{\omega_0^4 + \Omega_0^4 + 2\omega_0^2 \Omega_0^2 \cos(ka)} \tag{69}$$

(Note that the quantity under the square root is always positive since it lies between $(\omega_0^2 - \Omega_0^2)^2$ and $(\omega_0^2 + \Omega_0^2)^2$)

The spectrum is thus made of ($\Omega_0 < \omega_0$):

- The lower band called *acoustical branch*

$$\omega = \sqrt{\omega_0^2 + \Omega_0^2 - \sqrt{\omega_0^4 + \Omega_0^4 + 2\omega_0^2 \Omega_0^2 \cos(ka)}} \tag{70}$$

i.e., $0 < \omega < \Omega_0 \sqrt{2}$ when $0 < k < \pi/a$.

- The upper band called the *optical branch*

$$\omega = \sqrt{\omega_0^2 + \Omega_0^2 + \sqrt{\omega_0^4 + \Omega_0^4 + 2\omega_0^2 \Omega_0^2 \cos(ka)}} \tag{71}$$

i.e., $\sqrt{2(\omega_0^2 + \Omega_0^2)} > \omega > \omega_0 \sqrt{2}$ when $0 < k < \pi/a$.

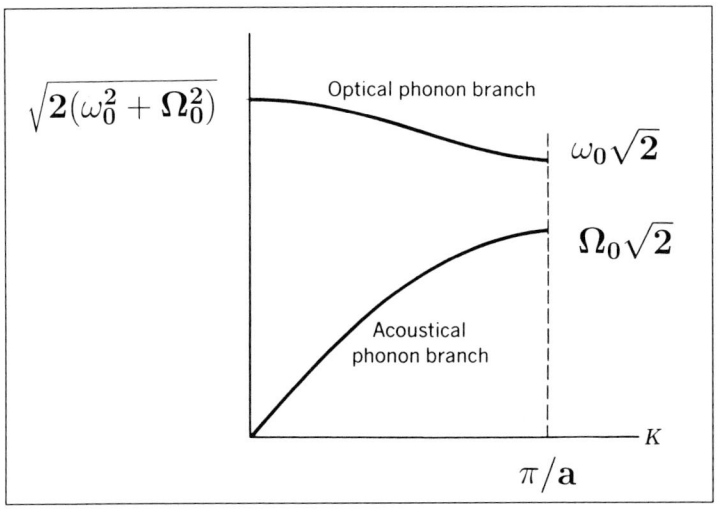

Figure 29. Dispersion curves of an infinite diatomic linear chain.

Thus there is a gap of width $\sqrt{2}(\omega_0 - \Omega_0)$ (Fig.29). Let us study the displacements at $k = 0$, the equations (67) and (68) and (64) lead to:
- for the acoustical branch: $u_0 = u_1$, i.e., the two atoms of the unit cell move in the same direction
- for the optical branch: $Mu_0 + mu_1 = 0$, i.e., the two atoms of the unit cell move in opposite directions in such a way that their center of mass is fixed. When the atoms of masses m and M carry equal charges with opposite signs such a vibration mode can be excited by the electric field of a light wave, so that the branch is called the optical branch.

b) *Semi-infinite chain*

Consider now a semi-infinite chain, the free end of which is occupied by an atom of mass m. The equation of motion of this atom is:

$$-m\omega^2 u_1 = \beta(u_2 - u_1) \tag{72}$$

or, using the same notations as above:

$$(\omega^2 - \omega_0^2)v_1 + \omega_0\Omega_0 v_2 = 0 \tag{73}$$

Let us look for a localized state at the free end of the chain, i.e., obeying the following conditions:

$$v_p = \exp(-ka)v_{p-2} = xv_{p-2} \tag{74}$$

with $|\exp(-ka)| < 1$, i.e., $Re(k) > 0$. In order to satisfy the equations for the atoms inside the chain, we must have (change k into ik in (69))

$$\omega^2 = \omega_0^2 + \Omega_0^2 \pm \sqrt{\omega_0^4 + \Omega_0^4 + 2\omega_0^2\Omega_0^2 \cosh(ka)} \tag{75}$$

The system (73) and (68) (for $j = 1$ and in which we have changed ik into $-k$ and replaced v_3 by $\exp(-ka)v_1$) allows to determine v_1 and v_2. The corresponding secular equation is, taking (74) into account and after some simple algebraic manipulations:

$$x^3\omega_0^2 + x^2\Omega_0^2 - x\omega_0^2 - \Omega_0^2 = 0 \tag{76}$$

which can be factorized into:

$$(x^2 - 1)(x\omega_0^2 + \Omega_0^2) = 0 \tag{77}$$

The only interesting solution (satisfying $|x| < 1$) is:

$$x = -\Omega_0^2/\omega_0^2 = -m/M \tag{78}$$

Substituting (78) for x into (75) we obtain the following frequency:

$$\omega_s^2 = \omega_0^2 + \Omega_0^2 = \beta \frac{M+m}{mM} \tag{79}$$

It is easy to see that $\sqrt{2}\Omega_0 < \omega_s < \sqrt{2}\omega_0$: the frequency of this localized mode belongs to the gap between the acoustical and optical branches.

Using (72) it is easily shown that:

$$u_2 = -\frac{m}{M}u_1 \tag{80}$$

and since

$$u_{2j+1} = (-1)^j \left(\frac{m}{M}\right)^j u_1 \tag{81}$$

and

$$u_{2j} = (-1)^{j-1} \left(\frac{m}{M}\right)^{j-1} u_2 \tag{82}$$

we get

$$u_{2j} = u_{2j+1} \tag{83}$$

In conclusion, there is always a surface mode in the gap between the optical and acoustical branches provided that the chain is ended by the lightest atom. This surface mode does not exist in the opposite case.

3.5.3. Extension to semi-infinite crystals

Let us limit ourselves, to simplify, to crystals with two atoms per unit cell. The phonon spectrum is then made of three acoustical branches and three optical branches. However, contrary to the one dimensional case, there is not always a gap between the band of frequencies of the acoustical modes and that of optical modes. This has important consequences on the existence of surface modes.

Let us take as an example the case of the (001) face of alkali halides which crystallize in the NaCl structure. We expect intuitively, and also from the above calculation on the diatomic linear chain, that the phonon spectrum is very sensitive to the ratio of masses of the two constituent atoms. We present here two extreme cases [28]: NaBr in which $M(Br)/m(Na) \approx 7$ and KCl in which $M(K)/m(Cl) \approx 1$:

- *Dispersion curves of NaBr(001)*: Fig.30 shows the phonon dispersion curves of a slab of 15 atomic layers of NaBr(001). As expected from the difference of masses, there is an absolute gap in the projected bulk band structure (dashed curves). There is a large variety of surface localized modes (full curves): acoustical modes (S_1, S_6, S_7) and optical modes (S_2, S_3, S_4, S_5). The main vibrational character of ions belonging to the surface plane is also indicated: $SP_\perp$ and $SP_{//}$ for vibrations in the sagittal plane (i.e., containing the surface normal and the wave vector) perpendicular and parallel to the surface, respectively, SH for a parallel shear mode. The indices + or − indicate that the positive or negative ion is the most involved in the surface vibration. Finally, hatched regions signal the presence of surface resonances.

- *Dispersion curves of KCl(001)*: In KCl, the corresponding masses being very similar, there is no absolute band gap. Thus there is an overlap between acoustical and optical branches and the phonon spectrum is more complicated. In Fig.30 it is seen that, in many surface modes, both types of ions are involved in the vibration and an optical Rayleigh mode S'_1 appears at $\bar{X}$.

REFERENCES

1. M.C. Desjonquères and D. Spanjaard, Concepts in Surface Physics, Springer-Verlag, Springer Series in Surface Science 30, 1993 2nd edition, 1996; A. Zangwill, Physics at Surfaces, Cambridge University Press, Cambridge, 1988; M. Lannoo and P. Friedel, Atomic and Electronic Structure of Surfaces, Springer-Verlag, Springer Series in Surface Science 16, 1991; H. Lüth, Surfaces and Interfaces of Solids, Springer-Verlag, Springer Series in Surface Science 15, 1993
2. J. Friedel and M.C. Desjonquères, in La Catalyse par les Métaux, G.A. Martin and A.J. Renouprez Eds., Edition du CNRS (1984) 28
3. For a compilation of experimental results see: J.M. Mac Laren, J.B. Pendry, P.J. Rous, D.K. Saldin, G.A. Somorjai, M.A. Van Hove and D.D. Vvedensky, Surface Crystallography Information Service, A Handbook of Surface Structures, 1987
4. J. Sokolov, F. Jona and P.M. Marcus, Sol.St.Comm. 49 (1984) 307
5. N.D. Lang and W. Kohn, Phys.Rev.B1 (1970) 4555; N.D. Lang in Theory of the Inhomogeneous Electron Gas, F. Lundquist and N.H. March Eds., Plenum Press, 1981
6. K.W. Jacobsen, Comm.Cond.Mat.Phys. 14 (1988) 129

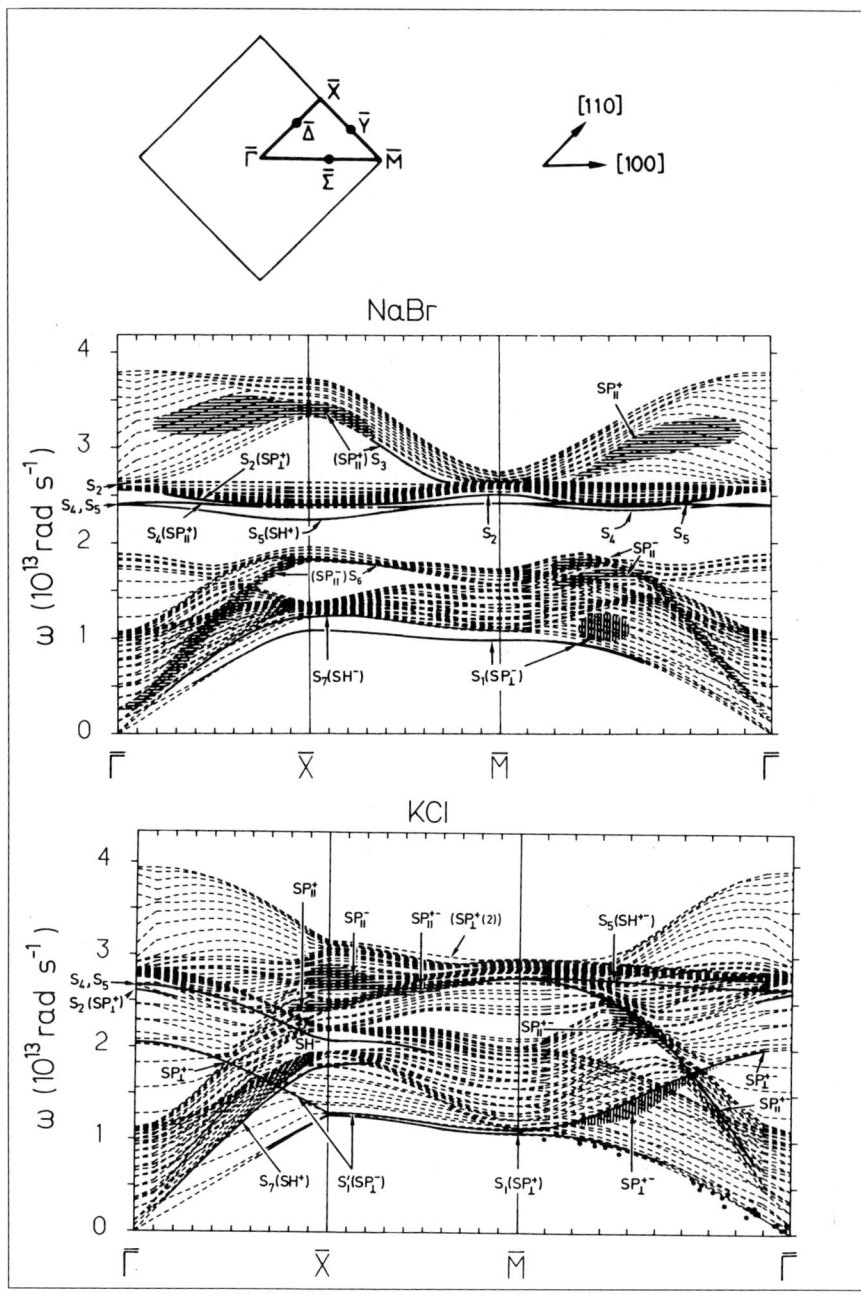

Figure 30. Phonon dispersion curves in NaBr(001) and KCl(001), the notations are explained in the main text.

7. M.C. Desjonquères and D. Spanjaard, J.Phys.C 15 (1982) 4007
8. B. Legrand, G. Tréglia, M.C. Desjonquères and D. Spanjaard, J.Phys.C 19 (1986) 4463 and references therein
9. M. Guillopé and B. Legrand, Surf.Sci. 215 (1989) 577
10. K.W. Jacobsen and J.K. Norskov, Phys.Rev.Lett. 59 (1987) 2764; A.B. Baddorf, I.W. Lyo, E.W. Plummer and H.L. Davis, J.Vac.Sci.Technol. 5 (1987) 782
11. J.F. Van der Veen, R.M. Tromp, R.G. Smeenk and F.W. Saris, Surf.Sci. 82 (1979) 468; R. Baudoing, Y. Gauthier and Y. Joly, J.Phys.C 18 (1985) 4061
12. J.K. Norskov, Surf.Sci. 299/300 (1994) 690
13. J.B. Pendry, Low Energy Electron Diffraction, Academic Press, 1974
14. D.P. Woodruff and T.A. Delchar, Modern Techniques of Surface Science, Cambridge University Press, 1986
15. P. Nozières et F. Gallet, J.de Physique 48 (1987) 353
16. J.C. Heyraud and J.J. Métois, Acta Met. 28 (1980) 1789, Surf.Sci. 128 (1983) 334, Surf.Sci. 177 (1986) 213, J.Cryst. Growth 82 (1987) 269
17. G. Tréglia and M.C. Desjonquères, J.de Physique 46 (1985) 987
18. D. Castiel, L. Dobrzynski and D. Spanjaard, Surf.Sci. 59 (1976) 252
19. R.E. Allen, G.P. Alldredge and F.W. de Wette, Phys.Rev.B4 (1971) 1661
20. Lord Rayleigh, Proc.London Mat.Soc. 17 (1887) 4
21. J. Szeftel, Surf.Sci. 152/153 (1985) 797; J.E. Black, D.A. Campbell and R.F. Wallis, Surf.Sci. 105 (1981) 629
22. A. Blandin, D. Castiel and L. Dobrzynski, Sol.St.Comm. 13 (1973) 1175
23. A. Fasolino, A. Santoro and E. Tossatti, Phys.Rev.Lett. 44 (1980) 1684
24. G. Allan and M. Lannoo, Phys.Rev.B37 (1988) 2678
25. D. Pines, Elementary Excitations in Solids, Benjamin, 1964
26. L. Dobrzynski and P. Masri, J.Phys.Chem.Solids 33 (1972) 1603
27. C. Kittel, Introduction à la Physique des Solides, Dunod, 1972
28. W. Kress, F.W. de Wette, A.D. Kulkarni, U. Schröder, Phys.Rev.B35 (1987) 5783

Dislocations and stress relaxation in heteroepitaxial films

L.P. Kubin
LEM, CNRS-ONERA

29 Av. de la Division Leclerc, BP 72, 92322 Châtillon Cedex

A few important elastic properties of dislocations that are relevant for a study of stress relaxation in thin layers are recalled at an elementary level. The elastic interactions of a dislocation with a free surface and an interface are then briefly discussed. The concept of critical thickness for the relaxation of misfit strains in heteroepitaxial films is introduced in a simplified manner. This notion, which makes use of an equilibrium argument, is critically discussed in the last part with emphasis on several kinetic effects and energy barriers, particularly those associated with the mechanisms of dislocation generation and mobility.

1. INTRODUCTION

The existing models for the relaxation of misfit stresses by dislocations in heteroepitaxial films have their advantages and limits. As they are based on the elastic properties of dislocations, they can be worked out in a formal manner at the mesoscopic scale. In what follows, however, only a simplified outline will be presented. More detailed analyses will be found in the reviews and articles quoted in reference. In practice, there are many competing mechanisms that can induce the relaxation of misfit stresses in thin films. Even when it seems possible to define a critical thickness for the relaxation of misfit strains, a check of the presence of accommodating dislocations is necessary. Finally, the available models being based on equilibrium considerations, the predicted values can only be achieved in ideal experimental conditions.

In part 2, a few elementary properties stemming from the elastic theory of dislocations are recalled and explained in the simple terms. The basic concepts are those of elastic energy, from which the line tension and curvature properties of a dislocation line can be deduced, and of the Peach-Koehler force on a dislocation. In part 3, the interaction of dislocations with free surfaces and interfaces are discussed. The current model for the critical thickness is introduced by considering the critical stress for the motion of a dislocation in a thin film. A discussion is proposed in part 4 of two dislocation properties that may lead to the occurrence of metastable non-equilibrium states, generation and mobility. Whereas the latter is reasonably known from investigations at different scales as well as experiment, the former can only be understood at the atomic scale and has not been, as yet, the object of predictive models.

2 - ELASTIC PROPERTIES OF DISLOCATIONS

2.1. Introduction to dislocations

The reader is referred to a recent review [1] for an elementary introduction to dislocation theory, to the classical text-book by Friedel [2] for a complete discussion of the properties of dislocations in physical terms and to the book by Hirth and Lothe [3] for a formal account of the elastic theory of dislocations.

Figure 1 shows the traditional scheme illustrating the motion of an edge dislocation, seen end-on, in a crystal. The dislocation line glides in the horizontal plane under the effect of an externally applied shear stress τ. At every moment the atomic displacements caused by this motion are localised in a small neighbourhood, the core region, which is materialised by a circle in Fig. 1-a. The dimension of this region is not larger than one or a few lattice parameters. This can be checked from the high-resolution electron micrograph of fig. 2-a, which shows the core structure of an edge dislocation in a thin foil of h.c.p. titanium. After several translations by unit lattice parameters (Fig. 1-c), the dislocation line vanishes at the surface of the crystal, leaving a step of atomic height. The amplitude and direction of this step define the Burgers vector b of the dislocation.

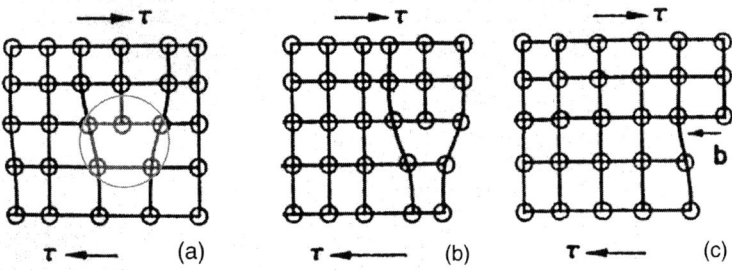

Figure 1. Schematic: An edge dislocation of Burgers vector b, seen end-on, shears a crystal (a, b) and vanishes leaving an atomic step at the surface (c).

In the dislocation theory, one makes a distinction between dislocation properties that stem from the atomic structure of the core and those that can be deduced by considering the volume external to the core. The core properties govern the nature of the slip systems as well as the intrinsic mobilities of the defects (cf. section 4.3.). These properties depend on the nature of the crystal considered and can be investigated with the help of atomistic simulations. Outside the core, the atomic distortions do not vanish but are small enough to allow performing estimates within a continuum frame. Within the elastic theory of dislocations, it is possible to show that the stress and strain fields surrounding a dislocation line are actually long-ranged. They decay as 1/r, where r is the distance to the geometrical centre of the core. A calculation of the elastic energy shows that most of the energy of a dislocation is actually stored outside the core. As a consequence, the elastic theory of dislocations is able to describe the energetic aspects of dislocation behaviour, including the forces applied to dislocations, at distances larger than one or a few lattice spacings. In the linear isotropic formulation, which is in many cases sufficient for practical purposes, the elastic theory of dislocations gives access to many properties that are fully generic and depend on the nature of the crystal considered via two elastic constants only. Fig. 2-b shows a dislocation microstructure, as observed in the electron microscope in diffraction contrast conditions. The defects appear as dark lines of width 10-20 nm that represent the region where electron diffraction is disturbed by the deformation of the lattice. At this scale, the dislocation lines do not appear as atomic configurations but as individual objects. This is the way they are usually treated in the continuum approach.

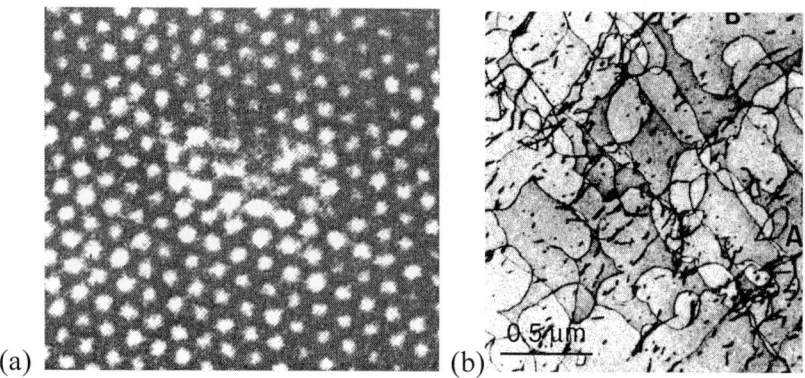

Figure 2. (a) - High-resolution electron micrograph of an edge dislocation, seen end-on, in a thin film of h.c.p. titanium (Courtesy of S. Naka). The missing half-plane is easily seen at glancing angle. (b) - Electron micrograph in diffraction contrast, showing a dislocation microstructure in a deformed Mo-Nb alloy.

As will be seen in section 2.3., dislocations are defects of such a high energy that their density (i.e., the total line length per unit volume) is practically zero at equilibrium, irrespective of the temperature. Dislocations are often generated by local stress fluctuations during crystal growth, from the liquid or vapour phase. It is very difficult to eliminate them subsequently because they are locally stabilised by their mutual interactions. During annealing treatments, they can mutually annihilate, but a substantial fraction of the total density often gets further stabilised in configurations of lower energy. Thus, just like grain boundaries, dislocations are metastable defects and real crystals are seldom found in the equilibrium configuration of zero density. As a consequence, the dislocation density present in a crystal at a given moment depends on its previous thermo-mechanical history.

2.3. Elastic energy

The concept of dislocation was proposed for the first time in 1936 in order to explain why crystals plastically yield under stresses much smaller than those necessary to shear a crystallographic plane as a whole. The elastic theory was developed by taking advantage of earlier calculations developed in order to explain the occurrence of internal stresses in crystals. Transmission electron microscopy observations of dislocations, which were initiated in 1956, confirmed all the expectations of the theory and, further, substantially contributed to its development. Whereas the elastic theory of dislocations is now practically in closed form, the focus has been in the recent years on the core properties, the collective behaviour of dislocation and the connection between the various scales of investigation.

A dislocation line is a defect that performs the accommodation between sheared and non-sheared areas in a crystal. Fig. 3 illustrates a virtual process, called a Volterra process, by which such a defect is constructed according to its definition. In a piece of crystal, a line delimiting a planar area is drawn that is either closed on itself or ends at a free surface. A cut of this area is performed up to the dislocation line and the upper part of the crystal is sheared along the cut by an amount b, where b is a lattice translation, with respect to the lower part. In the final relaxed configuration of Fig. 3, empty circles represent the atoms located in the plane immediately above the sheared area and small full circles the atoms immediately below. Along the direction of the line perpendicular to the shear, the accommodation involves an extra-half plane as three atomic rows in the upper part of the crystal correspond to two rows in the lower part (cf. Fig. 3). This portion of the line whose direction is perpendicular to the Burgers vector is, by definition, of edge character. In the lower-left part of the crystal, the line is parallel to the shear and is said to be of screw character. The horizontal planes are transformed into a helicoidal surface with vertical axis. At intermediate positions, the dislocation line is said to have a mixed character. The ability of dislocations of edge character (and of mixed character) to accommodate misfit stresses is obvious from Fig. 3. From this construction, it is also immediately seen that the motion of the line in its slip plane propagates a shear in the crystal.

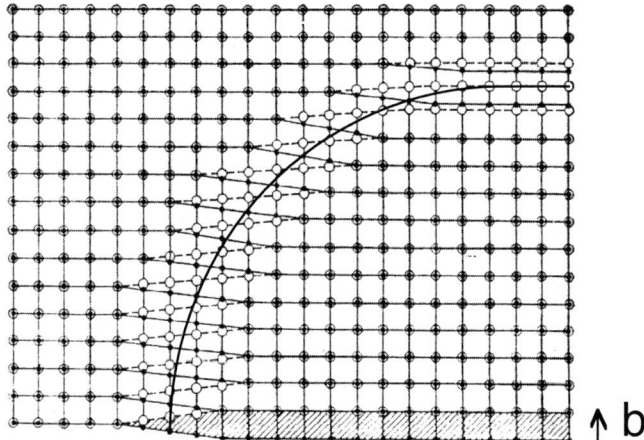

Figure 3. Construction of a dislocation line by a Volterra process.

In a continuum medium, the elastic energy of a dislocation line can be calculated by making use of the Volterra construction. A full discussion of the method for estimating rigorously the self-energy of a dislocation line of arbitrary shape will be found in [3]. In what follows we make use of the common approximation that yields simple solutions valid for infinite and straight dislocation lines [2]. The strain fields thus calculated are of the form $bf(\theta)/r$, where r is the distance to the line and θ an angular variable. In the case of the screw dislocation, these stress fields are radial and only include shear components. For the edge or mixed dislocation, both normal and shear components are involved. The divergence at r = 0 obviously results from the failure of the continuum approach in the dislocation core region. The stress fields are calculated from the strain fields with the help of Hooke's law and the local elastic energy density is obtained as a quadratic form of the strain components, as is always the case in linear elasticity. Integrating over the volume surrounding the core of the defect (of radius b) to remove the divergence, one obtains the total elastic energy per unit length of dislocation line, W_u:

$$W_u = \frac{\mu b^2}{4\pi K} \text{Ln}\left(\frac{D}{b}\right) \quad (1)$$

The integration over the volume naturally leads to a logarithmic dependence, whereas the quadratic dependence of the elastic energy on strain leads to a proportionality to the Burgers vector squared. K is a constant that depends on the dislocation character ($K_{screw} = 1$, $K_{edge} = 1-\nu$, where ν is the Poisson's ratio). The integration of the energy density is carried up to an outer cut-off distance D. The latter can be either the dimension of the crystal for an isolated dislocation or the distance to the next dislocation of opposite sign that screens the field of the considered one. For a dislocation loop, the reader may convince himself with the help of Fig. 3 that the portions of a dislocation loop opposite to each other have strain and stress fields of opposite sign, so that the screening distance is then the radius of the loop. The definitions of the inner (or core) and outer cut-off radii are particularly critical at the nanoscale. They may significantly influence the value of the energy and of all the quantities that derive from it.

The dislocations that are found in real crystals are those of lower energy. Hence, the Burgers vector is always the smallest possible lattice translation, at least in compact structures. For instance in the f.c.c. lattice and those deriving from it like the diamond cubic (d.c.) lattice, the preferential Burgers vector is a/2<110>, where a is the lattice parameter. This dislocation glides in the compact {111} planes, which exhibit the smallest resistance to shear. Such dislocations are called perfect dislocations because they carry a shear that restores the perfect crystal outside the core region (cf. Fig. 3). With smaller Burgers vectors, the elastic energy can be further reduced but the process depicted in Fig. 3 produces a planar fault since the upper and lower halves of the crystal no longer match after the shearing process. The corresponding linear defect is then called a partial dislocation. Configurations involving partial dislocations can be observed in crystals where the stacking fault energy is low enough so that the total energy is effectively reduced. In f.c.c. and d.c. crystals, Shockley partial dislocations have a Burgers vector a/6<112> and trail a stacking fault in the {111} planes.

An estimate of the line energy of a dislocation in a bulk crystal can be obtained by making use of typical values for the quantities involved (e.g., $b \approx 0.3$ nm and a dislocation density of $\rho = 10^{12}$ m^{-2} in a metal, hence a screening distance $D \approx \rho^{-1/2} \approx 10^{-6}$ m). W_u is then found of the order of μb^2. This energy is about μb^3 per atomic length of line, which is a quite large value since μb^3 is typically of the order of 5 eV for a metal and more in d.c. semiconductors). This explains why dislocations cannot spontaneously appear under the effect of thermal fluctuations. Another consequence of this high energy is that mechanisms involving substantial changes in the elastic energy of a dislocation configuration cannot be thermally activated. In comparison, the core energy, as measured from molecular dynamics simulations, is a small fraction of the total energy, not larger than a few percent. Thus, mechanisms involving changes in core configurations such as those occurring during dislocation motion can be thermally activated.

2.3. Force on a dislocation

The motion of a dislocation by glide, as depicted in Fig. 1, occurs in such a way as to minimise the total elastic energy stored in the crystal. Therefore, in a crystal of finite size, a single dislocation always sees a force tending to move it towards the closest free surface (cf. section 3.1.), in order to restore a perfect crystal. In the same way, in the presence of an internal or external stress field, a dislocation is submitted to a force tending to move it in such a way as to reduce the total energy of the crystal, i.e., in practice the total elastic energy W. Therefore, one has per unit length of line :

$$F = -\nabla W \qquad (2)$$

A direct estimate of this force can be obtained easily, even in complex situations. We consider first the simple case of the motion of a dislocation under the effect of an externally applied shear τ in its slip plane. The slip area is $S = \ell L$ (cf. Fig. 4). The mechanical work during the whole shearing of the crystal is $W_m = \tau b S$. One assumes that it is possible to define for the same process a force per unit length, F_u, acting on the dislocation line and normal to it in its slip plane. The total force on the line is $F_u \ell$ and the corresponding mechanical work is $W_d = F_u \ell L = F_u S$ at the dislocation scale. From these two estimates for the same energy, we draw :

$$F_u = \tau b \qquad (3)$$

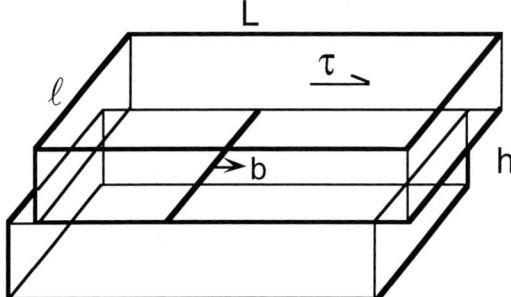

Fig. 4 : The shearing of a crystal by a dislocation line.

It must be noted that this force is not a material force but a configurational force, as the dislocation line is not a material object but an atomic configuration. Nevertheless, it is possible to define a fictitious force applied to the geometrical line, that will further allow treating the dynamics of this defect. This force results from the interaction of the elastic field induced by the external stress and the one associated with the dislocation line. Equation 3 has been generalised by Peach and Koehler to the interaction of a dislocation line with any arbitrary stress field, $\underline{\underline{\sigma}}$, external or internal, which is now written in tensor form :

$$\underline{F_u} = \underline{\underline{\sigma}} \cdot \underline{b} \times \underline{\ell}_u \qquad (4)$$

In this equation, ℓ_u is the unit vector parallel to the dislocation line. Because of the outer product involved, the configurational force is always normal to the line. It is important to notice that the Peach-Koehler force is not necessarily in the slip plane and also has a component normal to it. The stress tensor can be non-uniform in space, what matters is only its value at the position of the line. In particular, if $\underline{\underline{\sigma}}$ originates from another dislocation its components are inversely proportional to the distance. As a consequence, we see From eq. 4 that the interaction forces between (infinite) dislocations share the same property.

2.5. Curvature, multiplication

We consider a dislocation line pinned at its extremities by contact interactions with other defects (cf. Fig. 5), for instance other dislocation segments. Under an applied stress the line bows out and reaches an equilibrium shape. Its line length increases and the work of the applied stress must produce the corresponding line energy. An extremely useful approximation consists of considering the dislocation line as an elastic string. This allows defining a resistive force, the line tension, that opposes any increase in line length. In terms of forces, and within this approximation, the normal force on the dislocation line τb is equilibrated by the line tension force, T. The latter is defined as the derivative of the line energy with respect to the line length. For curved segments, the outer cut-off radius in the expression for the energy (eq. 1) is taken as the curvature radius. Then, we have

$$T = \frac{\mu b^2}{4\pi K} \text{Ln}\left(\frac{R}{b}\right) \tag{5}$$

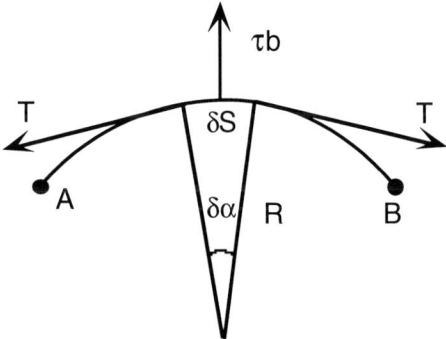

Figure 5. Curved equilibrium shape of a dislocation segment pinned at both ends under the combined effect of an applied stress and of the line tension.

The equilibrium condition is obtained by looking at the vertical projection of the forces on a small line element δs, the local curvature radius R being defined by $R = \delta s/\delta\alpha$ (cf. Fig. 5). One straightforwardly obtains:

$$\tau = \frac{T}{bR}, \tag{6}$$

which is reminiscent of the Laplace formula for an elastic membrane under internal pressure. For the sake of simplicity, it is often assumed that the curvature radius is constant along the line, although, strictly speaking, this is not the case. Indeed, the line tension depends on the character via the coefficient K in eq. 5.

The interaction of dislocation lines with fixed point or point-like obstacles in their slip planes can be described from the knowledge of the line tension and of the maximum interaction energy with the obstacles. For the present purpose, we consider the case where the obstacles are extremely strong ones, which means that they cannot be sheared by the dislocation lines. This leads to the Frank-Read mechanism of dislocation multiplication, as illustrated by Fig. 6, where we assume that the line is pinned by two other dislocations. Under an increasing applied stress, the dislocation segment AB bows out, achieving successive equilibrium configurations until it reaches a semicircular shape (2). At this step, the curvature radius is R= d/2 where d is the initial length of the segment. If the stress is further increased, the curvature radius can no longer decrease uniformly. Then, the line tension force cannot equilibrate the Peach-Koehler force and it is no longer possible to define an equilibrium shape. The dislocation line moves under the effect of the net resulting force applied to it and rotates around the pinning points (3, 4). In configuration (5), the two segments that meet each other after one rotation have stress fields of opposite sign and mutually annihilate. A single loop (6) is then produced, whereas the initial segment (1) is restored. The repetition of this process induces the emission of successive concentric loops, thus increasing the dislocation density.

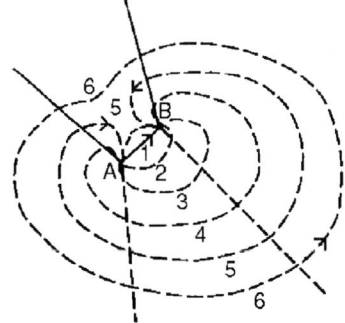

Figure 6. Dislocation multiplication by the Frank-Read mechanism.

The Frank-Read source mechanism is responsible for dislocation multiplication in crystals. It can operate in many different manners, particularly in thin films and layers, but it always requires a critical stress τ_{FR} that corresponds to the critical, approximatively semicircular shape:

$$\tau_{FR} = \frac{2T}{bd} \tag{7}$$

The type of scaling exhibited by eq. 7 has a quite general range of validity as far as the interaction of dislocations with localised obstacles is concerned. The critical stress to move dislocations over obstacles with an average spacing d is inversely proportional to d. This is the base of the strengthening of bulk crystals by impurity clusters, precipitates or other dislocations. At the nanoscale, the confinement of the dislocation lines to small dimensions induces a hardening for the same reason. However, as mentioned above, the characteristic length also enters the line tension in a logarithmic form via the value of the outer cut-off radius.

3. DISLOCATIONS IN THIN FILMS, CRITICAL THICKNESS

3.1. Interaction with free surfaces and interfaces

In epitaxial films, dislocations principally experience two types of internal forces. The first one, the "image force", stems from the difference in elastic constants between the two media on each side of the interface. The second one, that will be discussed in section 3.3., is the misfit stress that originates from the difference in lattice parameters. The general case where these two internal stresses occur simultaneoulsy is extremely difficult to treat and requires complex numerical procedures and simulations (cf. the recent works [4] and [5] that contain full continuum simulations on several problems related to internal stress relaxation by dislocation motion). We first consider the image force and assume that there is not misfit stress.

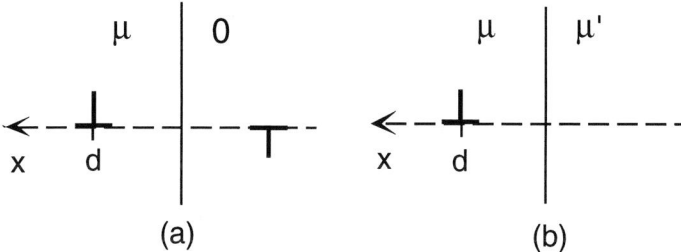

Figure 7. Interaction of a dislocation with (a) - a free surface and (b) - an interphase.

In the case of the interaction with a free surface, the force on the dislocation line is always such that the latter is attracted towards the surface in order to reduce the total elastic energy of the crystal. Simple calculations can be performed for a straight dislocation line parallel to either a free surface or an interface (cf. Fig. 7). Since there should be no normal stress on a free surface, one has to add a correction to the stress field of a dislocation in an infinite medium to account for this boundary condition. In the case of Fig. 7-a, it is sufficient to consider a mirror dislocation of opposite sign in a virtual medium with same elastic constants as the crystal considered. The normal stresses of the two dislocations cancel out at the free surface and the interaction force that attracts the dislocation towards the surface is the attractive interaction of two dislocations of unlike sign at a distance 2d. This force is be written :

$$F = -\frac{\mu b^2}{4\pi K d}, \tag{8}$$

where K depends on the dislocation character (cf. eq. 1). It must be noted that this approximation of the image force is valid for calculating the force on the dislocation but not for computing its stress field in a semi-infinite medium (except for pure screw dislocations). One immediate consequence of this attraction is that if a dislocation line is free to take its equilibrium shape, it will always tend to arrive normally at a surface. However, since the line energy depends on character and is minimum in screw orientation (cf. eq. 1), deviations from normality can be observed.

For an interface, separating two media with elastic constants μ and μ' (cf. Fig. 7-b), the same considerations apply but in a more complex manner [2], leading to :

$$F = k \frac{(\mu - \mu')b}{d}, \tag{9}$$

where k is a constant. In qualitative terms, the dislocation lines always tend to move towards the "soft" medium where they store less elastic energy.

A dislocation approaching a surface distorts it, as it tends to deform it in such a manner as to form an incipient step. The local increase in the surface energy induces a repulsive force that can be estimated in terms of the surface energy. One can show [6] that the repulsive force is of the form:

$$F = \beta \frac{\mu b}{d^2} \tag{10}$$

In the case of the interaction of a dislocation with a free surface, the equilibrium distance to the surface that results from the competing effects of the attractive image force (eq. 8) and the repulsive surface deformation (eq. 10) is smaller than one lattice parameter. This is clearly unphysical as linear elasticity does not apply at such small range. In crystals with clean surfaces, the image force is always predominant and the dislocations always emerge at the surface. For interphases, one may expect that in favourable situations the misfit dislocations are not located exactly at the interface but at a small distance from it, in the softer medium. The concept of image force has many applications. It allows discussing the effect of oxide layers, which are usually "strong" materials from the elastic point of view, as well as the complex situations that may arise in multilayers.

3.2. The capped layer approximation

Figure 8 shows a threading dislocation in a thin film of thickness h, grown on a substrate that we assume to have same elastic constants as the film. Under the effect of its interaction with the misfit stress, the dislocation bows out. If the misfit stress is large enough, a critical bowed-out configuration can be reached and the dislocation starts moving, trailing an interfacial segment that relaxes the misfit stress. The critical condition for motion can be approximated by constructing the image configuration. Then, the problem reduces to that of the motion of a dislocation confined between two walls (i.e., a capped layer) of thickness 2h.

The critical stress has, within a numerical coefficient, the same form as the Frank-Read stress defined above (cf. eq. 7) and is written :

$$\tau_c \approx \frac{T}{bh} \qquad (11)$$

This approximation of the capped layer is often used for simple estimates [7, 8].

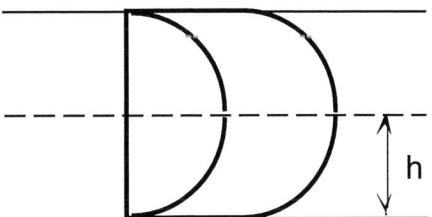

Figure 8. Motion of a threading dislocation in a thin epitaxial layer of thickness h and in a capped layer of thickness 2h.

3.3 Misfit dislocations

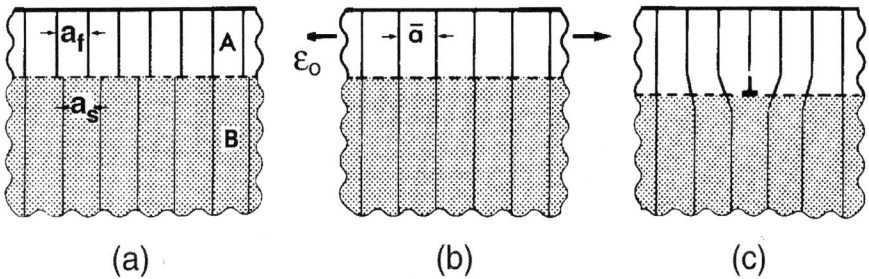

Figure 9. (a) - Non-interacting film and substrate with different lattice parameters. (b) - Elastic accommodation. (c) - Relaxation by misfit dislocations (After Jesser and van der Merwe [9], by courtesy of Elsevier).

In heteroepitaxial films, the elastic stresses can be relaxed in many different manners. Several mechanisms can occur at the interface like amorphisation, phase transformation, interdiffusion, reaction and the formation of point or planar defects, including stacking faults and microtwins. The relaxation by the formation of misfit dislocations is also a common mode of internal stresses relaxation that can be easily modelled at the mesoscopic scale. Fig. 9-a shows an incoherent interface between a thin film of lattice parameter a_f and a substrate of lattice parameter a_s. In Fig. 9-b, the interface is coherent and the film is elastically strained. When the film thickness is small compared to that of the substrate, the film becomes pseudomorphic, i.e., it adopts the lattice parameter of the substrate. The corresponding uniform strain, the misfit strain, is :

$$\varepsilon_o = \frac{a_s - a_f}{a_f} \tag{12}$$

The accommodation by misfit dislocations, through a vernier effect, is illustrated by Fig. 9-c. For the sake of simplicity, we assume here and in what follows that the Burgers vector of the dislocations is in the plane of the interface, i.e., $b = a_f$. Dislocations having a slip plane inclined with respect to the interface are less efficient for relaxing misfit stresses, which is accounted for by a projection factor. In the relaxed configuration of Fig. 9-c, one has (p+1) planes in the film for p planes in the substrate, hence $(p+1)a_f = pa_s$. As a consequence, the spacing p between the dislocations in the fully relaxed configuration is written

$$p = b/\varepsilon_o, \tag{13}$$

and the number of dislocations of same Burgers vector per unit length of the interface is

$$n_o = \varepsilon_o/b \tag{14}$$

3.4. The critical thickness

It is experimentally observed that misfit dislocations appear beyond a certain critical thickness value. To calculate this critical thickness, one assumes planar pseudomorphic growth. The earliest arguments, as given by Frank and van der Merwe in 1949 [10] (see also, [9, 11] for reviews of early work), are based on equilibrium considerations. The total elastic energy per unit area of interface in a strained layer of given thickness and containing an array of accommodating dislocations consists of a uniform strain energy W_ε plus the energy W_d of the interfacial dislocations. The equilibrium configuration is assumed either to minimise the total energy with respect to the dislocation spacing or to satisfy the critical condition $W_\varepsilon = W_d$. Thus, this type of model assumes that the array of interfacial dislocations appears and adjusts itself as a whole, ignoring the mechanisms by which the dislocations are formed and moved into place. A modification of these models by Matthews and Blakeslee has lead to the current standard model of the critical thickness [12] (see also [13-15] for discussion).

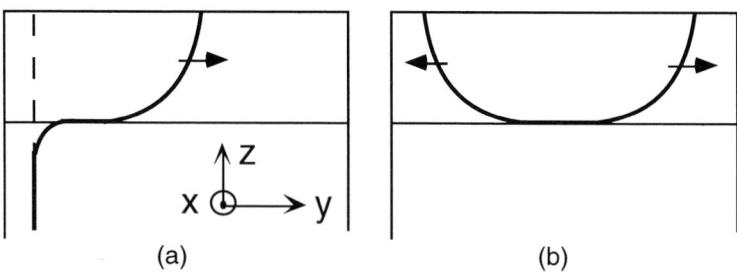

Figure 10. Formation of interfacial dislocations from the motion of a threading dislocation (a) or a dislocation loop nucleated at the surface (b).

It is assumed that dislocations pre-exist in the film that can be either threading dislocations or dislocation half-loops nucleated from the surface, as sketched in Fig. 10. The critical thickness for the onset of relaxation is achieved when the effective force on such dislocations reaches the critical value for irreversible motion.

It must be noted that when the critical condition is reached, the net force on a threading segment is zero, as the line is in equilibrium under the competing effects of the line tension force and the force applied by the misfit stress. In such conditions and according to eq. 2, the total elastic energy of the crystal is also minimum. Therefore, although the calculation of the critical thickness is usually presented in terms of forces, it is also based on an energy minimisation argument.

To estimate the critical thickness, one first calculates the stress field in the film within the assumptions of small film thickness and two-dimensional pseudomorphic growth. The coordinates (x, y) are taken in the interface plane and z along the normal to the interface (cf. Fig. 10). The film is subject to two biaxial stresses $\sigma_{xx} = \sigma_{yy}$ and since there is no normal stresss, $\sigma_{zz} = 0$. The corresponding strains are $\varepsilon_{xx} = \varepsilon_{yy} = \varepsilon_o$ and ε_{zz}. According to Hooke's law, we have:

$$\sigma_{xx} = \sigma_{yy} = \lambda(2\varepsilon_o + \varepsilon_{zz}) + 2\mu\varepsilon_o \tag{15-a}$$

$$\sigma_{zz} = 0 = \lambda(2\varepsilon_o + \varepsilon_{zz}) + 2\mu\varepsilon_{zz}, \tag{15-b}$$

where $\lambda = \mu v/(1-2v)$ is a Lamé coefficient. From eqs. 15, one draws easily $\sigma_{xx} = \sigma_{yy} = 2\mu\varepsilon_o(1+v)/(1-v)$.

Threading dislocations are often of screw character and their slip planes are inclined with respect to the interface, so that their Burgers vectors have a non-zero projection in the interface plane. For the sake of simplicity we consider here a simpler case that involves no geometrical factor. The vertical threading dislocation is assumed to have a Burgers vector b parallel to the x direction in the interface plane. Its glide plane is then the vertical (y, z) plane. According to the Peach-Koehler formula (eq. 4), the normal force on the dislocation line induced by the misfit stresses is along the negative y direction and its modulus is $\sigma_{xx}b$ per unit length of line. The corresponding shear in the dislocation slip plane (cf. eq. 3) is $\tau_d = \sigma_{xx} = 2\mu\varepsilon_o(1+v)/(1-v)$.

The dislocation can move over large distances when the Peach-Koehler force exceeds the resistive line tension force. Within the capped layer approximation, this is realised when $\tau_d = \tau_c = T/bh$ (cf. eq. 11). Taking the line tension from eq. 5, with an inner cut-off radius b and an outer cut-off D = h, one eventually obtains:

$$\varepsilon_o = \phi \frac{b}{h_c} \text{Ln}\left(\frac{h_c}{b}\right), \tag{16}$$

with $\phi = (1+v)/[8\pi K(1-v)]$. In the present case. This equation, which is often referred to as the Matthews criterion, provides a relation between the misfit strain and the critical thickness at which an isolated dislocation starts trailing an interface dislocation.

For thicknesses h* larger than h_c, additional interfacial dislocations are formed and further relax the residual elastic strain. For an array of interface dislocations with spacing p* larger than the minimum value p, the relaxed elastic strain is $\varepsilon = b/p^*$ (cf. eq. 13). The non-relaxed elastic strain is $\varepsilon_o - b/p^*$, which provides the residual driving force for further relaxation. Assuming that all the dislocations move independently of each other, eq. 16 is modified as follows

$$\varepsilon_o - \frac{b}{p^*} = \phi \frac{b}{h_c^*} \text{Ln}(\frac{h_c^*}{b}) \qquad (17)$$

This allows estimating the progressive densification of the array of interfacial dislocations as a function of the film thickness. At equilibrium, the driving force vanishes and eq. 17 yields an infinite thickness, corresponding to a vanishing line tension stress.

The standard model for the critical thickness has been the object of many developments and extensions (cf. e.g. [9, 14-16] that also discuss the limitations). Improved versions basically attempt to calculate more precisely the self-energy of the dislocation lines in a semi-infinite medium composed of two materials with different elastic constants.

4. DISCUSSION

4.1. Equilibrium and metastability.

The very first occurrence of misfitting dislocations is not easy to detect and this has been taken as an argument to explain some discrepancies between theory and experiment. When the relaxation occurs by a dislocation mechanism, a general trend is that the critical thickness model seems to work reasonably well in metallic systems exhibiting two-dimensional growth. The reason is that the presence of threading dislocations with high intrinsic mobility is almost unavoidable in metals. It is not always so in semiconducting materials where, as will be discussed below, the mechanisms of nucleation and motion of the dislocations are strongly affected by the covalent nature of the bonding.

The criterion for the relaxation of residual stresses beyond the critical thickness (eq. 17) usually yields, values smaller than the experimental ones. One possible reason is that the threading dislocations interact during their motion with other dislocation of same or different Burgers vectors. A simple model, [7, 13], considers the blocking of moving threading dislocations by the stress field of interfacial dislocations. This induces a local increase of the critical height.

There are, however, fundamental limitations to any equilibrium argument. Figure 11 shows a comparison between experimental and theoretical values of the critical thickness obtained by Houghton[17] in SiGe/Si(100), which yields a fairly good agreement after annealing at a sufficiently high temperature. However, in the same system, the measured critical thickness is found to increase with decreasing annealing temperature and the agreement becomes much less satisfactory. This indicates that the dislocations cannot reach their equilibrium configurations because they have to overcome energy barriers related to their mobility (cf. section 4.4.) or, if they were not initially present in the film, to their nucleation (cf. section 4.3.). In such conditions, given the growth conditions and temperature, it takes the dislocations a characteristic time to overcome these barriers with the help of thermal fluctuations. This time has to be compared with other characteristic times, for instance the ones related to the growth velocity and to temperature changes when annealing experiments are performed. Thus, dislocation configurations can be frozen in metastable states, particularly at low temperature or when the driving force is small (for instance when the misfit strain is partly relaxed). In addition, the growth conditions also influence the result in a manner that is not always predictable. Apart from the static model outlined above, there is no kinematic model of general value that can take all these features into account.

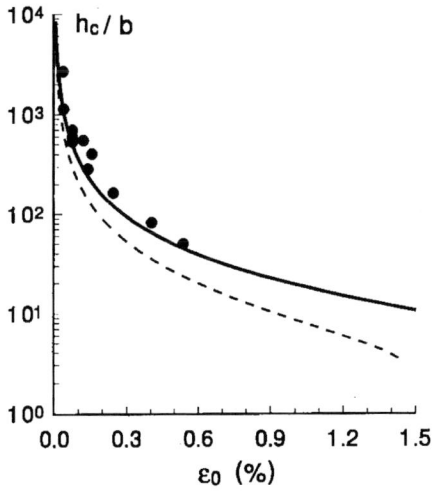

Figure 11. Comparison between theoretical predictions and experimental values (filled circles) of the critical thickness in annealed SiGe/Si(100). The prediction of a simplified Matthew criterion (eq. 16) and of the fully developed version of the model are shown in the dashed and full curves, respectively (Experimental results by D.C. Houghton [17], as reproduced in [13], by Courtesy of the Materials Research Society).

As a consequence, the main outcome of the equilibrium models is that they provide an absolute minimum value for the critical thickness below which no relaxation by misfitting dislocations can occur. To discuss the energetic barriers associated with mestastable dislocation states, it is necessary to examine the conditions that prevail when the film thickness reaches its theoretical critical value. If there are no threading dislocations, as may be preferentially the case in semiconducting materials, the nucleation of dislocations constitutes a first critical step. Once dislocations have been nucleated, or in the case more frequent in metals and alloys where there are grown-in threading dislocations, another energy barrier is associated with the dislocation mobility. Finally, when dislocations move in a thin film, they may interact and multiply according to various mechanisms (cf. [18] for a review and [19] for a mesoscopic simulation). Schematically, when the initial dislocation density is small, multiplication can occur in a spatially correlated manner through mechanisms akin to that of the Frank-Read source (cf. section 2.5.). This usually leads to the formation of interfacial arrays of misfitting dislocations. If the density is large, one may expect numerous interactions preventing the dislocations to reach the interface. This last step is, therefore, largely pre-determined by the two previous ones, on which we focus now.

4.2. Dislocation generation

Owing to the very large energy of dislocations, the homogeneous nucleation of a small critical loop is practically not thermally activated and requires very large athermal stresses. Approximate estimates performed within a continuum frame indicate that the latter must be produced by misfit strains of about two percent, which is not consistent with observation. In such conditions, nucleation is necessarily heterogeneous. It is assisted by the local stress concentrations induced by imperfections (precipitates, steps, growth defects etc....) at the interface, in the bulk or at the surface. The nature of these defects strongly depends on the growth mode, the experimental conditions and the composition of the film and the substrate.

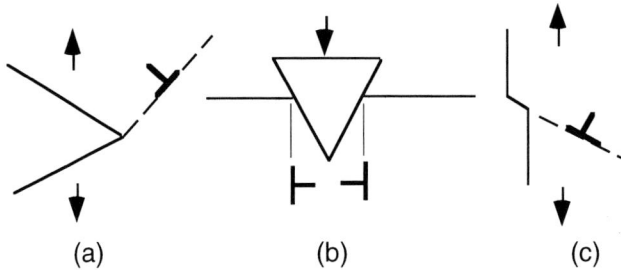

Figure 12. Three situations involving dislocation nucleation under stress. (a) - Nucleation at the tip of a loaded crack. (b) - Nucleation under an indenter. (c) - Nucleation at a surface step in a loaded crystal.

It must be emphasised that there is at present no general model for the critical conditions at which heterogeneous dislocation generation can occur, either in thin films or in other situations. Fig. 12 illustrates three cases, among many others, where dislocation generation is a critical step. One is the nucleation of dislocations at a crack tip (Fig. 12-a), in relation with the brittle-to-ductile transition in materials (cf. the continuum model [20]). The second (Fig. 12-b) deals with the nucleation of dislocations under a nanoindenter and the third one is a model configuration in which dislocations are nucleated at a surface step in a stressed crystal (Fig. 12-c). All these situations have recently been the object of molecular dynamics simulations. Depending on the mode of loading and the crystal orientation, one observes the nucleation perfect or partial dislocations or microtwins (cf. e.g. [6, 21]).

In epitaxial layers, nucleation at the surface becomes a significant mechanism as soon as the growth is no longer two-dimensional. As surface instabilities and island growth occur with "medium" and "large" misfit values, there are many sites for stress concentration from which dislocations and other defects can be nucleated. In the $Si_{1-x}Ge_x/Si$ system, where the misfit can be varied in the range 0-4% by modifying the germanium content, a "medium" misfit value is about 2%. In this compound, as well as in other similar systems, very clear evidences for heterogeneous dislocation nucleation at the surface have been obtained (cf. the review [22] and other articles in the same issue).

4.3. Dislocation mobility

The situation is more favourable as far as dislocation mobilities are concerned, since the latter are reasonably well known from studies on bulk single crystals. This leads to the possibility of modelling the relaxation kinetics of threading dislocations. We consider here the intrinsic mobility, i.e., the motion of isolated dislocations. The effective mass of a dislocation is equivalent to that of a line of atoms [2] and inertial effects are negligible at the mesoscopic scale. As a consequence, steady state velocities are considered, which depend on the net (or "effective") stress on a dislocation, i.e., on the local applied stress minus resistive contributions like the line tension stress. Temperature dependencies occur when the interaction with the lattice, the lattice friction, is significant.

In metallic crystals and at temperatures larger than 0.15 T_m, where T_m is the melting temperature, the lattice friction has disappeared under the effect of thermal activation. The dislocation mobility is governed by the interaction of the moving strain field with electrons and phonons. Under an effective stress τ^*, the steady-state velocity v is written:

$$v = \tau^* b / B, \qquad (18)$$

where B is a weakly temperature-dependent damping constant of the order of 10^{-5}-10^{-4} Pa/s. Thus, with moderate stresses of the order of a few MPa, dislocations can easily reach velocities of the order of 10 m/s. If dislocation motion is not restricted by the interaction with other dislocations, there is practically no kinetic barrier.

In covalent crystals, the mobility is reduced due the interaction of the dislocations with the crystal lattice itself. The lattice friction stems from the distortion of the covalent bonds during dislocation motion. It is responsible for brittleness at low temperature and its influence can persist up to rather high homologous temperatures. We consider here the case of silicon, which is particularly well documented. Below about 850 °C, bulk silicon crystals are brittle. In the temperature range between about 850 °C and 1200 °C (0.6-0.85 T_m), the dislocation mobility is governed by the lattice friction. The influence of the latter progressively disappears with increasing temperature and above 1200 °C the dislocation mobility becomes similar to that of metallic crystals. These critical temperatures are sensitive to the mode of loading and the magnitude of the dislocation velocity. A criterion allowing to check the existence of a lattice friction is that the dislocation lines are no longer smoothly curved under stress but follow crystallographic directions.

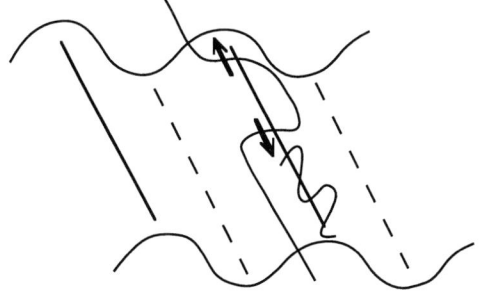

Figure 13. Saddle-point configuration for the motion of a dislocation line over the periodic Peierls barrier. The configuration consists of two kinks whose lateral motion over the secondary Peierls stress transfers the line to its next stable position in the lattice.

The lattice friction can be visualised as a periodic energy barrier, the Peierls barrier, of same periodicity as the lattice, that the dislocations have to overcome under the combined influence of stress and temperature. At equilibrium under zero stress, the dislocation lines tend to stay along the <110> directions, as far as possible of the dense atomic rows in the {111} slip planes in order to minimise the local lattice distortions. At the absolute zero of temperature, they move in a rigid manner over the Peierls barrier when a critical stress value is reached. At a non-zero temperature, this motion is thermally activated and occurs via the formation of a critical saddle-point configuration that is depicted in Fig. 13. This critical configuration consists of a kink pair that expands via the sideways motion of the two kinks, thus transferring the line to the next stable site. During their motion, the kinks also experience a secondary Peierls force, so that the total activation enthalpy for the motion of the line is the sum of the enthalpies for kink pair nucleation and kink propagation. The kink-diffusion model by Hirth and Lothe [3] yields a realistic picture of this process. An equivalent simplified formulation of the dislocation velocities due to Alexander and Haasen [23] is often used for the comparison with experimental results. In both cases, the velocity is of the form:

$$v \propto \left(\frac{\tau^*}{\mu}\right)^m \exp\left(-\frac{\Delta H}{kT}\right), \tag{19}$$

where $m \approx 1$, ΔH is the total activation enthalpy and kT the Boltzmann factor.

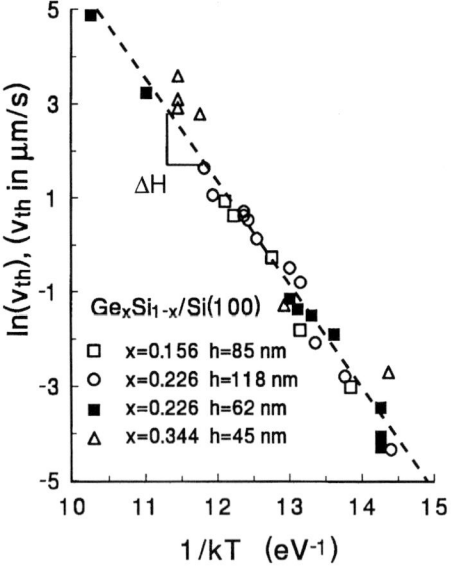

Figure 14. In situ measurement of dislocation velocities in $Ge_xSi_{1-x}/Si(100)$ for different germanium concentrations. The slope of this Arrhenius plot yields the activation enthalpy ΔH for dislocation motion (after Nix et al. [24], reproduced from [13] by Courtesy of the Materials Research Society).

Figure 14 shows results of in situ measurement of this activation energy performed by Nix et al. [24] in $Ge_xSi_{1-x}/Si(100)$ epitaxial layers. The value of the activation energy, 2.2 eV, is the same as the one determined from numerous studies on bulk silicon crystals. Due to the high value of the energy barrier, the mobility strongly decreases with decreasing temperature.

5. CONCLUDING REMARKS

The static Matthews criterion predicts the minimum thickness below which no interfacial dislocations can appear in strained epitaxial layers. Apart from this simple model, there is no consistent kinetic frame of general value and one can only perform case studies. The discrepancies between measured and theoretical values of the critical thickness are due to the occurrence of metastable dislocation states. The latter originate, in order of decreasing importance: - From the thermally activated nature of heterogeneous dislocation nucleation, a mechanism that can only be investigated at atomic scale. - From the thermally activated nature of the dislocation mobility when it is governed by a lattice friction. This mechanism is now being studied at the atomic scale (see e.g. [25]), but it is reasonably well known at the mesoscale. - From athermal dislocation interactions.

Depending on the type of application considered, one may wish to obtain either a dislocation-free film with non-relaxed elastic strains, or a relaxed film with a minimum dislocation density outside the interfacial region. There are many practical ways to approach such objectives. A film of thickness $h < h_c$ contains no misfit dislocations. When one wishes to obtain dislocation-free films of thickness $h > h_c$, it may be dangerous to perform the growth in conditions where the microstructure is metastable. Indeed, such layers may relax in a rather abrupt manner during further thermal treatments. An effective way to control the quality of an epitaxial layer consists of reducing the number of generation sites available when the film thickness reaches the critical value. For two-dimensional growth, intermediate compliant or graded films can accommodate part of the misfit strain, either elastically or by dislocation mechanisms, thereby increasing the critical thickness of the upper layer. For three-dimensional growth, the use of a monolayer of surfactant blocks surface diffusion and induces planar growth. Many other possibilities exist (see [26] for the case of SiGe/Ge) that are beyond the scope of the present paper.

REFERENCES

1. L. Kubin in Quasycristals: current topics, E. Belin-Ferré et al. (eds.), World Scientific, Singapour (2000) 342.
2. J. Friedel, Dislocations, Pergamon Press, Oxford, 1967.
3. J. Hirth and J. Lothe, Theory of Dislocations, Krieger Publishing Company, Malabar, Florida, 1982.
4. R.V. Kukta, Ph. D. Thesis, Brown University, Providence (RI), USA, 1998.
5. C. Lemarchand, B. Devincre, L.P. Kubin and J.L. Chaboche in Multiscale Phenomena in Materials, I.M. Robertson et al. (eds.), Materials Research Society, Warrendale PA, Vol. 578 (2000) 87.
6. S. Brochard, Doctoral Thesis (In French), University of Poitiers, France, 1998.

7. L. Freund, J. Appl. Phys., 68 (1990) 2073.
8. L. Freund, L., Adv. Appl. Mechanics, 30 (1994) 1.
9. W.A. Jesser and J.H. van der Merwe, in Dislocations in Solids, F.R.N. Nabarro (ed.), North-Holland (Amsterdam), Vol. 8 (1989) 421.
10. F.C. Frank and J.H. van der Merwe, Proc. R. Soc. London, Ser. A, 198 (1949) 205.
11. J. van der Merwe, J. Woltersdorf and W.A. Jesser, Mat. Sci. Eng., 81 (1986) 1.
12. J.W. Matthews and A.E. Blakeslee, J. Cryst. Growth, 27 (1974) 78, 29 (1975) 273, 32 (1976) 265.
13. L.B. Freund, MRS Bulletin, 17 (1992) 52.
14. E.A. Fitzgerald, Materials Science Reports, 7 (1991) 87.
15. L.B. Freund, Advances in Applied Mechanics, 30 (1994) 1.
16. M. Putero, Doctoral Thesis (In French), University Aix-marseille III, Chap. 1, 1999.
17. D.C. Houghton, J. Appl. Phys., 70 (1991) 2136.
18. F.K. LeGoues, MRS Bulletin, 21 (1996) 38.
19. K.W. Schwarz and F.K. LeGoues, Phys. Rev. Lett., 79 (1997) 1877.
20. J.R. Rice, J. Mech. Phys. Solids, 40 (1992) 239.
21. S. Brochard, P. Beauchamp and J. Grilhé, Phil. Mag A, 80 (2000) 503.
22. A.G. Cullis, MRS Bulletin, 21 (1996) 21.
23. H. Alexander and P. Haasen, Solid St. Phys., 22 (1968) 27.
24. W.D. Nix, D.B. Noble and J.F. Turlo, in Thin Films: Stresses and Mechanical Properties II, M.F. Doerner et al. (eds.), Mater. Res. Symp. Proc., Pittsburgh, PA, Vol. 188 (1990) 315.
25. A. Valladares, J.A. White and A.P. Sutton, Phys. Rev. Lett., 81 (1998) 4903.
26. M. Mooney, Mat. Sci. and Eng. Reports, R17 (1996) 105.

An atomistic approach for stress relaxation in materials

Guy Tréglia

Centre de Recherche sur les Mécanismes de la Croissance Cristalline (CNRS)[*]
Campus de Luminy, Case 913, 13288 Marseille Cedex 9, France

Abstract : One finds in realistic materials different sites with environments which differ from their own pure bulk environment. It is the case as well for pure materials, due to defects such as grain boundaries or surfaces, as for multicomponent systems (alloys, heteroepitaxial films) in particular in presence of a strong size mismatch. This induces a stress which can be either tensile or compressive. In order to properly account for the asymmetry between these two characters, the stress needs to be modelled at the atomic scale. This allows us to predict the various ways a system has to release (partially) its stress, would it be by structural (reconstructions, superstructures) or chemical (segregation, interfacial alloy formation) rearrangements. A particular emphasis is put on the possible use of local pressure maps as tool to predict the (atomic, chemical) rearrangements for stress release.

1. INTRODUCTION

The purpose of this lecture is to give some tools to study stress release in complex materials at the atomic scale. This can be achieved in the framework of approaches which can be either semi-phenomenological or grounded on the electronic structure. These are the later ones which will be presented here. More precisely, we will show how the essential quantities, defined in the lecture of P. Müller concerning elastic effects in crystal growth [1], can be modelled thanks to the general electronic structure background given in the lecture by M.C. Desjonquères [2] and to the mechanisms described by L. Kubin [3].

We first give in section 2 a qualitative outlook of the various situations in which a material undergoes some stress. This already occurs for a pure element when the periodicity of the infinite bulk material is broken in some way, either along one direction by an interphase (grain boundary, surface) or in all directions as in finite clusters. Obviously the sources of stress increase when different elements have to coexist (alloy, heteroepitaxial growth), particularly if they present a strong size-mismatch. In that case, the coupling between chemical and morphological aspects brings new situations of stress, and also new ways to relax them, by the introduction of point or extended (structural, chemical) defects.

Then, in section 3 we describe the general theoretical framework in which stress relaxation can be modelled at the atomic scale, taking properly into account the asymmetry between the tensile or compressive character. We first define the criteria to be used to relax a system, depending on whether one favours the energetic or the stress point of view. Then we discuss the various energetic models which can be derived from the electronic structure : many-body interatomic potentials or effective Ising-like models.

Finally, the section 4 is devoted to the illustration of the previous sections in some specific cases. Thus, going from the simplest to the most complicated cases, we first treat the

---

[*] associé aux Universités Aix-Marseille II et III

case of pure metals by a comparative study of the defect energies versus pressure profiles. Then we show for bulk alloys how chemical ordering intimately couples to atomic relaxation in case of strong size mismatch. The introduction of a surface makes things still more complex, segregation phenomena interacting with atomic rearrangements. The formation of superstructures at alloy surfaces or during heteroepitaxial growth gives rise to a wide variety of accommodation modes. In all cases one put some particular emphasis on the caution to be taken before using local pressure maps as tool to predict the various ways a system has to release (partially) its stress, would it be by structural or chemical (segregation, interfacial alloy formation) rearrangements.

2. ORIGIN OF STRESS AND ACCOMMODATION MODES

In the most general way, one can say that a material undergoes some stress each time it contains sites the environments of which differ from what they would be in the infinite ideal pure bulk. There are many possible reasons for such situations, but for the sake of simplicity, let us classify them as due to either geometrical or chemical defects.

2.1. Pure elements

A stress occurs as soon as the perfect infinite bulk periodicity is broken by a defect. The double price to pay is the energy of this defect and the stress it induces. This can occur in the bulk in the vicinity of a grain boundary or of a surface, or for a finite cluster. In each case, the environment being different from that in the bulk, due to bond breaking, and/or distortions, the atoms in the defect region are no longer in equilibrium if kept at their bulk positions. This leads to the existence of a defect energy and of a stress, even after relaxation since the the remaining bonds are constrained by the remaining bulk periodicity.

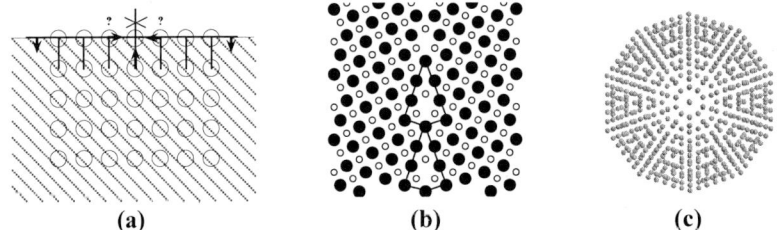

Figure 1. Schematic atomic structures of a surface (a), a grain boundary (b), and an icosahedral cluster (c).

Thus for a surface, breaking bonds costs an excess energy on the one hand, and a tensile stress on the other hand. Both are damped when going into the bulk, in general in an oscillatory manner. Due to the degrees of freedom perpendicular to the surface, the atoms can move vertically towards more favourable positions (surface relaxation). In general, these relaxations are inwards for metallic or covalent systems. Unfortunately, the remaining 2D periodicity forbids them to act similarly inside the surface, leading to some residual excess energy and stress (see Fig.1a). This is the addition of these excess energies which lead to the so-called surface energy [4]. However, the atoms can try to move laterally, either by small or collective displacements. These are the so-called surface reconstructions, which are more difficult to predict in a qualitative way than surface relaxations. Indeed, even though the tensile character of surface stress could suggest that the system should try to compensate it by shortening the remaining bonds, then over close-packing the surface, the actual behaviour is more complex. For example, in fcc metals, the low index surfaces which reconstruct present opposite behaviours : an increased close-packing for the (100) (hexagonal reconstruction) and (111) (long period herring-bone reconstruction) orientations but a lowering close-packing for the (110) one (missing row reconstruction) [4].

In the case of grain boundaries, the region in the vicinity of the defect is also the place for strong modifications of the environment of the atoms : variations of both the number of neighbours and of the interatomic distances, even before any relaxation (see Fig.1b). Moreover, due to the coexistence of both shorter and longer distances, it is difficult to describe in a qualitative way the stress (one expects some sites to be tensile and others compressive). As a consequence, a wide variety of atomic rearrangements is possible, which can lead to some polymorphism of the grain boundary [5].

Finally, the situation of small clusters is different from the above ones since then, due to the lack of any periodicity, the system is completely free to relax in all directions and then to adopt structures which have nothing to do with the bulk ones. Such structures present better energies while keeping a stress distribution. This is the case for instance for small clusters of fcc elements which can present icosahedral shapes with forbidden fivefold symmetries [6] (see Fig.1c).

2.2. Alloys

In an alloy, the coupling between chemical and geometrical effects is a new source of stress, especially when the size-mismatch beween the different components is large. Once again, the situation varies from the bulk to the surface. In the former case, for a dilute alloy, one expects local rearrangements around the impurity which would differ depending on the sign of the size-mismatch (dilation vs compression). For more concentrated systems, the way to accommodate this size-mismatch can even induce contrary behaviours. Indeed, beyond a critical value, intermixing becomes difficult and the system can choose to release its stress by either a precipitation phenomenon (Fig.2a) or reversely (Fig.2b) the formation of well ordered compounds (Hume Rothery phases) [7]. Obviously, in the case of precipitation, there might exist some critical size below which the precipitate is constrained at the parameter of the lattice, and beyond which it could release its stress, with a complex interface behaviour between solvent and solute (see Fig.2a).

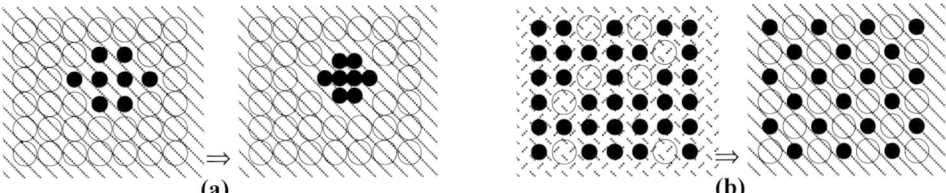

Figure 2. Schematic atomic structures of an alloy presenting a tendency to either phase separation (a) or chemical ordering (b). In both cases we illustrate a possible way for the system to relax its stress.

The situation is still more complex at the surface of a mixed system, would it be obtained as the surface of an alloy or from heteroepitaxial growth. In the former case, surface segregation can be considered as a partial way to release the stress induced by size-mismatch

[8]. In the other case, one can stabilize strained films which release their stress beyond a given critical thickness [9]. Obviously, the way to release the stress is different depending on whether it is tensile or compressive. Schematically, various accommodation modes [3] can be found, from a periodic corrugation of the surface as in Fig.3a to the introduction of extended (continuous matching) or localized defects (misfit dislocations) as in Fig.3b.

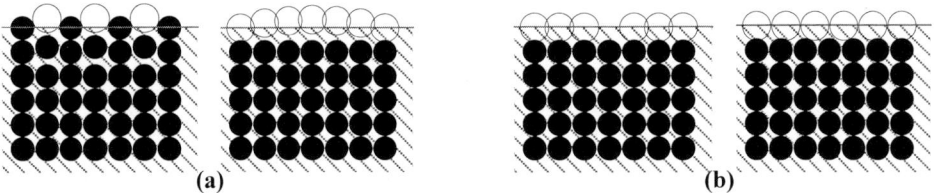

Figure 3. Schematic atomic structures of an alloy surface, illustrating the various ways it can accommodate a strong size-mismatch either by some corrugation (a) or by the introduction of defects (b).

3. MODELLING

3.1. Criteria to be used ?

3.1.1. Energetic criterion

The essential questions are: once granted that some stress is undergone within defective regions, in which case will it be partially released, what is the driving force for this relaxation and what is the best accommodation mode ? From a thermodynamical point of view, the only unambiguous criterion is to minimize the total energy of the system, and more precisely the cost in energy induced by the presence of the defect. This defect energy can be defined in rather similar ways for the various situations described above, roughly as the energy balance involved in its creation. More precisely it is the difference in energy between an initial state where the material is perfect and a final one in which the defect has been introduced, normalised by either the unit cell area or the number of atoms of the defect :

$$E_{defect} = \frac{(E_{final} - E_{initial})}{N_{norm}} \quad (1)$$

in which N_{norm} is the suited normalisation factor. Thus for a pure surface (Fig.4a) this energy balance is the surface energy E_{surf} defined as the energy required to cut an infinite bulk material (initial state) into two pieces (final state), then forming two surfaces each one with $N_{surf} = N_{norm}/2$ atoms. For an alloy, the energy which has to be minimized has something to do with a mixing energy E_{alloy} (or a solution energy E_{sol} for a single impurity) in which the initial state is made of atoms of the two species in their own pure bulk and the final state the alloy for a given configuration (Fig.4b). This energy is normalized with respect to the number of atoms of the minority species $N_{norm} = N_{imp}$. The extension to the case of heteroepitaxial growth of this definition leads to the adsorption energy E_{ads}, in which $N_{norm} = N_{ads}$ atoms of a given species are adsorbed on the surface of a semi-infinite material of an other species, the final state being the covered substrate and the initial one the bare substrate, the other species being in its vapour phase (Fig.4c).

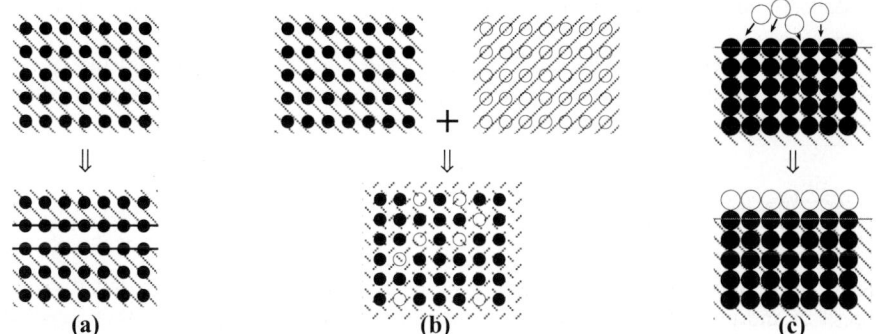

Figure 4. Schematic atomic structures of the various initial and final states involved in the calculation of the defect energy (1) : a pure surface (a), an alloy (b), an heteroepitaxial grown film (c).

It is possible to restrict the definition (1) to the only atoms affected by the existence of the defect if one is able to develop the total energy in terms of local site contributions :

$$E_{tot} = \sum_{i \text{ sites}} E_i \qquad (2)$$

in which case (1) can also be written as :

$$E_{defect} = \frac{\sum_{i \in defect}(E_i - E_i^0)}{N_{norm}} \qquad (3)$$

in which are only counted the sites belonging to the defect region, for which the energy E_i has been perturbed with respect to its value E_i^0 in the initial perfect state. In the previous case, this defect region involves the first layers for the surface, all the sites for the alloy, and the deposit and first substrate layers for the heteroepitaxial growth. Obviously definition (3) can be used (and minimized), as it is, if one aims at determining the equilibrium atomic structure of a grain boundary or of a cluster.

3.1.2. Stress criterion

The major interest of the energetic criterion is that it is completely unambiguous : at equilibrium, the stablest system is the one with the lowest energy. Its major drawback is that it requires the essential knowledge of the final state to be used. More precisely, once assumed a given structure for the defective material (surface reconstruction, ordered phase, deposit superstructure, grain boundary polymorphic structure, cluster shape, ...) one is able to optimize it and to say if it is stabler than an other one. As a consequence the procedure requires an exhaustivity which is difficult to achieve. It is then tempting to ask whether an other criterion could give information on the stability of a structure eventhough one does not know towards what other it could evolve. Hence the idea to calculate the stress profile, with the subsidiary question to know if a structure will try to minimize this stress and if a critical stress has to be not exceeded ! In other words, does a system relax or reconstruct to minimize its stress ?

A first tool which can be used to characterize the stress distribution in a material containing a high number of inequivalent sites is the local hydrostatic pressure which is defined at T=0 K by [10] :

$$P_i = -\frac{dE_i}{dLog(V)} \qquad (4)$$

where V is the atomic volume. The sign of P_i gives that of the stress : positive for a compression and negative for a tension. A map of the local pressure could then give some indication on the tendency of the various sites to relax in one or the other way. This local pressure being isotropic, it is sometimes better suited to get some anisotropic information on the stress. This is for instance the case for the surface stress [1] which is related to the variation in energy induced by a deformation ε_{ij} (i,j=x,y,z). This surface stress is then a tensor (3x3), which for a well suited choice of the axis, can be put under a diagonal form :

$$\bar{\sigma} \equiv \begin{pmatrix} \sigma_{xx} & 0 & 0 \\ 0 & \sigma_{yy} & 0 \\ 0 & 0 & \sigma_{zz} \end{pmatrix} \qquad (5)$$

but reduces to (2x2) after vertical relaxation perpendicular to the surface (along z direction) which leads to $\sigma_{zz} = 0$, whereas this cannot be achieved for the other two components (σ_{xx}, σ_{yy}) which are bound horizontally by bulk infinite periodicity along x and y axis. This surface stress is obtained as a derivative of surface energy, which writes (in the Shuttleworth notation) as [1] :

$$\sigma_{ij} = \frac{1}{A}\frac{\partial(AE_{surf})}{\partial \varepsilon_{ij}} = E_{surf}\delta_{ij} + \frac{\partial E_{surf}}{\partial \varepsilon_{ij}} \qquad (6)$$

where A is the surface area. Note that the surface local pressure given by (4) is closely related to surface stress (5-6) since it is nothing but the trace of the tensor :

$$P_{surf} = -\frac{1}{3}\text{Tr}[\bar{\sigma}] \qquad (7)$$

3.1.3. Questions

Once given the above definitions, the questions we have to answer are the followings. Does a material tend to minimize its stress, and is there some critical value of this stress beyond which one could consider the system as unstable ? How could we predict the stablest state from the single knowledge of the unstable one, and from this point of view is a pressure map a good indicator of the (atomic, chemical) rearrangements to come ? More precisely, is it justified to predict that compressive (tensile) sites could be sources for constitutive vacancies (additional atoms) in pure metals and/or be substituted by smaller (bigger) atoms in bimetallic systems ? If so, such tendencies could serve as guides for the possible reconstruction (surfaces, grain boundaries, clusters) and segregation (surface, grain boundaries) phenomena. Answering these questions first requires to calculate the above quantities (E_i, P_i, σ_{ij}) and therefore to get a reliable energetic model at the atomic scale.

3.2. Energetic model

3.2.1. Atomistic approach

A reliable energetic model at the atomistic scale requires to define it from the electronic structure. It is obviously impossible at the present time to derive an equilibrium configuration (chemical and/or geometrical) directly from ab initio electronic calculations in view of the complexity of the systems we want to describe. It is therefore necessary to build semi-empirical descriptions of the electronic structure, sufficiently simple to be used in the numerical simulations (Molecular Dynamics, Monte Carlo) performed to determine the equilibrium state. In the case of transition and noble metals or of covalent systems, the Tight-Binding scheme provides a well suited framework to do that [11]. However, it is still too heavy to be used directly in the simulations so that other approximations are required, which do not have to fulfill the same criteria depending on the addressed problem. Thus, if interested in atomic rearrangements, and at least in the case of metals, a crude description of the local density of states (*ldos*) is only required so that it is possible, from the single knowledge of its second moment (SMA : Second Moment Approximation [12]), to derive many-body potentials which can then be used in molecular dynamics simulation at fixed chemistry. On the other hand, if interested in chemical rearrangements, it is necessary to go beyond second moment (up to at least four), in which case it is possible to derive an Ising type model on a fixed rigid lattice (TBIM : Tight-Binding Ising Model [13]) which can then be used in mean-field calculations or Monte Carlo simulations. Obviously, both degrees of freedom should be coupled (for instance in mixed Monte Carlo), which requires to use the SMA beyond its strict range of validity, taking lessons of TBIM. Finally, once the equilibrium configuration is reached, it is possible either to determine its electronic structure in a detailed manner if interested in specific applications (but this is not the purpose here) or to calculate local energies and related quantities such as pressure and stress.

3.2.2. Many-body potential in the Second Moment Approximation (SMA)

Let us first recall the expression of the Tight-Binding hamiltonian for a pure metal, written in the basis of atomic orbitals λ at sites i $|i,\lambda\rangle$, as [11] :

$$H = \sum_{i,\lambda} |i,\lambda\rangle \varepsilon_{at}^{\lambda} \langle i,\lambda| + \sum_{i,j,\lambda,\mu} |j,\mu\rangle \beta_{ij}^{\lambda\mu} \langle i,\lambda| \qquad (8)$$

which involves two parameters, the effective d atomic level $\varepsilon_{at}^{\lambda}$ and the hopping integrals $\beta_{ij}^{\lambda\mu}$. From this hamiltonian, it is possible to calculate the *ldos* at site i as [2] :

$$n_i(E) = -\frac{\text{Im}}{\pi} \sum_{\lambda} \langle i,\lambda | (E\ \hat{1} - \hat{H})^{-1} | i,\lambda \rangle \qquad (9)$$

and from it the total energy $E_{tot} = \sum_i E_i$, with site energies E_i given by [14]:

$$E_i = \sum_\lambda \int^{E_F} (E - \varepsilon_{at}^\lambda) n_i^\lambda(E) dE + \frac{1}{2} \sum_{j \neq i} \iint dr_1 dr_2 \frac{Q_i(r_1) Q_j(r_2)}{|r_1 - r_2|} \qquad (10)$$

where the first term is a negative (attractive) contribution due to the band formation from the atomic level, and the second one a repulsive interaction between spheres with charge Q_i. Unfortunately the latter contribution is too weak to treat correctly the repulsive interaction between the ions at short distance so that it has to be replaced by a pairwise empirical contribution. In what concerns the first term, it reduces to a simple analytic form if one replaces the actual ldos by a schematic one having the same second moment. It leads to a many-body contribution which cannot be written as a sum of pair interactions, but instead as the square-root of such a sum [12]:

$$E_i = -\sqrt{\sum_{j, r_{ij} < r_c} \left\{ \beta^2 \exp\left[-2q\left(\frac{r_{ij}}{r_0} - 1\right)\right]\right\}} + \sum_{j, r_{ij} < r_c} \left\{ A \exp\left[-p\left(\frac{r_{ij}}{r_0} - 1\right)\right]\right\} \qquad (11)$$

where r_{ij} is the distance between sites i and j, r_c is the cut-off distance beyond which the interaction vanishes and r_0 the nearest neighbour distance in the bulk. Under this form, this potential depends on four parameters: A, β, p and q, which are fitted for a pure metal to experimental values of the cohesive energy (E_{coh}), the lattice parameter, the bulk modulus (B_{mod}) and shear elastic constants (C_{44} and C'), or alternately to the universal equation of state [15]. This fitting procedure of the potential parameters on bulk properties allows us to transfer them to more complex cases. In particular, for surfaces, the SMA potential leads to an inwards relaxation of the surface layer, in agreement with experiments, contrary to simple pairwise ones which predict an outwards relaxation [2]. Note however that the corresponding surface energies (E_{surf}) are generally found twice too small. Such behaviours are also recovered by semi-empirical potentials of the same type, such as the embedded atom method (EAM [16]) or glue model [17].

It is then simple to use this potential in a Molecular Dynamics simulation by solving the Newton movement equation [18]:

$$\vec{F}_i(t) = m \frac{d\vec{v}_i(t)}{dt} = -\frac{dE_{tot}}{d\vec{r}_i} \qquad (12)$$

in which F_i, but also P_i and σ_{ij}, are obtained in an analytic way in terms of A, β, p and q thanks to the expression of SMA. The Molecular Dynamics simulations can be performed either at finite temperature if interested in diffusion processes, or at 0 K if only searching for the stablest structure minimizing the internal energy. In the latter case, the procedure is to quench the system by cancelling the velocity (v_i) of an atom i when it goes against the corresponding force (F_i): $\vec{F}_i(t) \cdot \vec{v}_i(t) < 0 \Rightarrow \vec{v}_i(t) = 0$.

3.2.3. Effective Ising Model (TBIM)

In presence of two different species A,B, the hamiltonian (8) has to be extended to take into account the effect of chemical order on its two parameters [14]. First the diagonal one $\varepsilon_{at}^{\lambda}$ has to be replaced by $\varepsilon_i^{A,\lambda}$ or $\varepsilon_i^{B,\lambda}$ depending on the corresponding site i to be occupied by an A or B atom, and the off-diagonal parameter $\beta_{ij}^{\lambda\mu}$ by $\beta_{ij}^{\lambda\mu}(A,A)$, $\beta_{ij}^{\lambda\mu}(B,B)$ or $\beta_{ij}^{\lambda\mu}(A,B)$ depending on the respective occupancies of sites i and j. In general, one can neglect the dependence of the hopping integrals with respect to chemical ordering (no off-diagonal disorder) which is weak compared to that of the diagonal parameter : indeed the bandwidth, related to β, varies a few whereas the atomic level decreases by 1eV from an element to the other along a transition series. Then, neglecting also the (weak) dependency of the atomic level with respect to the d orbital degeneracy λ, the hamiltonian writes :

$$H(\{p_i^X\}) = \sum_{i,\lambda,X} p_i^X |i,\lambda\rangle \varepsilon^X \langle i,\lambda| + \sum_{i,j,\lambda,\mu} |j,\mu\rangle \beta_{ij}^{\lambda\mu} \langle i,\lambda| \qquad (13)$$

where $\{p_i^X\}$, with $p_i^X=1$ or 0 depending on that the site i is occupied or not by an atom of type X (=A,B), denotes the chemical configuration, i.e. the localization of A and B atoms on the lattice.

It is obvious from this hamiltonian that a description of the *ldos* is only affected by chemical ordering if it includes at least four moments (since one has to count closed paths of four hoppings on the lattice to explore the chemical nature of a site and of its neighbours) [19]. This means that a *second moment* approximation should not be used to describe chemical ordering. The alternative is to make a *N-moment* calculation and then to treat the energy of the system for a given chemical configuration by a perturbation treatment [20] with respect to a completely disordered state treated within an adequate approximation (here the Coherent Potential Approximation, CPA [21], in which the actual on-site energies $\varepsilon_i^{A,\lambda}$ and $\varepsilon_i^{B,\lambda}$ are replaced by an average complex effective potential σ). One obtains :

$$E_b = \sum_{i,X} p_i^X \int^{E_F} n_i^X(E).E.dE - E_F \int^{E_F} n_i^X(E).dE = \overline{E}(c) + H^{mix}(\{p_i^X\})$$

$$H^{mix}(\{p_i^X\}) = \sum_{i,X} p_i^X h_i^X + \frac{1}{2} \sum_{i,j,X,Y} p_i^X p_j^Y V_{ij}^{XY} \qquad (14)$$

in which the two energetic parameters (h_i^X, V_{ij}^{XY}), for X,Y=A,B, are calculated from the electronic structure of the disordered alloy in the CPA approximation.

$$h_i^X = \frac{\text{Im}}{\pi} \int^{E_F} dE \sum_{\lambda} \log[1 - (\varepsilon^X - \sigma_i)\overline{G}_{ii}^{\lambda\lambda}(E)] \qquad (15a)$$

$$V_{ij}^{XY} = -\frac{\text{Im}}{\pi} \int^{E_F} dE \ t_i^X(E) t_j^Y(E) \sum_{\lambda,\mu} \overline{G}_{ij}^{\lambda\mu}(E) \overline{G}_{ji}^{\mu\lambda}(E) \qquad (15b)$$

in which $\overline{G}_{ij}^{\lambda\mu}(E)$ is the Green operator in the disordered state and t_i^X the so-called t-matrix :

$$\overline{G}_{ij}^{\lambda\mu}(E) = \langle i,\lambda |\left([E-\sigma]\hat{I}-\hat{H}\right)^{-1}|j,\mu\rangle \qquad t_i^X = \frac{\varepsilon^X - \sigma_i}{1-(\varepsilon^X - \sigma_i)\sum_\lambda \overline{G}_{ii}^{\lambda\lambda}(E)} \qquad (15c)$$

This means that, although the total energy cannot be written as a sum of pair interactions, the single small part of it which depends on chemical configuration can nevertheless be written in an Ising-like form. Moreover, for all problems which reduce to exchanges between A and B atoms (chemical ordering, phase separation, interfacial segregation, interdiffusion), it is easy to see that the energy balances only involve differences of the parameters h_i^X and V_{ij}^{XY} both between A and B atoms and between the sites affected by the exchange. An important example is that of surface segregation for which the segregation energy writes [22] :

$$\Delta H_p^{TBIM} = \Delta h_p + (1-2c)Z^{tot}V - \sum_{n=-q}^{n=+q}(1-2c_{p+n})Z_{p,p+n}V_{p,p+n} \qquad (16)$$

$$\Delta h_p = (h_p^A - h_p^B) - (h_{bulk}^A - h_{bulk}^B) \qquad (17)$$

$$V_{ij} = \frac{\left(V_{ij}^{AA} + V_{ij}^{BB} - 2V_{ij}^{AB}\right)}{2} \qquad (18)$$

in which it can be shown that Δh_p differs from zero at the surface only (p=0) in which case it is closely related to the difference in surface energies between A and B species ($\Delta h_0 \approx E_{surf}^A - E_{surf}^B$) and V_{ij} is the effective pairwise term between sites i and j [13]. For a given lattice it decreases with distance and can be restricted to first neighbours [23], in which case $V=V_{ij}$ drives the tendency of the bulk alloy to order (V > 0) or to phase separate (V < 0). This interaction is enhanced if n or m is a surface site ($|V_0| > |V|$) [13].

Obviously, such a description developed on a rigid lattice only accounts for the chemical stress but not for the one induced by size-mismatch. This can be corrected in an approximate way by adding a contribution to this energy balance [24], in the form of a local term :

$$\Delta H_p = \Delta H_p^{TBIM} + \Delta H_p^{size}(c) \qquad (19)$$

where the additional term is also essentially non vanishing at the surface, in which case it is calculated in the dilute limits, $\Delta H_0^{size}(c \to 0,1)$, by means of an *ad-hoc* SMA potential fitted within the assumption that A and B elements only differ by their lattice parameter [25]. As will be seen in §4.3.1 the result is a tendency to segregation of the impurity, for the biggest atoms only in the case of close-packed surfaces, and for both species in the case of open ones thanks to atomic relaxations.

3.2.4. Consistency SMA-TBIM

In spite of the above discussion concerning 4th moment vs 2nd moment prescriptions, one can be tempted to study atomic rearrangements for a fixed chemistry configuration by using a SMA potential with mixed A-B parameter playing the role of an effective off-diagonal disorder replacing the actual diagonal one. This implies to fit them to the essential characteristics of the alloy, i.e. its phase diagram, in order to reproduce at least the actual tendency to order or to phase separate. In practice, this can be achieved by fitting the solution energies and/or the ordering ones. Nevertheless the coupling between ordering processes and size-mismatch induced stress is far from being simple, so that it is useful to check *a posteriori* the consistency between what we guess to reproduce from the chemical trends, and what we really get. In particular, one can ask what would be effective ordering pair interactions calculated in this extended SMA model before using it to treat simultaneously chemical and atomic rearrangements. Some partial answers will be given in sections 4.2 and 4.3.

4. ILLUSTRATIONS

The purpose of this section is to illustrate the possible link between the energy of a defect and the pressure or the stress it induces. More precisely, we will show up to what limit a pressure map can be used as a guide for possible atomic and/or chemical rearrangements in a material. In pure metals a naïve idea could be that compressive (resp. tensile) regions would be privileged places for the introduction of vacancies (resp. additional atoms) leading to a decrease (resp. increase) of close-packing. For alloys, other solutions are available to release the stress induced by size mismatch, among which replacing the big (resp. small) atoms by atoms of the other species in compressive (resp. tensile) regions [26]. The question is to know up to what limit stress can be considered as the driving force for atomic reconstructions or formation of superstructures, the energetic criterion being ultimately the only unambiguous rule.

4.1. Pure elements

4.1.1. Surfaces

It has been previously mentioned that breaking bonds always leads to a tensile stress, at least when using realistic potentials such as the SMA one. As a consequence, a surface, even vertically relaxed, remains under tension while presenting an energetic cost (surface energy). Although all the surfaces are found experimentally and theoretically to relax, only some of them reconstruct [2]. In general, the latter obey some well identified trends along the periodic table. Thus the (100) surface of bcc metals reconstructs (zig-zag reconstruction or dimerization) for the column of W but not for that of Ta, indicating a role of the d-band filling which can indeed be put in evidence in terms of Peierls-like distortion [27]. The case of fcc metals in the last three columns of the transition series is a little bit more puzzling since now the tendency to reconstruct or not varies from one series to the other. More precisely, the elements of the 5d series reconstruct whereas those of the 3d and 4d ones do not, the reconstruction being very dependent on the crystallographic orientation. Thus, the (100) and even (111) faces reconstruct towards more compact structures (hexagonal for the (100) [28], long period herring-bone one for the (111) [29]) whereas the (110) face tends to be less compact (missing row reconstruction [30]). One can then wonder how does the nature of the element under consideration act ? A possible lead could be the existence of a critical value of the energetic cost of the defect. One can see in the figure 5a, in which the surface energy is scaled to the bulk cohesive energy in order to exhibit the intrinsic property of the defect (decoupled to tendencies due to the bulk such as the parabolic behaviour of the cohesive energy), that no tendency appears (in spite of a peculiar behaviour of the 5d series).

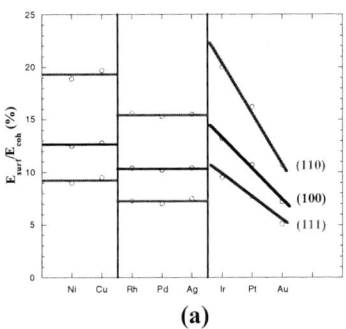
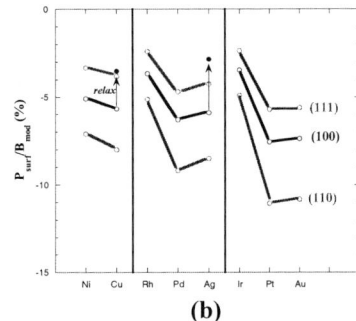

(a) (b)

Figure 5. Calculated variation of the surface energy normalized to the cohesive one (a) and of the surface pressure normalised to the bulk modulus (b) for fcc metals along the three transition series, for the three low index surface orientations. The empty (full) dots are for the unrelaxed (relaxed) calculations performed with SMA interatomic potentials.

The second lead could be the existence of a critical stress. One has then plotted in a similar way the ratio of the (tensile) surface pressure [31] to the bulk modulus, but here again no clear effect can be seen. This is not so surprising in view of the different characters of the reconstructions which do not act in all cases in favour of an increase of close-packing at the surface. It is therefore difficult to get from Fig.5b any critical stress beyond which a surface should reconstruct, even though elements of the 5d series seem to be over stressed with respect to others.

In order to make such arguments more quantitative, it is necessary to check, at least for a given reconstruction, that one is able to reproduce a decrease (increase) of the energy of the systems under reconstruction for those which are known to (not) reconstruct experimentally, and to identify what are the driving forces for this trend. Then, if it works, one can calculate the surface stress before and after a reconstruction in order to see if the tendency to reconstruct derived from the energy calculation is followed by a decrease of the stress (as guessed) or not ? This is illustrated here in the case of the missing row reconstruction [31]. In that case, as can be seen in figure 6, the SMA potential gives the overall experimental energetic trend from the 3d series to the 5d one, Pd being a little bit apart. Moreover, this trend can be shown to be driven by the increase from the 3d series to the 5d one of the q parameter which characterizes the decay of the hopping integral with distance [30]. What we can ask to a stress criterion is to be at least as efficient as this energetic criterion. Unfortunately, as can be seen in Fig.6 [31], the situation is far from being so clear for the surface stress. Indeed, not only the stress is not reduced in all cases, but in addition it is highly anisotropic depending on whether it is perpendicular or parallel to a close-packed direction. An alternative would be to use as an indicator not the stress but instead the variation of the surface energy with strain which can be derived from the second way to write the Schuttleworth relation (6):

$$\frac{\partial E_{surf}}{\partial \varepsilon_{ij}} = \sigma_{ij} - E_{surf}\delta_{ij} \qquad (20)$$

As can be seen in Fig.6, the correlation between the sign of this slope and the energetic criterion along a close-packed row is almost perfect, so that this could be a better guide, but it is now necessary to go beyond the single observation of a correlation [31] ! Another important result is that, although relaxation does not change qualitatively the trends for the surface energy, it is essential in the case of stress calculations [31].

	Ni	Cu	Rh	Pd	Ag	Ir	Pt	Au
experiment	(1x1)					(1x2)		
ΔE_{surf} (1x2) - (1x1)	> 0	≤ 0	≈ 0			< 0		
$\Delta\sigma_{yy}$ (1x2) - (1x1)	> 0		≈ 0			> 0		
$\Delta\sigma_{xx}$ (1x2) - (1x1)	< 0		≈ 0			> 0		
$dE_{surf} / d\varepsilon_{yy}$ (1x1)	> 0	< 0	≈ 0			< 0		

Figure 6. On the left-hand side are shown the various possible criteria to be used for predicting the (1x2) missing row reconstruction of the fcc (110) surface [31]. Within the grey scale, a black (low grey) box means that the criterium favours (unfavours) the reconstruction, while the intermediate grey gives no indication. The reconstruction is illustrated on the right-hand side, in which the dots are for surface atoms and the crosses for the first underlayer ones, the possible missing rows being indicated by empty dots.

Finally, it is worth noticing that the missing row reconstruction is perhaps a little bit apart from the others which in general tend to favour an increase of close-packing at the surface by addition of atoms (hexagonal-type reconstructions), or at least some trial for the surface atoms to shorten bonds (dimerization of bcc or covalent materials). This peculiarity of the (110) face could be at the origin of the complex behaviour of stress in this case. Similar studies have been performed for other types of reconstructions, among which the hexagonal one for fcc metals of the 5d series, including ab initio calculations [32], which tend to generalize the results that reconstruction does not lower the tensile stress in all cases, in agreement with some experimental data [33].

4.1.2. Grain boundary

The case of a grain boundary is also illustrative of what can be learnt from a stress profile concerning the relaxation of the defect. This is described here in the particular case of a Σ5 (210) [001], for which different poymorphic forms can exist which are indeed found with comparable stabilities from the energetic calculations performed within SMA [34]. The main two structures are shown in figure 7a. As can be seen, in a somewhat similar way to the case of surfaces, these competitive structures can be associated to either an increase or a decrease of the close-packing in the grain boundary region. Note however that the situation is still more complex than that of surfaces, since in the latter case all sites present tensile characters whereas now one gets an inhomogeneous pressure profile presenting both compressive and tensile sites (see figure 7c). In particular, the most close-packed structure exhibits a site with a strong compressive character which does not exist in the other, and which can be shown to be also strongly penalizing from the energetic point of view (Fig.7b). However one has to be carefull before concluding that the structure should present an enhanced stability under introduction of a vacancy at this site. Indeed, as can be shown from the comparison with the competitive less close-packed structure, the energetic profile is indeed less marked in the latter case but also it is less damped as a function of the distance to the defect plane. This means that one has to prefer an overall view when comparing different structures and to take some caution before using a too much local analysis of the stress to predict the addition or the suppression of individual atoms to relax the defective structure, collective rearrangements being sometimes more efficient. One will see later that, in the case of alloys, changing the chemical species of the atoms at given sites can be also a more efficient way to reduce the local stress, which can be at the origin of segregation phenomena.

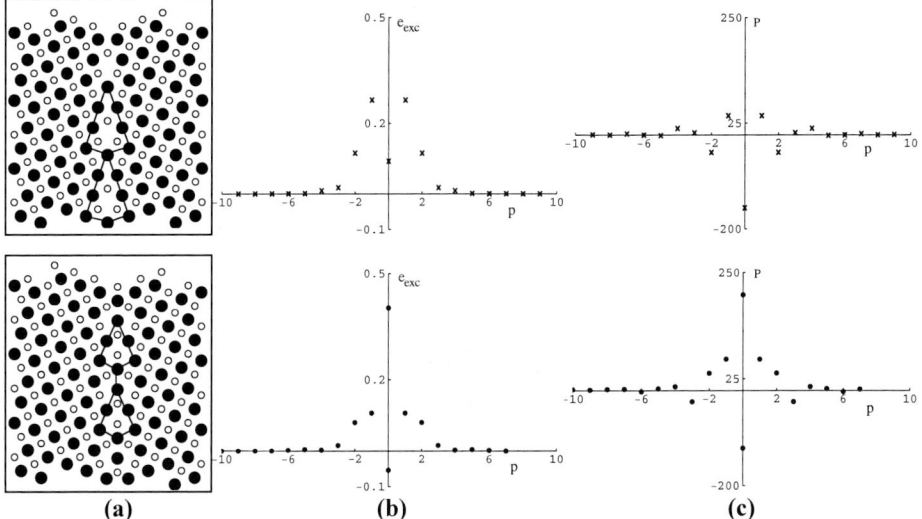

Figure 7. Polymorphism of a Σ5 (210) [001] grain boundary in pure Cu (upper and lower part) from SMA calculations [34] : (a) relaxed atomic structure, (b) excess energy (in eV) and (c) pressure profiles (in kbar) as a function of the index of the layers (p) parallel to the boundary plane (p=0).

4.1.3. Clusters

Let us finish our atomistic study of stress in pure materials by the case where it is induced by the finite character of the object under study. Thus it can be shown, using realistic potentials such as the SMA or other equivalent ones, that small clusters of a fcc metal can exist under shapes which exhibit forbidden fivefold symmetries. More precisely, the icosahedral structure can be preferred to the fcc one (cubo-octahedron or Wulff polyhedron) up to a critical size beyond which a structural phase transition to the bulk structure occurs (Fig.8a). This transition, which is metal dependent, is driven by the ratio surface/bulk which decreases with increasing size [35]. Naïvely speaking, an icosahedron can be viewed as a cluster presenting only close-packed (111)-type facets whereas a fcc cluster also presents more open (100) facets which cost more energy. On the contrary, the five-fold symmetry penalizes the core of the icosahedron, hence the transition beyond a given size. A more detailed analysis can be made from the stress profile inside each cluster (see figure 8b). Indeed, although the fcc cluster presents essentially trends which are similar to those of extended surfaces (tensile facets and a rapidly damped profile inside), the icosahedron exhibits a very inhomogeneous profile in which pressure re-increases in the core of the cluster leading to a strongly compressive core region, which is also penalized from the energetic point of view [35]. Contrary to the previous case of grain boundary, the monotonous aspect of this profile allows us in the present case to use it as a guide for introducing vacancies in the compressive region which are expected to play a stabilizing role. This is confirmed by energetic calculations (Fig.8c) which show that indeed there exists constitutive vacancy clusters (the size of which depends on the nature of the metal but which can go up to tetrahedron) which stabilize the icosahedron and give new magic numbers lower than the usual ones [36]. Although the existence of such hollowed 3D clusters is difficult to evidence experimentally, 2D ones have indeed been observed during growth [37].

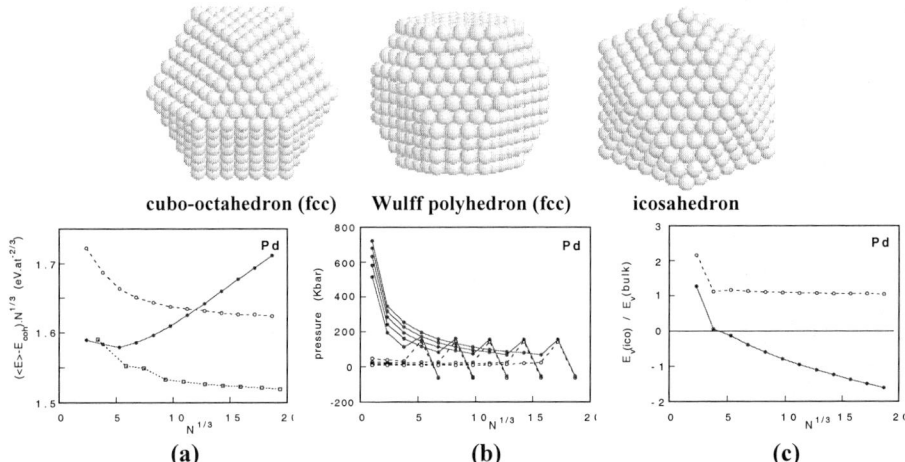

Figure 8. (a) Comparison between the energies (SMA) of the different shapes illustrated in the upper part (empty squares : Wulff polyhedron, empty dots : cubo-octahedron, full dots : icosahedron), as a function of the reversal of their size (cubic root of the number of atoms), in the case of Pd [35]. The pressure profiles as a function of the successive shells around the central atom for clusters of different sizes (empty dots : fcc polyhedron, full dots : icosahedron) are shown in (b) [35]. The formation energy of a monovacancy (empty dots : fcc polyhedron, full dots : icosahedron) scaled to its value in the bulk is plotted in (c) for clusters with increasing size [36].

4.2 Bulk alloy

The purpose here is to show that atomic relaxations induced by a strong size-mismatch can modify the local chemical order in a material. This can lead to some segregation in the defective region (grain boundary) or even to a change in the intimate tendency of the system to favour the formation of heteroatomic bonds (tendency to chemical ordering) or homoatomic ones (tendency to phase separation).

4.2.1. Coupling order-relaxation (impurity)

Our aim is to define the limits of a possible consistency between a rigid latice approach (TBIM) suited to a chemical configuration study and a complete effective SMA interatomic potential extended to alloys in spite of the problems evoked above. The first step, in order to identify the influence of atomic relaxations on chemical ordering, is to calculate the ordering pair interaction V_{ij} between sites i and j (see eq.18), which can be achieved along two different procedures. The first one is to calculate the solution energy, i.e. the energy balance involved during the dissolution of a single impurity from its own bulk structure :

$$\Delta E_{sol} = \sum_n Z_n \left(V_n^{AB} - \frac{V_n^{AA} + V_n^{BB}}{2} \right) = -\sum_n Z_n V_n \qquad (21)$$

where the summation acts on the different shells (n = 1, 2, 3, ...) of neighbours of the impurity (Z_n being the corresponding coordination numbers). A positive (negative) value of ΔE_{sol} then reveals a tendency to phase separation (ordering). The second method is to bring two initially isolated impurities towards positions of nth neighbours, in which case the energy balance per impurity is :

$$\Delta E_n = \frac{V_n^{AA} + V_n^{BB}}{2} - V_n^{AB} = V_n \qquad (22)$$

From these two calculations, performed within SMA before and after atomic relaxations, one will learn what is the influence of the stress induced by size-mismatch on chemical ordering. This will bring us useful information on the consistency between Ising-like and many-body potentials, and on a possible renormalization of V to account for size effect in the former case. The calculations have been performed for two systems presenting a similar strong size mismatch (>10%), Cu-Ag and Cu-Pd, but opposite bulk ordering tendencies (phase separation in the former case and ordered phases in the latter one) [38].

The first calculation concerning the dissolution of a single impurity does not show any pronounced effect, which was expected since the mixed parameters of the SMA potential have been fitted to solution energies. The essential result is that atomic relaxation is more important when the impurity is the largest (due to the asymmetry of the potential with respect to tensile or compressive strain), and for the system presenting a tendency to phase separation (Cu-Ag). In both cases relaxation decreases the solution energy, which means that it favours order with respect to phase separation, but never leads to any change of sign. A local analysis of the local contribution on the sites of the impurity and of its neighbours give a more detailed information. It is found in the Ag-Cu case that an Ag atom « likes » to be surrounded by Cu ones, but that a Cu atom « does not like » at all to be surrounded by Ag ones, while in the Cu-Pd case the Cu atoms « like » to be surrounded by Pd ones and the reversal does not matter. In other words, for ordering systems the less cohesive element « likes » to be surrounded by the other whereas for phase separation ones the most cohesive element « does not like » to do that [38]. Here, Cu plays alternately one role and the other !

The situation is more complex when bringing closer two isolated impurities. Indeed the calculation on a rigid lattice (no relaxation) reveals a tendency to bulk ordering for both systems (see figure 9) ! The atomic relaxation of the system decreases this tendency, up to changing the sign and then recovering the good tendency for Cu-Ag or leading to more realistic values for CuPd. This tendency is fully confirmed by *ab initio* calculations [39]. This means that now relaxation favours phase separation, contrary to what occured for a single impurity. Another surprising result is that relaxation enhances the interaction between second neighbours (V_2) for Cu-Ag, in contradistinction with simple sterical arguments. Less spectacular, one recovers that relaxation plays a more important role when the impurity is the largest atom and one confirms the previous argument relating the respective cohesive energies of the two species and the sign of V.

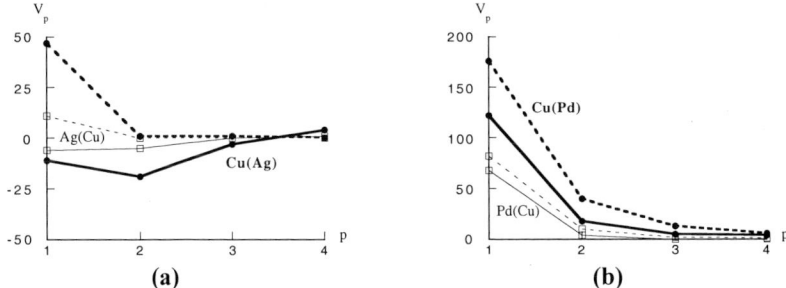

Figure 9. Variation of the effective pair interaction V_p (in meV), calculated within SMA, as a function of distance (p=1 : 1st neighbours, p=2 : 2nd neighbours ...) in the Cu-Ag (a) and Cu-Pd (b) systems [38]. Both dilute limits are shown : thick lines for Cu(Ag) and Cu(Pd) and thin lines for Ag(Cu) and Pd(Cu). The full (dotted) lines refer to the relaxed (unrelaxed) situation.

Finally, it is tempting to reconstruct ΔE_{sol} (Eq. 21) from the values of V derived using Eq. (22). This leads (obviously) to a total disagreement in the non relaxed situation, for which the problem can be solved by calculating ΔE_{sol} only once the impurity has been prepared at the lattice parameter of the matrix. Unfortunately, such a procedure does not hold in the relaxed case, for which the inconsistency between Eq. (21) and Eq. (22) remain [38].

4.2.2. Coupling order-relaxation (clusters)

One can wonder how this influence of the atomic relaxation on the ordering tendency evolves when going to less dilute systems. It is studied here in the case of the Ni-Ag system, which presents the same characteristics as the Cu-Ag one (Ni playing the role of Cu and being smaller than Ag), but for which experimental data are available at least close to the surface [40]. More precisely one has calculated the relaxation of a precipitate (here Ni) in a matrix (here Ag), as a function of the size of the cluster in three cases : on a rigid lattice (tensile stressed Ni), on a relaxed one at constant number of Ni atoms (N_{at}^{Ni}) and finally after a complete relaxation including the optimization of N_{at}^{Ni}. Two main results are obtained [41].

The first one is that, contrary to the case of free clusters, the relaxation of an icosahedron changes it into a fcc cubo-octahedron, due to the stress imposed by the interface with the fcc matrix. This means that the interfacial size-mismatch prevents from forbidden symmetries, even at small sizes. The second one is that the solution energy varies in a completely different manner in the three cases, as shown in figure 10.

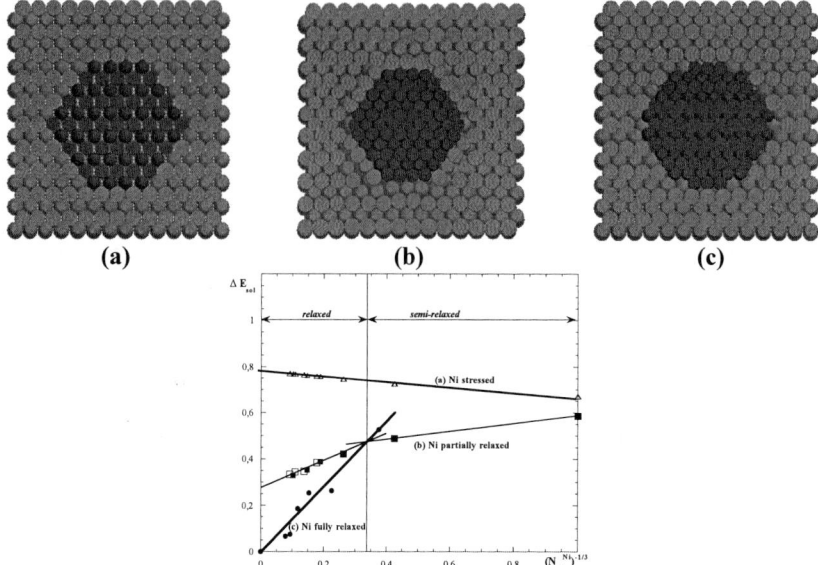

Figure 10. The upper part illustrates the relaxation of a Ni cluster in a bulk Ag matrix within SMA calculations [41]. The tensile stressed situation is shown in (a) whereas the Ni cluster is relaxed either by keeping constant the number of Ni atoms (b) or by varying it (c). The corresponding solution energies (in eV) are plotted in the lower part as a function of the reversal of this number, the curves being linearly fitted according to Eq. (23).

First, one can see that the Ni cluster tends to relax to its own parameter beyond a critical size of about 30 atoms, in agreement with SXRD measurements [40]. However, below this size, a partial stress is preferred from the energetic point of view. It is is tempting herealso to analyze this result in terms of effective pair interactions, which can be done by schematically separating in the energy balance ΔE_{sol} the interfacial contribution from that of the core of the precipitate. This leads to a linear dependency :

$$\Delta E_{sol} \approx \Delta E_b^{Ni} + \alpha \ V_1 \left(N_{at}^{Ni} \right)^{-\frac{1}{3}} \qquad (23)$$

in which α is a structure factor depending on the shape of the cluster (here $\alpha \approx 20$ for an isotropic shape) and ΔE_b^{Ni} is the difference between the average energy of a Ni atom in the core of the cluster and the bulk Ni cohesive energy. Fitting linearly the dependence of the solution energy with respect to $\left(N_{at}^{Ni} \right)^{-\frac{1}{3}}$ gives then access to the variation of the Ni core compared to bulk (constant term) and to the pair interaction (slope). The case of Ni stressed at the Ag lattice parameter confirms the conclusion derived for a single impurity, namely that the unrelaxed system presents a tendency to ordering instead of phase separation since the slope is negative. Moreover, the non vanishing value of the constant term is due to the fact that even for large clusters, the Ni core is kept at the Ag volume. The two different relaxed situations present both similarities and differences. For the totally relaxed case (with respect to both the number of atoms and their distances), it is nice to find that one indeed recovers in the large size regime both a positive slope giving the right value for V_1 (≈ -0.05 eV) and a straight line crossing the origin [41].

4.2.3. Coupling stress-segregation (grain boundary)

A conclusion of the section devoted to the case of the pure Σ5 (210) [001] grain boundary was the possible influence of stress on chemical segregation for an alloy. This is confirmed here in the case of the Cu-Ag system in both dilute limits, as can be seen in figure 11 in which the influence of the inhomogeneous pressure profile on the concentration one is exhibited [42]. The main conclusion derived from these curves is that the segregation driving force here is essentially the size-mismatch (which differs from the case of surfaces as will be seen in §4.3.1). This implies that there is always segregation of the impurity at some of the grain boundary sites since the coexistence there of tensile and compressive sites lead to favourable sites for segregation of either the largest (Ag) or smallest (Cu) atoms in order to release the stress. A consequence of this segregation of the impurity is a systematic reversal of segregation with bulk concentration, which does not exist at the surface for which the site energy generally prevails on size effect as will be seen in the next section.

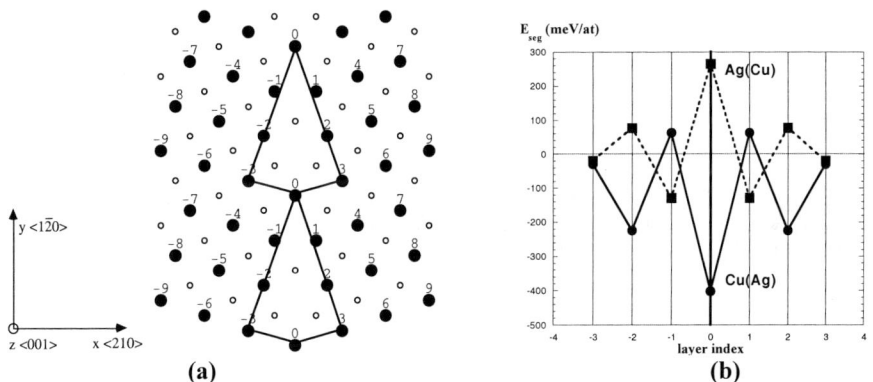

Figure 11. Segregation at a Σ5 (210) [001] grain boundary in a Cu-Ag alloy [42]. The segregation energy of a single impurity in a given layer p parallel to the boundary, as defined in (a), is shown in (b). Both dilute limits are shown : full line for Cu(Ag) and dotted line for Ag(Cu).

4.3. Alloy surfaces and/or heteroepitaxial growth

4.3.1. Coupling between relaxation and surface segregation

Similar to what occurs at a grain boundary, the presence of a surface allows a system to release a part of the stress induced by size-mismatch thanks to chemical segregation, i.e. a variation of concentration in the surface selvedge [8,22]. Nevertheless, this size effect competes with site (surface) energies which are now stronger due to the existence of broken bonds, alloying terms playing a less important role. We have already mentioned that this size effect could be treated in an Ising like description as an additive local term in the segregation energy, see Eq. (19). This was already the case in phenomenological models [8], in which this contribution was calculated in the dilute limit within elasticity theory by assuming that an impurity (with atomic radius r_B and bulk modulus K_B) can release all its bulk elastic energy at the surface of a matrix (with atomic radius r_A and shear modulus G_A) [43] :

$$\Delta E_0^{elast} = \pm \frac{24\pi K_B G_A r_B r_A (r_B - r_A)^2}{3 K_B r_B + 4 G_A r_A} \qquad (24)$$

Such an elastic treatment implies that the impurity should always segregate, whatever the sign of size-mismatch. This completely neglects the asymmetry of a realistic potential between tension and compression. A more realistic treatment is then to calculate this contribution within an *ad-hoc* SMA potential fitted to account for differences between the two species limited to their size-mismatch. Some results are shown below in the case of a PtNi alloy (Fig.12a) [24].

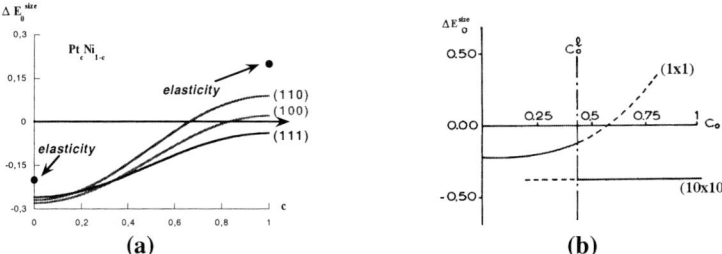

Figure 12. Size effect contribution (in eV) to the surface segregation energy calculated within an *ad hoc* SMA potential. The calculation is performed as a function of the bulk concentration only for PtNi in (a) for the three low index surfaces [24]. It is plotted as a function of the Ag surface concentration c_0 for the dilute Cu(Ag) system in (b) [44], in the case of both a (1x1) and a (10x10) surface structure (see later).

As can be seen, the major effect of the asymmetry of the potential is that only the largest impurity segregates, no effect being observed for the smallest one (except for open surfaces thanks to surface relaxation). This behaviour is at the origin of the profile transition observed in the PtNi system as a function of crystallographic orientation [45]. It is worth noticing that such a calculation, in which the size effect only depends on the bulk concentration, has to be changed if surface segregation is sufficiently strong to lead to a pure plane of impurities, in which case one can expect the stress to be less and less released at the surface. This implies that this driving force will decrease down to a limit where the surface could even change its atomic structure to relax this stress by forming a superstructure as the (10x10) one illustrated here in the case of CuAg (see figure 12b) [44].

However, the previous procedure to account for size-mismatch in surface segregation is simplified by the assumption of decoupling (in an additive form) geometrical and chemical rearrangements to be able to model the phenomenon on a rigid lattice. It has then to be checked at least once that a full calculation, using a complete SMA potential (accounting for both chemical and geometrical differences) to simulate segregation on a non rigid lattice, keeps the essential trends, in order to relax simultaneously atomic and chemical stress. This can be achieved within a mixed Monte Carlo simulation which includes both atomic exchanges and displacements. This leads in the case of Cu(Ag) (100) [46] to a segregation isotherm which exhibits a strong coupling between a Fowler type chemical surface transition (stepped isotherm step) from an almost pure Cu surface layer to an almost pure Ag one, and an atomic structural transition from a square surface layer towards a quasi- hexagonal one.

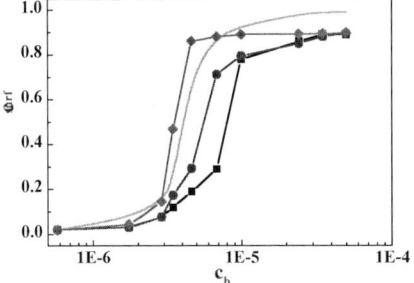 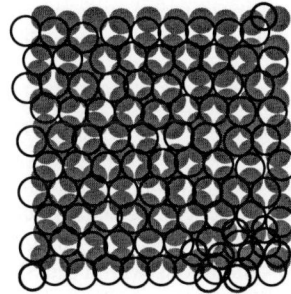

Figure 13. Left hand side : segregation isotherm in the $Cu_{1-c}Ag_c$ (100) system using a complete SMA potential [46]. The thin line results from a mean-field calculation whereas the ones with symbols have been obtained from mixed Monte Carlo simulations performed either by increasing the bulk concentration (■) or by decreasing it with (●) and without (♦) the expelled Ag adatoms included. The hexagonal reconstruction of the surface layer induced by Ag segregation is illustrated on the right hand side (full circles : Cu, empty circles : Ag).

4.3.2. Coupling between relaxation and chemical ordering at the surface

However, the above description still assumes a 2nd moment description which can be criticized. It is then tempting to see how atomic and chemical rearrangements couple at the surface by extending the calculation previously performed in the bulk of the effective pair interactions. This is performed here for the same CuAg and CuPd systems as a function of the orientation of the surface in both dilute limits [38].

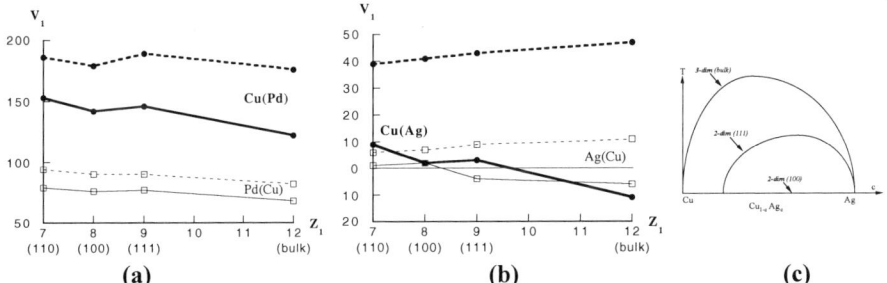

Figure 14. Variation of the effective pair interaction V_1 (in meV) between nearest neighbours, calculated within SMA, as a function of the the number of nearest neighbours of a surface atom, in the Cu-Pd (a) and Cu-Ag (b) systems [38]. Both dilute limits are shown : thick lines for Cu(Ag) and Cu(Pd) and thin lines for Ag(Cu) and Pd(Cu). The full (dotted) lines refer to the relaxed (unrelaxed) situation. The schematic deformation of the phase diagram for Cu-Ag from the three-dimensional to two-dimensional is shown in (c): the (100) surface presents a solid solution in the whole range of concentration.

The main result is that breaking bonds tends to decrease the tendency of a system to phase separate (or equivalently increase its tendency to chemical order). This enhancement of V_1 for ordering systems leads at the surface to a ratio $V_0=1.2\ V$ in good agreement with TBIM prescription. The situation is more complex for phase separation systems since in that case increasing V can even change its sign. It is indeed what is observed beyond a critical miscut (which depends on the dilute limit) indicating that a surface can be chemically ordered on top of a phase separating system. The dependence of this reversal of the ordering tendency with both concentration and surface orientation has been shown experimentally for Cu-Ag [47]. This leads to the schematic two-dimensional phase diagram shown in figure 14.

4.3.3. Crystallographic anisotropy of superstructures: Ag/Cu

In presence of a strong size-mismatch between a surface layer of a given species on top of a substrate of another one, would it be obtained by segregation or deposition, a large interfacial stress is expected in the pseudomorphic state, which can be relieved by various mechanisms, depending on surface orientation. The prediction of the most efficient way to relax interfacial stress as a function of the crystallographic orientation can be obtained by atomistic simulations within the SMA potential. This is illustrated here in the particular case of the Ag/Cu system which has been the subject of many experimental studies [48]. Indeed, it can be considered as a prototype one for systems presenting a large size-mismatch, a strong tendency to phase separation in the bulk and for which the deposit (or segregated) element has the lowest surface energy. All these driving forces (strong Ag surface segregation) indeed lead to the « stability » of a full Ag layer on top of the Cu substrate at completion of the first monolayer under heteroepitaxial growth or surface segregation.

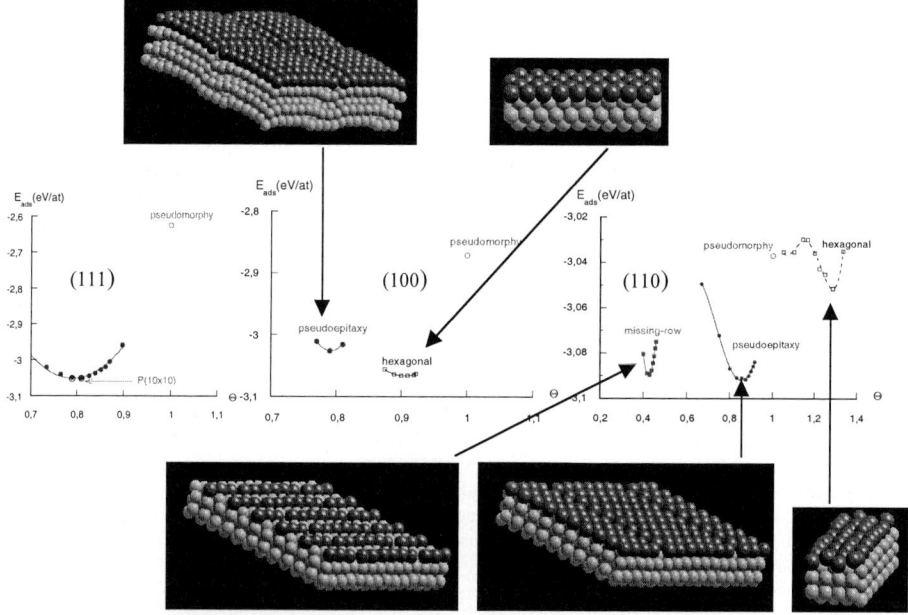

Figure 15. Variation of the adsorption energy with coverage in the Ag/Cu system, for the three low index surfaces and the various superstructures exhibited here for the (100) and (110) orientations, and the Moiré structure exhibited later for the (111) one [49].

As can be seen in Fig.15, a strongly anisotropic character is evidenced, which reveals the existence of several possible reconstructions to release the surface stress which, as for pure elements, can change the epitaxial layer into a more or less close-packed state [49]. Moreover, some surfaces which do not reconstruct in their own semi-infinite pure phase can reconstruct under heteroepitaxial stress. This is illustrated here for the three low index faces of Ag/Cu. Thus, many superstructures (nxm) appear which have to be optimized as a function of their periodicity (n,m), and which can either all appear successively for different coverages as in the (110) case, or be limited to the single stablest one (reconstructed islands) from the lowest coverage as in the (100) case. This behaviour is driven by the shape of the dependence of the adsorption energy as a function of coverage, which is characteristic of either repulsive (for the (110) surface) or attractive (for the (100) and (111) surfaces) interactions. The case of the close-packed (111) surface which seems to be the simplest is in fact more complex when we look at it into details as we will see now.

4.3.4. Various accommodation modes for stress in Ag/Cu(111)

As said above, the large interfacial stress expected in the pseudomorphic state can be relieved by various mechanisms : atomic relaxations, formation of misfit dislocations or point defects such as vacancies. We analyze within the SMA method the relative efficiency of atomic relaxations on one hand and of the formation of vacancies on the other hand to relax interfacial stress when one Ag monolayer covers a Cu (111) substrate [50]. LEED, surface X-ray diffraction (SXRD) and STM observations have shown that two structures can exist, with a periodicity between (9 x 9) and (10 x 10) [51], as expected from the ratio of lattice parameters of Ag and Cu (about 13 %). Although the periodicity of the superstructure is unambiguous, the unit cell is more complex. In fact, two superstructures are found, both theoretically [50] and experimentally [51,52], with equivalent stabilities. The first one is called hereafter the Moiré structure and corresponds to a continuous taking up of the misfit (see figure 16a). The corrugation of this first Ag layer is very high, about 1.1 Å, the low regions corresponding to the near on-top sites whereas the hcp sites are the higher ones. A very similar map is obtained for the first Cu underlayer, the amplitude being a little bit larger (1.5 Å). This corrugation is actually damped beyond the tenth layer [49]. The second structure (Fig.16c) is characterized by a periodic network of triangles as observed in STM and is referred to as the triangular structure. Interpretation of these triangles in terms of corrugation due to partial dislocation loops in the first Cu plane under the deposit has been proposed [50], in strong analogy with similar observations for Au / Ni (111) [53]. It is suggested that the stability of this triangular structure comes from the fact that the triangles, which form localized stacking-faults, allow to avoid the on-top or near on-top position of some atoms in the Ag layer imposed in the Moiré-structure. Thus these Ag atoms, which are coordinated with only one Cu atom in the Moiré structure, become threefold coordinated in the triangular structure. The theoretical study, taking into account the different mechanisms to relax the interfacial stress (Ag and Cu vacancy formation, partial dislocation loop), indicates that the most efficient relaxation mechanism is the formation of partial dislocation loops in the first Cu substrate layer, requiring the formation of four or five Cu vacancies per unit cell in this plane. The introduction of the dislocation loop in the substrate has two main consequences compared to the previously proposed structure without Cu vacancies. First, it strongly reduces the extension towards the substrate of the perturbation, characterized by the corrugation and the stress induced by the adlayer. Second, the calculated morphology is now in almost perfect agreement with the STM images.

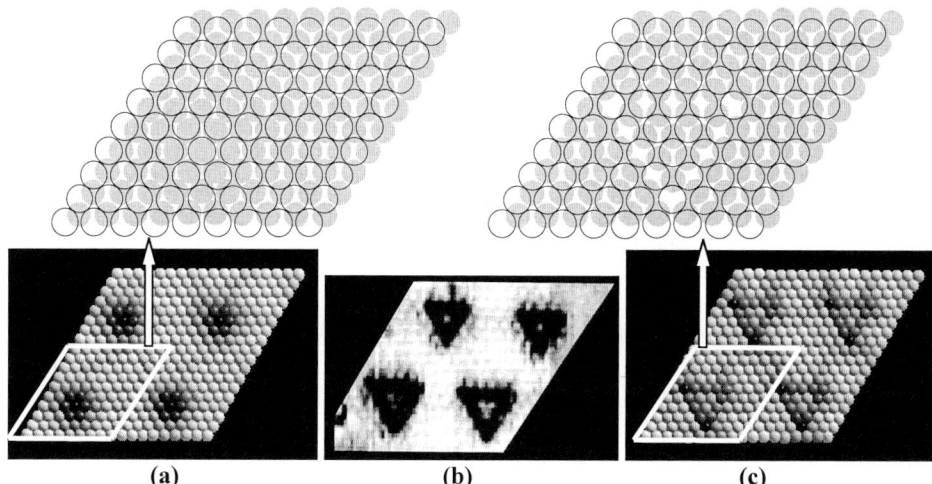

Figure 16. Ag/Cu (111) (10x10) competing Moiré (a) and triangular (c) superstructures issued from SMA quenched molecular dynamics simulations [50] compared to STM observations (b). In the lower part are shown elevation maps (the grey scale going from darkest for hollow regions to white for elevated ones). In the upper part are shown the coincidence lattices for both structures (full circles : Cu atoms, empty circles : Ag atoms).

From the experimental point of view, STM studies have shown that the Moiré structure appears for low temperature deposits, whereas the triangular structure is observed at temperatures higher than room temperature [52]. One can then ask whether the transition between the two superstructures is an equilibrium phase transition (characterized by a change of sign of the free energy difference between the two configurations at about 300 K) or if it simply results from a kinetic evolution. In the latter case the high temperature structure, i.e. the triangular one, could be the stablest one at any temperature, but the creation of vacancies in the first Cu underlayer would require to overcome some thermal activation barriers, which should only be efficient at temperatures higher than 300 K. In order to elucidate the physical nature of the transition one has to calculate the free energy of the two superstructures as a function of temperature. The influence of temperature on the internal energy is obtained by Monte Carlo simulations, whereas the vibrational entropy is calculated by the recursion method [54]. One can show that increasing temperature slightly favours the Moiré structure along a free energy criterion, as a result of a complex interplay between internal energy and vibrational entropy [55]. This indicates that the experimentally observed transition between the Moiré structure at low temperature and the triangular one at higher temperature has probably a kinetic origin and is not a true equilibrium phase transition.

4.3.5. Stress profile: a guide for rearrangements to come ?

Here we adopt a rather different point of view [56]. Instead of considering the optimization of the number of coordination as the main driving force, we will focus on the efficiency of the vacancy formation to relax local stress, as it was predicted in icosahedral clusters [36]. We show that this approach is more general than the previous one. Moreover it allows to discuss the relation between constitutive interfacial vacancies which lower the internal energy of the interface and vacancy segregation on specific sites, which is the enhancement of the vacancy concentration on these sites relatively to the bulk one at finite temperature. We will also illustrate that the same driving force, i.e. the reduction of the local stress, is responsible of the chemical segregation in the first underlayer.

From this new point of view, a useful guide to predict the most efficient relaxation mechanism is to characterize the regions which undergo the largest stress. Thus the local pressure map in the first Cu underlayer allows to predict a complementarity between the segregation of vacancies in compressive sites and the Ag enrichment, due to the larger atomic radius of the Ag atoms, in the tensile ones. Segregation energy maps, both for vacancies and Ag atoms, confirm this prediction and lead to several perspectives for the kinetics of dissolution of an Ag deposit over a Cu (111) substrate.

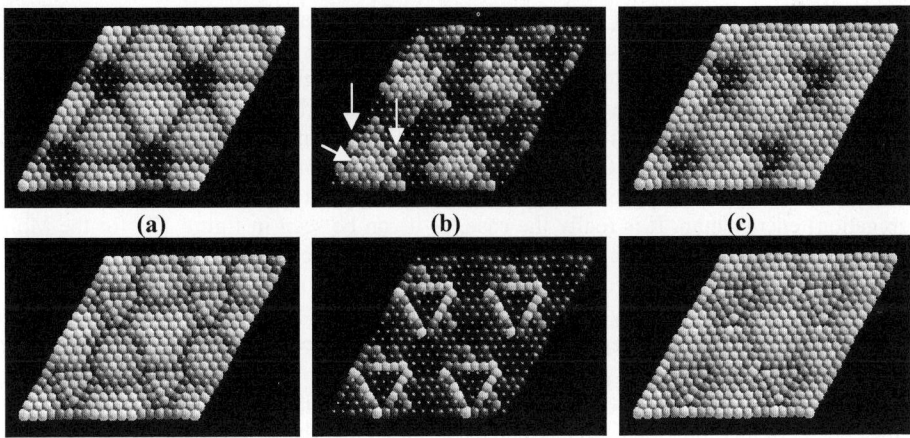

Figure 17. First Cu layer in the Ag/Cu (111) (10x10) competing Moiré (upper part) and triangular (lower part) superstructures [56] : local pressure (a), monovacancy formation (b) and Ag impurity segregation energy maps. The grey scale is chosen so that in (a) the pressure goes from -75 kbar (white) to 85 kbar (black), in (b) the energy goes from -0.1 eV (white) to 1.2 eV (black) and in (c) from -0.4 eV (white) to 0.2 eV (black).

For the Moiré structure, the local pressure map in the first Cu underlayer shown in figure 17a is strongly correlated with the corrugation one, the hollow sites (which are below the near on-top sites of the Ag plane) undergoing a rather large compressive pressure. From this map, it is expected that the vacancy formation energy will be low in the compressive sites and high in the tensile ones. The map of these formation energies, shown in figure 17b, confirms totally this simple idea : the average value in the region corresponding to the near on-top positions of the Ag adlayer is of about 0.15 eV, drastically lower than the formation energy in the bulk : 1.24 eV. As a consequence, a very large vacancy segregation is expected on these sites at finite temperature. Moreover, a few sites (three), corresponding to the white dots marked by arrows in figure 17, are found to exhibit negative formation energies, which means that these monovacancies are constitutive defects of the (10x10) Moiré structure. It is then tempting to see whether creating simultaneously these three vacancies leads to a still more efficient stabilizing defect. This is not the case, which means that one has to be careful when using the map of figure 17b as a practical guide to remove the Cu atoms. A more efficient way is to remove those which are in the close vicinity of the perfectly on-top Ag atom, starting from the one which is just below and removing the followings along a close-packed row passing through this on-top position. In that case these on-top positions are now avoided, the vacancies having formed a partial dislocation loop around a localized stacking-fault. Therefore, instead of speaking in terms of constitutive defects of the Moiré structure, we will describe the triangular structure as a new structure *per se*.

Considering now the triangular structure, we can ask about the characteristics of the local pressure map in the first Cu underlayer. This map, shown in figure 17a, indicates that some compressive sites still exit near the edges of the triangle described in figure 16. It is then tempting to calculate the map for the vacancy formation energies. In figure 17b, we can see that the formation energy is very low near the triangle edges. However, this formation energy remains positive for these sites, even though very close to zero. It means that the monovacancy in the triangular (10x10) structure is no longer a constitutive defect but instead a thermal one with a very large concentration.

We have seen that the local pressure map is a good guide to locate constitutive and segregated vacancies in the compressive sites of the first underlayer. We can ask whether such efficiency is achieved to predict Ag segregation energy map in the first underlayer. Since Ag has a larger atomic radius than Cu, we can predict that Ag atoms will segregate in the tensile sites, i.e. that the Ag segregation energy map is quite opposite to the vacancy formation (or segregation) energy map. It is really the case, as it can be seen in figure 17c for the Moiré structure and the triangular structure. An important consequence is that the segregation energy map in the first underlayer strongly depends on the superstructure adopted in the surface plane. The map is rather homogeneous for the triangular structure, whereas in the Moiré one Ag atoms will avoid the regions below the on-top positions, leading to a rather inhomogeneous Ag segregation profile.

Up to now we do not have considered vacancy formation and Ag segregation in the surface plane. In fact, previous calculations have shown that the vacancy formation energy in the Ag surface plane is always positive in the Moiré structure, meaning that Ag advacancies are not constitutive surface defects. This result, which is rather surprising, comes from the competition between the coordination effect and the local pressure one. Actually the Ag on-top site, which is the less coordinated one and thus the most favourable site for vacancy formation if we consider only the coordination effect, is also the most tensile site and then the less favourable site for vacancy formation due to local pressure effect. Such competition prevents the formation of constitutive defects in the surface plane.

4.4. Heteroepitaxial growth and surface alloy formation

4.4.1. Two-dimensional layers : asymmetry tension-compression in Co-Pt system

A peculiar effect of realistic potential is its asymmetry between compressive and tensile stress. This is illustrated here in the case of the growth modes of the (Pt, Co) system. In the compressive case (Pt/Co) a stress driven alloy is formed at the coalescence of strained platinum islands. In the tensile case (Co/Pt) a dislocation network develops itself allowing an over-closepacking of the film, as for Au(111) reconstruction.

The two elements present a large size-mismatch (11%), have almost the same surface energy which means that growth is mostly governed by the structure. The structural difference between compressive and tensile situations comes from the fact that the dilation has a lower energy cost than the contraction. Therfore, when depositing big atoms onto small ones, pseudomorphy is highly unfavourable. Moreover the stress induced by the presence of unfavourable on-top positions can be partially relaxed by removing atoms from the substrate into adlayer (here) or kinks (Ag/Cu). The latter effect is important near the coalescence leading to a stress driven alloy formation [57].

On the contrary when depositing small atoms on big ones, pseudomorphy is the most favourable situation in the first stages of growth. The lattice dilation has however a limited spatial extent which means that the strain at completion of the first layer would be too large to get a perfect pseudomorphy. The strain induced by size-misfit is partially released through the introduction of contractive reconstruction dislocations, separating the regions of coherent epitaxy, but with alternately fcc and hcp stacking with respect to the substrate [58]. This mechanism allows an overclose-packing of the film like for 5d metal surface reconstruction. This mechanism is generally valid for tensile strain relaxation in hexagonaly close-packed metal-metal interfaces.

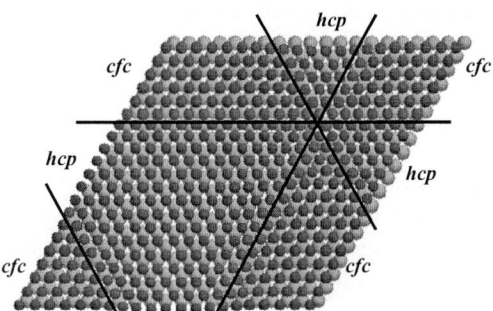

Figure 18. Co / Pt (111) superstructure from SMA calculations, exhibiting pseudomorphic fcc and hcp regions, separated by compressive discommensuration lines [58]. The Co (Pt) atoms appear as small (large) balls.

4.4.2. Three-dimensional clusters, adsorbed (Pd / MgO) or buried (Ni-Ag)

Once characterized the 2D interfacial relaxation of stress due to size-mismatch in presence of a surface, one can wonder about that of 3D interfaces (which have been studied up to now only with vacuum or in the bulk). Different situations occur depending on whether the cluster remains adsorbed on top of the substrate or is buried below.

The former case is observed during 3D heteroepitaxial growth, which is the common case for oxide/metal deposition. Its theoretical treatment needs to extend the previous SMA potential to also account for metal-oxide interactions, which can be achieved by fitting an analytic form to *ab initio* calculations performed for well defined rigid configurations of adsorbed layers [59]. This is illustrated here in the case of the Pd/MgO(100) system which has been studied a lot experimentally [6] and which presents atomic rearrangements at both the surface of the cluster and at its interface with the oxide substrate. The rearrangements at the interface are driven by the stress induced by the size-mismatch between Pd and MgO. More precisely, a pseudomorphy would lead to a 8 % tensile stress for the Pd interface atoms. The situation then looks a little bit as that of Co/Pt, except that now the Pd interfacial layer is also strained by the above Pd atoms. This leads to both similarities and differences since one observe at equilibrium, for sufficiently large clusters, anisotropic pseudomorphic domains, which release their tensile stress along dislocations, thanks to movements of Pd atoms in the 3D cluster along (111) slip planes. This leads once again to hcp-fcc stacking faults presenting a low defect energy.

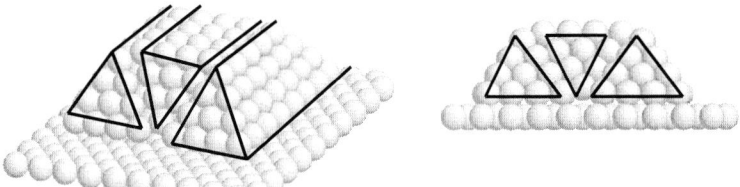

Figure 19. Pd / MgO (001) cluster with a dislocation line in the interfacial layer, issued from calculation [59].

The case of clusters buried near the surface can occur under annealing of a thin film deposited on a substrate which tends to segregate, for two almost immiscible elements. In that case, one expects the substrate atoms to climb, burying the deposited ones which can either remain under 2D shape or adopt 3D ones. This has been experimentally observed for many systems among which Ni/Ag (100) for which AES studies predict a dissolution of the Ni layer limited to the first layers below the surface [60] and SXRD the formation of precipitates in which Ni has relaxed at its bulk parameter [40].

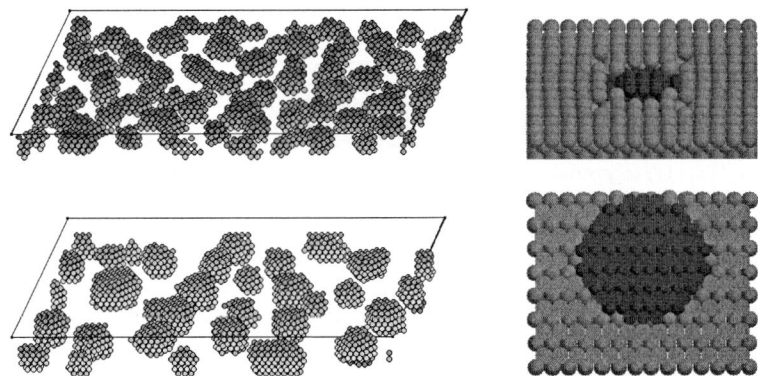

Figure 20. Buried Ni clusters issued from dissolution of one Ni monolayer initially deposited on top of a (100) Ag surface. In the left part are shown two snapshots of the kinetic Monte Carlo dissolution performed using a KTBIM model on a rigid lattice [62]. In the right one is shown the relaxation of such a cluster viewed from above and along a perpendicular cut [41].

From the theoretical point of view, a kinetic extension of the previous TBIM model (KTBIM [61]) allows to study the dissolution, on a rigid lattice, of a thin film, with a proper treatment of the main energetic driving forces responsible of surface segregation and bulk ordering. This leads to a dissolution which blocks on the first ten layers with formation of Ni buried clusters obeying to an Ostwald ripening behaviour (see figure 20) [62]. The next question is then whether allowing relaxation of the system would lead to changes in the result, concerning the preference for 3D buried precipitates instead of 2D buried layers, and in each case, does Ni relax or not ? The answer is that, although 2D buried layer prefer to keep their tensile stress, the 3D clusters are relaxed [41], inducing a corrugation of the surface which can be compared to that observed by STM (see figure 20) [63].

5. CONCLUSION

This lecture intended to show that an atomistic treatment of stress was necessary to account properly for the rich variety of behaviours occuring in materials such as surfaces or grain boundaries in pure elements, free, deposited and buried clusters, alloy surfaces or heteroepitaxially grown thin films. For multicomponent systems, a particular emphasis has been put on the complex coupling between atomic and chemical degrees of freedom. In all cases, it has been shown that it was essential to account properly for the asymmetry between the tensile or compressive characters of stress to predict the various ways a system has to release (partially) its stress, would it be by structural (reconstructions, superstructures) or chemical (segregation, interfacial alloy formation) rearrangements. Finally, a particular emphasis has been put on the caution to be taken when using local pressure maps as tools to predict the (atomic, chemical) rearrangements for stress release.

ACKNOWLEDGEMENTS

This lecture is based on a common work performed with Bernard Legrand during the last fifteen years. It would not exist without him and I am very indebted to him for that and also for his critical reading of the manuscript. All the illustrations used here are drawn from original works by various authors who are gratefully acknowledged here :
- Isabelle Meunier has studied the coupling between chemical ordering and atomic relaxation, and also the various accommodation modes of stress in heteroepitaxial Ag/Cu and Ni/Ag systems.
- Christine Goyhenex has evidenced the role of the asymmetry between tensile and compressive stress in the Pt-Co system.
- Christine Mottet has performed the work devoted to clusters, free in a first time and then deposited on an oxide substrate with Jacek Goniakowski and Wilfried Vervisch. She has also studied the anisotropy of superstructures in the Ag/Cu system.
- Fabienne Berthier, Robert Tétot and Jérôme Creuze achieved the studies concerning both the grain boundaries and the coupling between superstructure and segregation at alloy surfaces.
- Andrés Saúl is responsible, on the one hand with Jean-Marc Roussel of the study of dissolution kinetics in Ni/Ag, and on the other hand with Stéphane Olivier of that devoted to surface stress.
- François Ducastelle is at the origin of the development of TBIM.

REFERENCES

1. P. Müller and R. Kern, contribution to the present issue.
2. M.-C. Desjonquères and D. Spanjaard, contribution to the present issue.
3. L. Kubin, contribution to the present issue.
4. M.-C. Desjonquères and D. Spanjaard, *Concepts in Surface Physics*, (Springer-Verlag Berlin Heidelberg New York 1993, 1996).
5. L. Priester, J. Thibault and V. Pontikis, Sol. State Phenom. **59-60** (1998) 1.
6. C.R. Henry, Surf. Sci. Rep. **31** (1998) 231.
7. M. Gerl and J.-P. Issi, *Physique des Matériaux*, (Presses Polytechniques et Universitaires Romandes, 1997).
8. P. Wynblatt and R.C. Ku, *Interfacial Segregation*, eds. W.C. Johnson and J.M. Blakely (AMS, Metals Park, OH, 1979), p. 115.
   Y. Gauthier and R. Baudoing, *Surface Segregation Phenomena*, eds. P.A. Dowben and A. Miller (CRC Press, Florida 1990), p. 169.
9. J.W. Matthews, *Epitaxial Growth*, Part 2, ed. J.W. Matthews (Academic Press, New York, 1975), p. 559.
   J.H. van der Merwe, *Chemistry and Physics of Solid Surfaces*, eds. R. Vanselow and R. Howe (Springer, Berlin, 1984), p. 365.
10. V. Vitek and T. Egami, Phys. Status Solidi B **144** (1987) 145.
    P.C. Kelires and J. Tersoff, Phys. Rev. Lett. **63** (1989) 1164.
11. J. Friedel, *The Physics of Metals*, ed. J.M. Ziman (Cambridge University Press, 1969), p. 340.
12. V. Rosato, M. Guillopé and B. Legrand, Phil. Mag. A **59** (1989) 321.
13. G. Tréglia, B. Legrand and F. Ducastelle, Europhys. Lett. **7** (1988) 575.
14. F. Ducastelle, *Order and Phase Stability in Alloys*, (North Holland, 1991).
15. J.H. Rose, J. Ferrante and J.R. Smith, Phys. Rev. Lett. **47** (1981) 675.
    J.H. Rose, J.R. Smith, F. Guinea and J. Ferrante, Phys. Rev. B **29** (1984) 2963.
    D. Spanjaard and M.-C. Desjonquères, Phys. Rev. B **30** (1984) 4822.
16. M.S. Daw and M.I. Baskes, Phys. Rev. Lett. **50** (1983) 1285.
    M.S. Daw and M.I. Baskes, Phys. Rev. B **29** (1984) 6443.
17. F. Ercolessi, E. Tosatti and M. Parrinello, Phys. Rev. Lett. **57** (1986) 719.
    F. Ercolessi, M. Parrinello and E. Tosatti, Phil. Mag. A. **58** (1988) 213.
18. L. Verlet, Phys. Rev. **159** (1967) 98
    C.H. Bennett, *Diffusion in Solids, Recent Developments*, eds. A.S. Nowick and J.J. Burton (Academic, New York, 1975), p. 73.
    D. Frenkel and B. Smit, *Understanding Molecular Simulation*, (Academic Press, 1996).
19. F. Ducastelle, Computer Simulation in Materials Science, NATO-Asi (1991).
20. F. Ducastelle and F. Gautier, J. Phys. F **6** (1976) 2039.
21. B. Velicky, S. Kirkpatrick and H. Ehrenreich, Phys. Rev. **175** (1968) 747.
22. G. Tréglia, B. Legrand, F. Ducastelle, A. Saúl, C. Gallis, I. Meunier, C. Mottet and A. Senhaji, Comput. Mat. Sci. **15** (1999) 196.
23. A. Bieber, F. Gautier, G. Tréglia and F. Ducastelle, Solid State Commun. **39** (1981) 149.
24. G. Tréglia and B. Legrand, Phys. Rev. B **35** (1987) 4338.
25. D. Tomanek, A.A. Aligia and C.A. Balseiro, Phys. Rev. B **32** (1985) 5051.
26. J. Tersoff, Phys. Rev. Lett. **74** (1995) 434.

27. G. Tréglia, M.C. Desjonquères and D. Spanjaard, J. Phys. C **16** (1983) 2407, *and references therein.*
    B. Legrand, G. Tréglia, M.C. Desjonquères and D. Spanjaard, J. Phys. C **19** (1986) 4463.
28. M.A. Van Hove, R.J. Koestner, P.C. Stair, J.P. Bibérian, L.L. Kesmodel, I. Bartos and G.A. Somorjai, Surf. Sci. **103** (1981) 189, 218.
29. D.D. Chambliss, R.J. Wilson and S. Chiang, Phys. Rev. Lett. **66** (1991) 1721.
30. M. Guillopé and B. Legrand, Surf. Sci. **215** (1989) 577, *and references therein.*
31. S. Olivier, A. Saúl and G. Tréglia, to be published.
32. A. Filippetti and V. Fiorentini, Phys. Rev. B **60** (1999) 14366.
    A. Filippetti and V. Fiorentini, Phys. Rev. B **61** (2000) 8433.
33. H. Ibach, Surf. Sci. Rep. **29** (1997) 193.
    C.E. Bach, M. Giesen, H. Ibach and T.L. Einstein, Phys. Rev. Lett. **78** (1997) 4225.
34. F. Berthier, B. Legrand and G. Tréglia, Interface Science **8** (2000) 55.
35. A. Khoutami, B. Legrand, C. Mottet and G. Tréglia, Surf. Sci. **307-309** (1994) 735.
36. C. Mottet, G. Tréglia and B. Legrand, Surf. Sci. Lett. **383** (1997) L719.
37. C.L. Chen, T.T. Tsong, S. Liang and L. Zhang, Phil. Mag. Lett. **71** (1995) 357.
38. I. Meunier, G. Tréglia and B. Legrand, Surf. Sci. **441** (1999) 225.
39. S. Sawaya, thèse Université Aix-Marseille II (1999).
40. I. Meunier, J.-M. Gay, A. Barbier, B. Aufray, to be published.
41. I. Meunier, G. Tréglia and B. Legrand, to be published.
42. F. Berthier, B. Legrand and G. Tréglia, Acta Materialia **47** (1999) 2705.
43. J.D. Eshelby, Adv. Sol. State Phys. **3** (1956) 79.
    J. Friedel, Adv. Phys. **3** (1954) 446.
44. G. Tréglia, B. Legrand, J. Eugène, B. Aufray and F. Cabané, Phys. Rev. B **44** (1991) 5842.
    A Saúl, B. Legrand and G. Tréglia, Phys. Rev. B **50** (1994) 1912.
45. B. Legrand, G. Tréglia and F. Ducastelle, Phys. Rev. B **41** (1990) 4422.
46. J. Creuze, Thesis, Université Paris XI-Orsay, France (2000).
47. P.T. Sprunger, E. Laesgaard, F. Besenbacher, Phys. Rev. B **54** (1996) 8163.
48. J. Eugène, B. Aufray and F. Cabané, Surf. Sci. **241** (1991) 1.
    Y. Liu and P. Wynblatt, Surf. Sci. **241** (1991) L21.
    Y. Liu and P. Wynblatt, Surf. Sci. **310** (1994) 27.
49. C. Mottet, G. Tréglia and B. Legrand, Phys. Rev. B **46** (1992) 16018.
50. I. Meunier, G. Tréglia, J.-M. Gay, B. Aufray and B. Legrand, Phys. Rev. B **59** (1999) 10910.
51. B. Aufray, M. Göthelid, J.-M. Gay, C. Mottet, E. Landemark, G. Falkenberg, L. Lottermoser, L. Seehofer and R.L. Johnson, Microsc. Microanal. Microstruct. **8** (1997) 167.
52. F. Besenbacher, L. Pleth Nielsen and P.T. Sprunger, *The Chemical Physics of Solids and Heterogeneous Catalysis*, D.A. King and D.P. Woodruff, Eds (Elsevier, Amsterdam 1997) **8**, chap. 10.
53. J. Jacobsen, L. Pleth Nielsen, F. Besenbacher, I. Stensgaard, E. Laegsgaard, T. Rasmussen, K.W. Jacobsen and J.K. Norskov, Phys. Rev. Lett. **75** (1995) 489
54. D.A. Maradudin, E.W. Montroll, G.H. Weiss and I.P. Ipavota, *Theory of Lattice Dynamics in the Harmonic Approximation*, (Academic Press, New York, 1971).
    G. Tréglia and M.C. Desjonquères, J. Physique **46** (1985) 987.

55. I. Meunier, R. Tétot, G. Tréglia, B. Legrand, Appl. Surf. Sci (2001) *in press*.
56. I. Meunier, G. Tréglia, B. Legrand, R. Tétot, B. Aufray and J.-M. Gay, Appl. Surf. Sci. **162-163** (2000) 219.
57. C. Goyhenex, H. Bulou, J.-P. Deville and G. Tréglia, Phys. Rev. B **60** (1999) 2781.
58. C. Goyhenex and G. Tréglia, Surf. Sci. **446** (2000) 272.
59. C. Mottet, J. Goniakowski and W. Vervich, to be published.
60. A. Rolland and B. Aufray, Thin Solid Films **76** 51981) 45
    B. Aufray, H. Giordano, B. Legrand and G. Tréglia, Surf. Sci. **307-309** (1994) 531.
61. A. Senhaji, G. Tréglia, B. Legrand, N.T. Barrett, C. Guillot and B. Villette, Surf. Sci. **274** (1992) 297
    B. Legrand, A. Saúl and G. Tréglia, Material Science Forum, (Trans. Tech. Publi. Switzerland) **155-156** (1994) 165.
62. J.-M. Roussel, A. Saúl, G. Tréglia and B. Legrand, Phys. Rev. B **55** (1997) 10931.
63. D.A. Hite, O. Kizilkaya, P.T. Sprunger, M.M. Howard, C.A. Ventrice Jr., H. Geisler and D.M. Zehner, J. Vac. Sci. Technol. A **18** (2000) 1950.

Ab initio study of structural stability of thin films

Alain Pasturel

Laboratoire de Physique et Modélisation des Milieux Condensés, UMR 5493 , BP 166 CNRS, 38042 GRENOBLE-Cedex09, France

A modern electron theory which is relevant to alloy design has to do with the calculations of electron densities and total energies. I am talking about the energies of structures which range from perfectly crystalline, through heavily deformed to completely amorphous. This discussion in this presentation is mainly about the energy of a system when the electrons are in their ground state, which means the equilibrium state of electrons at zero temperature for a given atomic configuration. Fortunately, the ground state theory is of wider applicability than might be expected and the effects of temperature can often be taken into account.

I shall focus on how calculations of the total energy can be done from first principles, that is without any parameters which have to be adjusted to fit experimental data. Simplifications can be made along the way, and these are discussed in Treglia 's notes. In principle, the behavior of electrons and ions in a solid is obtained from solving the Schrödinger equation (or its relativistic analogue, the Dirac equation). The challenge is to find this solution for complex systems in a variety of fields, ranging from structural materials over epitaxial growth to mechanical properties. The development of efficient algorithms combined with the availability of cheaper and faster computers has turned density Functional Theory (DFT) into a reliable and feasible tool to study many complex processes in materials science and condensed matter physics. It will be the purpose of the first part of this presentation.

The total energy of a system together with the electron density are the two fundamental quantities calculated by theoreticians. However, many related quantities of more practical interest can be coaxed out of the same theory, such as elastic properties, energies of point and planar defects, epitaxial strain energies, … etc. For thin films, which are important in catalysis, magneto-optic storage media and interconnects in microelectronics, it is crucial to predict and control their morphologies. Indeed, the quality and structure of these systems is of paramount importance for these applications. These systems are usually strained due to film/substrate lattice mismatch. On would like to understand and predict the stability of these types of strained materials. It will be discussed in Part II.

Furthermore, with the addition of some statistical mechanisms, calculations from first principles are beginning to shed light on the growth phenomena and the origin of phase transition in surface alloys in relation to stress relaxation mechanisms. It will be presented in Part III.

1. DENSITY FUNCTIONAL THEORY:

- very good review articles

S. Lundqvist and N.H. March, Theory of Inhomogeneous Electron gas, Plenum Press, New York 1983.
R.O. Jones and O. Gunnarsson, Rev. Mod. Phys. **61** (1989) 689.
R.G. Parr and W. Yang, Density Functional Theory of Atoms and Molecules, Oxford University Press, New York 1989.

- A clear and clever book

D.J. Singh, Planewaves, Pseudopotentials and the LAPW Method, Kluwer Academic Publishers 1994.

- About the methods

H.L. Skriver, The LMTO Method, Springer, Heidelberg 1983.
M.C. Payne, M.P. Teter, D.C Allan, T.A Arias and J.D. Joannopoulos, Rev. Mod. Phys. **64** (1992) 1045.
S. Goedecker, Rev. Mod. Phys. **71** (1999) 1085.

- Application to materials science

E. Wimmer, J. Phys. IV France, **C6** (1997) 75
J. Hafner, Acta Mater. **48** (2000) 71.

2. ELASTIC CONSTANTS AND EPITAXIAL STRAIN ENERGIES:

2.1 Elastic constants:
J.M. Wills, O. Eriksson, P. Söderlind, A.M. Boring, Phys. Rev. Lett. **68** (1992) 2802.
P. Söderlind, O. Eriksson, J.M. Wills and A.M. Boring, Phys. Rev. B **48** (1993) 5844.
T. Kraft, P.M. Marcus, M. Methfessel and M. Scheffler, Phys. Rev.B **48** (1993) 5886.
P. Söderlind, R. Ahuja, O. Eriksson, J.M. Wills, B. Johansson, Phys. Rev. B **50** (1994) 5918.
G.Y Guo and H.H. Wang, Phys. Rev. B 62 (2000) 5136.
J. Bouchet, B. Siberchicot, F. Jollet and A. Pasturel, J. Phys. Cond. Matter **12** (2000) 172.

2.2 Bain paths:
P.J. Craievich, M. Weinert, J.M. Sanchez and R.E Watson, Phys. Rev. Lett. **72** (1994) 3076.
P.J. Craievich, J.M. Sanchez, R.E Watson and M. Weinert, Phys. Rev. B **55** (1007) 787.

2.3 Epitaxial strain energies:
T. Kraft, P.M. Marcus, M. Methfessel and M. Scheffler, Phys. Rev. B **48** (1993) 5886.
P. Alippi, P.M. Marcus and M. Scheffler, Phys. Rev. Lett. **78** (1997) 3892.
V. Ozolins, C. Wolverton and A. Zunger, Phys. Rev. B **57** (1998) 4816.
W. Li and T. Wang, Phys. Rev. B **60** (1999) 11954.
S.L. Qiu, P.M. Marcus and Hong Ma, Phys. Rev. B **62** (2000) 3292.
D. Spisak and J. Hafner, J. Phys. Cond. Matter **12** (2000) L139.

2.4 Surface relaxation and surface stress:
P.J. Feibelman, Phys. Rev. B **50** (1994) 1908.
P.J. Feibelman, Phys. Rev. B **51** (1995) 17867.
P.J. Feibelman, Phys. Rev. B **53** (1996) 13740.
P.J. Feibelman, Phys. Rev. B **56** (1997) 2175.
N. Moll, M. Scheffler and E. Pehlke, Phys. Rev. B **58** (1998) 4566.
G. Jomard, T. Petit, L. Magaud and A. Pasturel, Phys. Rev. B **60** (1999) 15624.
P.M. Marcus, X. Qian and W. Hübner, J. Phys. Cond. Matter **12** (2000) 5541.
G.E Thayer, V. Ozolins, A.K. Schmid, N.C. Bartelt, M. Asta, J.J. Hoyt, S. Chiang and R.Q. Hwang, Phys. Rev. Lett. **86** (2001) 660.

3. TEMPERATURE EFFECTS:

3.1 Thermal expansion:
J. Xie and M. Scheffler, Phys. Rev. B 57 (1998) 4768.
J. Xie, S. de Gironcoli, S. Baroni, M. Scheffler, Phys. Rev. B **59** (1999) 970.

3.2 Epitaxial growth:
C. Ratsch, P. Ruggerone and M. Scheffler, Morphological Organization in Epitaxial Growth and Removal, Vol 14, Eds Z. Zhang, M.G. Lagally, World Scientific, Singapore 1998, 3-29.
K. Schroeder, A. Antons, R. Berger, WI Kromen and S. Blugel, Int. Symp. On Structure and dynamics of Heterogeneous Systems, Eds P. Entel and S.E. Wolf, World Scientific, Singapore 2000.
C. Stampfl, H.J. Kreuzer, S.H. Payne, H. Pfnur and M. Scheffler, Phys. Rev. Lett. **83** (1999) 2993.

3.3 Surface alloys:
V. Drchal, J. Kudrnovsky, A. Pasturel, I. Turek and P. Weinberger, Phys. Rev. B **54** (1996) 8202.
R. Tetot, J. Kudrnovsky, A. Pasturel, V. Drchal and P. Weinberger, Phys. Rev. B **51** (1995) 17910.

Stress, strain and chemical reactivity: a theoretical analysis

Philippe Sautet

Institut de Recherches sur la Catalyse, CNRS, Theory and modeling group, 2 Av. A. Einstein, 69626 Villeurbanne Cedex, France
And Laboratoire de Chimie Théorique et des matériaux hybrides, Ecole Normale Supérieure, 46 Allée d'Italie, 69364 Lyon Cedex 07, France

The relation between stress, strain and chemical reactivity is described on the basis of several examples ranging from organic chemistry to surface chemistry and heterogeneous catalysis. A quantum chemical approach is used to understand the influence of strain on the energy of the initial and transition states for the chemical reaction. Reactivity is shown to correlate with the strain relief between these two states. In the case of surface chemistry on metals, both compressive or tensile strain can lead to an enhance reactivity, whether the influence of strain is dominant on the initial or transition state respectively.

1. INTRODUCTION

The concept of stress originates from the mechanics of solid systems, but has spread in our everyday life. A solid or a person is under stress when it or he is submitted to opposite forces, which compensate. Is a person under stress more reactive? The stress generally arises from an external action, such as a pressure, and a chemical system deforms to exert an opposite force and compensate for this action. One example is the case of a metal dimer, in epitaxial interaction with a surface of the same metal. The equilibrium distance of the dimer is generally shorter than that of the metal in the bulk. When interacting with the surface, epitaxial forces tend to elongate the dimer, in order to restore a bulk like distance. The equilibrium distance of the dimer on the surface results from a compromise between the dimer bond energy and the epitaxial energy. If the energy of the total system is stabilized from the gain in epitaxial forces, the stressed dimer is strained and destabilized in energy. When a system is destabilized, this makes it more reactive in many cases. In the present text, the relation between stress, strain and chemical reactivity will be detailed.

2. CHEMICAL REACTIVITY

As illustrated in figure 1, chemical reactivity can be characterized by a profile of energy, between the initial state, the reactants, and the final state, the products [1]. In the most general case, the energy profile must cross a reaction barrier, climbing to the transition state, for the reaction to occur. The chemical reaction rate is directly linked to this reaction barrier, by the Arrhenius exponential law. Some reactions proceed without energy barrier, especially in the case where the initial state corresponds to unstable species such as radical molecules. Stress will deform the system, and the strain will affect all states in the profile. The perturbation will modify the energy of the initial and transition state, hence changing the reaction barrier. The deformation is generally small so that the nature of the reaction pathway will not be modified. To first order, the geometry of the transition state will not be changed.

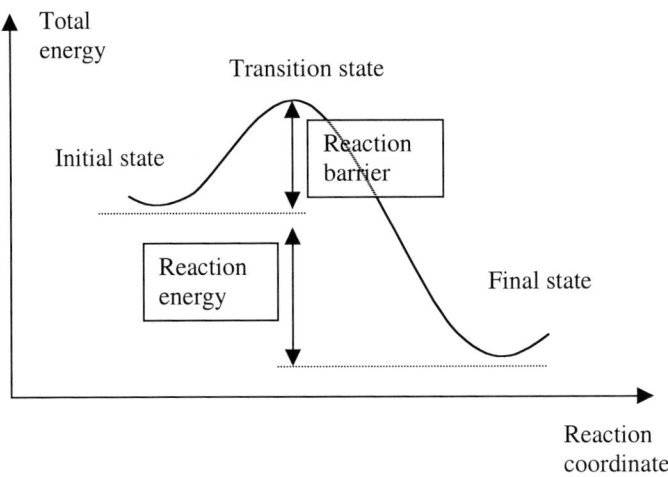

Figure 1: Energy profile of a chemical reaction.

It is not strait forward to fully characterize a reaction pathway, so that a more indirect criterion could be helpful. A chemical system can be characterized by a set of Molecular Orbitals [2]. The orbitals with low energy are occupied by two electrons, while those with high energy are vacant. The energy difference between the highest occupied MO (HOMO) and the lowest unoccupied MO (LUMO) can be seen as a qualitative reactivity criterion for the molecule. If strain reduces this energy difference, by destabilizing the HOMO or stabilizing the LUMO, then the reactivity might be enhanced.

The electronic structure of a chemical system results from interaction between atomic orbitals. These interactions will be affected by changes in the geometry of the system. Hence, for a modulation of reactivity, the strain is the correct parameter. The stress itself has no influence, other than being responsible for the strain.

2.1 A text book example from chemistry : butadiene

The butadiene molecule can exist in two forms: the linear molecule C_4H_6 and the cyclic form cyclobutadiene C_4H_4. The reactivity of these unsaturated molecules can be studied from the four π orbitals, which as schematically shown from a simple Hückel approach in figure 2 [2]. In this approach, the on site energy of the p_z atomic orbital of carbon is α, and the coupling between these p orbitals is β. Since β is negative, the π system of the open molecule is more stable (E= $4\alpha+4.4\beta$) than that of the cyclic form (E= $4\alpha+4\beta$). Indeed, the second molecular orbital is strongly destabilized in energy in the cyclic form: the π system is stressed by the σ molecular system. It is destabilized and more reactive, since it has a higher HOMO and a lower LUMO orbital.

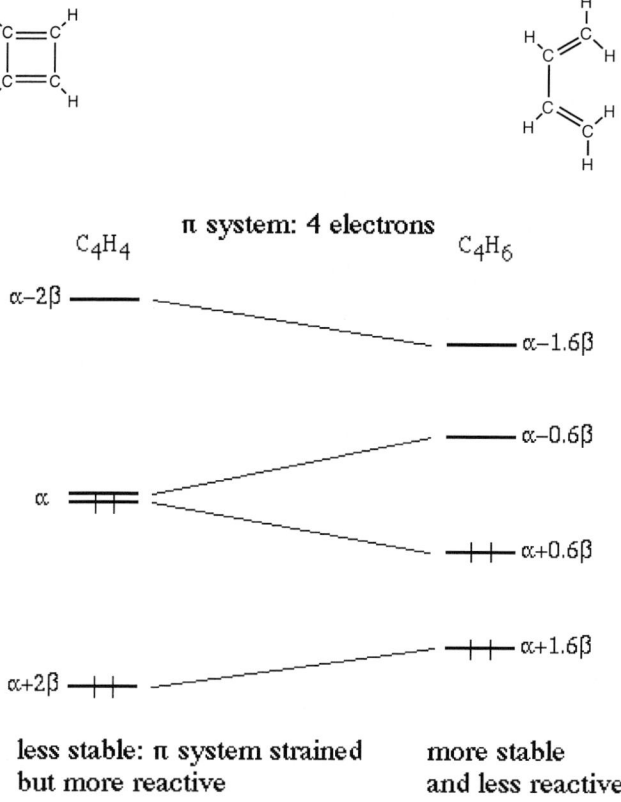

Figure 2: A comparison of the molecular orbitals of cyclobutadiene and butadiene. The energies of the orbitals obtained with the Hückel method are indicated.

2.2 Strain-reactivity relationship: an example from organic chemistry

The relation between strain and reactivity has been studied several years ago already in the domain of organic chemistry. One example is the addition of azide on various strained alkenes [3]. The basic reaction is the addition of a C=C bond on a N_3 unit.

Strain can be applied to the C=C bond by including it in a cyclic molecule. The constraint imposed by the cyclic structure yields a torsion deformation of the double bond. The magnitude of the strain can be controlled by the length and the nature of the cycle. The rate of addition has been measured for a series of strained C=C bonds, and the result is recalled in table 1. This rate shows very strong variations upon the chosen alkene, E-cyclooctene being for example 54 000 times more reactive than the reference cyclohexene. A first attempt has been to correlate the reactivity with the steric energy in the reactant molecule. This energy is evaluated with a molecular mechanics approach, each bond, bond angle and diedral angle being associated to a spring. The equilibrium value of each parameter, and the associated spring constants are fitted or obtained from quantum chemistry calculations. If all spring are at equilibrium, the elastic energy is zero. This fully relaxed situation can not be reached in a cyclic system, due to the constraint in the cycle, the total equivalent elastic energy is positive and is called the steric energy.

alkene	structure	Relative rate	Steric energy of alkene	Steric energy of product	Δ
Cyclopentene		42	8.1	11.4	-3.3
Cyclohexene		1	4.1	10.4	-6.3
cycloheptene		53	9.9	12.9	-3
cyclooctene		65	13.4	18.9	-5.5
norbornene		1.5×10^4	25.4	22.4	3
(E)-cyclononene		1×10^4	21.5	21.6	-0.1
(E)-cyclooctene		5.4×10^5	23.2	17.7	5.5

Table 1: relative reaction rate and steric energies (kcal.mol^{-1}) for the reactant alkene and the product triazoline. The difference Δ (steric energy of alkene – steric energy of triazoline) is given. Reprinted from reference 3.

The correlation between the reaction rate (in logarithmic scale) and the steric energy in the reactant is only succesful, since trans-cyclooctene is more reactive than norbornene, but has a lower steric energy. If the steric energy difference between the reactants and the products of the reaction is now calculated, the correlation becomes excellent. Hence the reaction barrier correlates with the strain relief upon reaction. If the stain relief is large (as it is the case for (E)-cyclooctene), the barrier will be lower and the reaction faster.

2.3 Straining a single metal atom complex: an example from coordination chemistry

Another example for the reaction between strain and chemical reactivity can be found in organo-metallic chemistry, where a single metal atom is interacting with molecular ligands. The complex between Nickel, ethene, and a di-phosphine ligand is a classical case [4] (see figure 3a). The two phosphorus based centers are here linked through a bridge, which length can be modulated, hence straining the P-Ni-P angle. If we consider a fragment where the ethene has been removed, the energies of the fragment molecular orbitals, especially of the HOMO, are modified if the angle is changed, as illustrated in figure 3b. This HOMO of the fragment shows an antibonding interaction between the phosphine ligand orbital and a metallic d type orbital. If the angle is decreased to 90°, the overlap between the ligand and the d orbital is optimal, and hence the anti-bonding orbital is destabilized in energy. On the contrary, if the angle is opened to 180°, the interaction between the d orbitals and the ligands is cancelled, hence lowering the HOMO energy. This HOMO orbital of the P-Ni-P fragment is the key orbital for the interaction with the ethene molecule. It interacts with the vacant π* orbital on ethene. If the angle is large, the metal fragment HOMO is low in energy, so that the reactivity toward ethene is small. However, if the angle is strained to a small value, the HOMO is higher in energy and the reactivity increases.

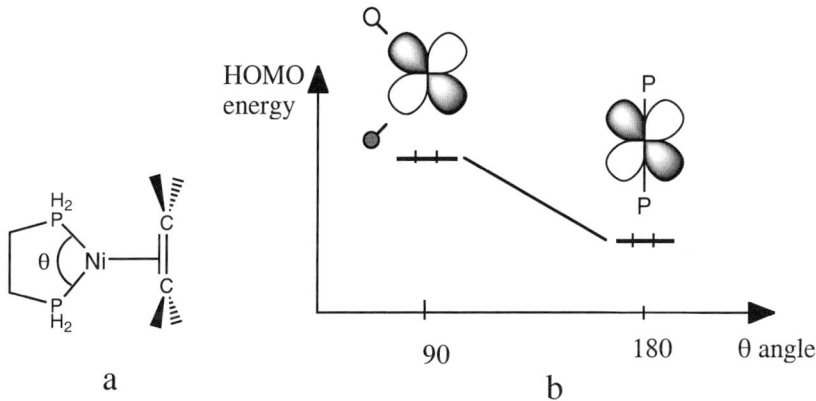

Figure 3: a) structure of the bridge di-phosphine Ni complex; b) Change in the HOMO energy of the P-Ni-P fragment as a function of the strain in the θ angle.

3. STRAIN AND REACTIVITY AT METALLIC SURFACES

Metals and alloys have a great importance in heterogeneous catalysis. Hydrogenation, hydrogenolysis or Fisher-Tropsch reactions are performed on transition metal particles [5]. There are various ways in catalysis to modify the reaction properties of metal surfaces, such as adding a promoter (an alcali atom for example) or a poison (sulfur or chlorine), forming a surface alloy or changing the surface structure with steps and defects. Among these ways to change the catalyst, applying a stress can be a very effective one [6].

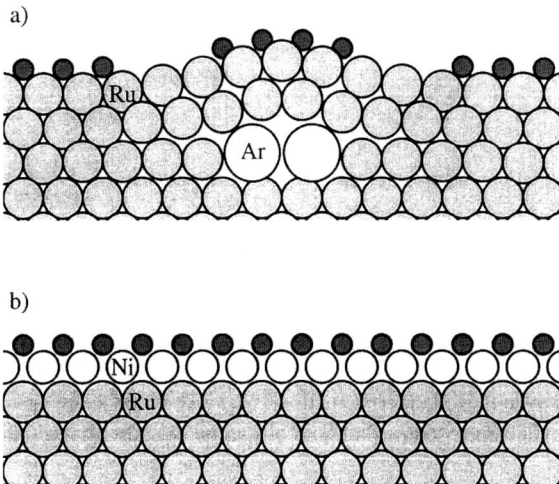

Figure 4: two experimental ways to induce a strain on a metallic surface: a) by the implantation of Ar bubbles, b) by pseudomorphic growth of a thin layer of a metal on another metal. Adsorbates are indicated by a small red ball. Adsorption is favored in the regions of tensile strain. Reprinted from reference 6.

Figure 4 shows two experimental ways to apply a stress and create a strain on a metallic surface. Menzel and coworkers modified a Ru surface by the implantation in subsurface of noble gas atoms [7]. These argon "bubbles" below the surface create an outward deformation of some areas of the surface. The top part of the bumps is a region where the surface is submitted to a tensile strain, while the periphery of the regions is subject to a compressive strain. Hence heterogeneity in strain can be induced by such surface modifications. Another way is to evaporate one metal on another one with a different atomic radius. If the chosen combination gives a pseudomorphic growth of a monolayer of one metal on top of the other, surfaces with important stress and strain can be obtained [8-13]. In each case the cell parameter and interatomic distances are different from the case of the pure surface. The question addressed in the following is how this strain in the surface modifies the chemisorption and reactivity properties of the catalyst.

There are several experimental evidences of the modified chemisorption at strained surfaces [7, 14-20]. Ru surfaces strained by implantation of Ar bubbles were exposed to oxygen chemisorption [7]. STM imaging showed a clear preference in chemisorption for areas corresponding to an expanded Ru lattice at the top of protrusions. CO chemisorption was studied with vibrational spectroscopy on copper layers where various amount of compressive or tensile strain was imposed from deposition on different substrates [14]. The CO stretching frequency showed a linear variation as a function of copper strain, a tensile strain corresponding to an increased frequency. Pd layers deposited on a Ni(111) and on a Ni(110) surface showed a greatly enhanced activity for butadiene C_4H_6 hydrogenation, keeping the excellent selectivity of Pd in partial hydrogenation to butene [18-21]. A similar result was also found for the hydrogenation a single olefin (butene). Therefore, the adsorbate-surface chemical bond and the molecular activation can be significantly modified even with a small strain in the overlayer.

Several effects are involved in the modification of the electronic properties of deposited metal layers on a substrate of a different metal. One of them is the above described modification of the metal-metal distance in the layer due to epitaxial growth. However the electronic interaction with the substrate atoms also has an influence. For example, the Fermi level of the system is fixed by the substrate, and not by the deposited layer. This electronic interaction is very important for sub-monolayer or monolayer deposits [12], since surface atoms bind directly with substrate atoms.

This influence of strain on chemisorption and reactivity can be modeled by total energy calculations.

3.1 Influence of strain from quantum calculations

The surface is generally described by a periodic slab under uniform lateral strain, or by a deposit of one metal on another metal, with the surface unit cell determined by the substrate metal. In the chosen examples, density functional theory (DFT) calculations [22-25] are performed using a plane wave basis set and ultrasoft atomic pseudopotentials.

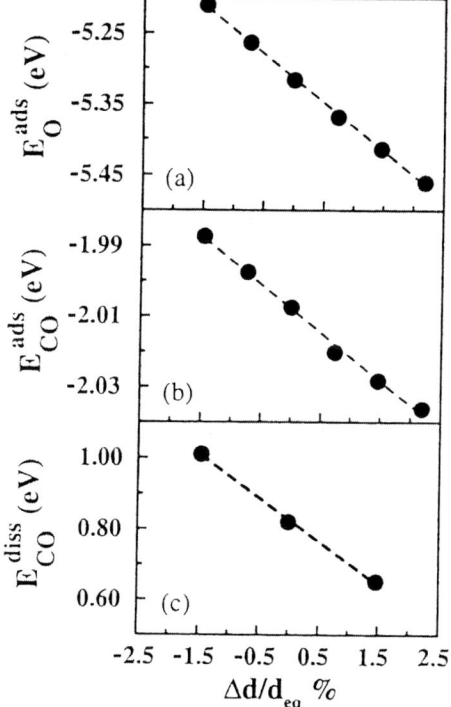

Figure 5: influence of strain on a Ru(0001) surface, expressed as the relative change in surface lattice constant, on the a) binding energy of an oxygen atom, b) binding energy of a CO molecule, c) dissociation barrier of CO. Reprinted from Reference 26

The exchange correlation functional is one form or another of the generalized gradient approximation GGA. The error in absolute adsorption energy compared to experimental values can be as large as 0.5 eV. However, energy differences and variations are usually much better described, since the errors tend to cancel for similar systems. Staining the surface is a moderate perturbation where a good description of energy trends can be obtained with DFT.

Figure 5 shows the calculated binding energy for an oxygen atom (a) and a CO molecule (b) as a function of applied strain on a Ru(0001) surface [26]. The influence on CO dissociation barrier is also shown (c). The substrate deformation is expressed in relative change in surface lattice constant (tensile and compressive strain associated to positive and negative values respectively). The binding energies of O and CO increase in absolute value for an expanded Ru lattice, while they decrease under compression. The variation is strong for O but much more subtle for CO. Expanded Ru surfaces show a stronger interaction with the CO molecule, and also a clear reduction of the CO dissociation barrier: they are more reactive. Compressed surfaces, on the contrary, appear to be less reactive.

Insight in this behavior can be obtained from a simple analysis of the change of the surface electronic structure under strain, as illustrated in figure 6 [6, 26]. The d-density of states at the surface is sketched. If the surface is submitted to a compressive or tensile strain, bond distances between metal atoms decrease or increase. The spatial overlap between d atomic orbitals on the metal atoms will hence increase or decrease, and so will the width of the d-band. The case of tensile strain and the associated narrowing of the d band is shown in figure 6. If the center of the d band was not changed, this narrowing would result in a charge transfer toward the d states, since they move below the Fermi level. Self-consistent field effect will prevent this charging, and the d states will shift up in energy in order to give a quasi charge neutrality. Hence the d- band center is shifted up upon tensile strain, and it would be shifted down if a compressive strain is applied.

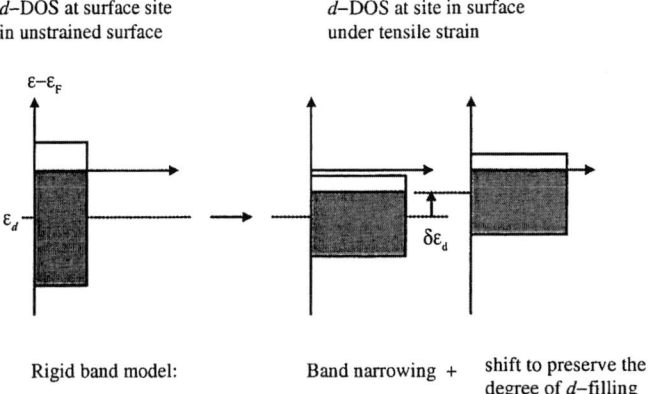

Figure 6: Change in the electronic structure of a metallic surface as a function of an applied tensile strain. The d band is represented by a box, the blue part being occupied. The arrow shows the position of the Fermi level. Reprinted from reference 6.

This displacement of the energy of the d-band center has important implications for the molecular interaction and reactivity. Figure 7 shows again the adsorption energy of O and

CO, and the dissociation barrier of CO on strained Ru surfaces, but this time expressed as a function of the calculated surface d-band center [26]. The correlation is clear: a higher d-band center results in a more reactive surface, associated with stronger binding energies and lower dissociation barriers. In the case of CO, a higher d-band center yields a more efficient back-bonding from the d orbitals to the π* CO orbitals. This population of the CO antibonding orbitals weakens the CO bond and makes dissociation easier. A similar trend of lower dissociation barriers with expanded lattices was obtained in the case of N_2 dissociation on Fe surfaces. Hence tensile strain of the surface seems to be associated with improved reactivity. Is this a general rule?

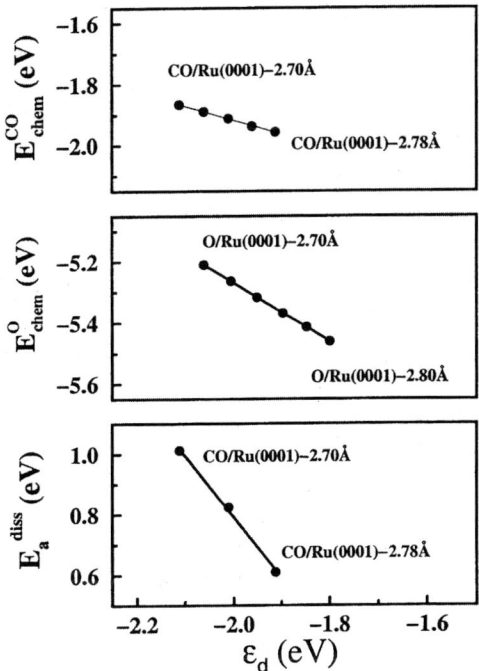

Figure 7: Adsorption energies for oxygen (a) and CO (b), and dissociation barrier for CO on strained surface. The data is presented as a function of the energy position of the center of the d-band, while the range of lattice constant variation is given by the labels. A clear correlation is shown. Reprinted from reference 26.

3.2 The compressed deposits of Pd on Ni(110): improved reactivity for hydrogenation of olefins and di-olefins

Deposits of Pd atoms up to 4 ML have been created on a Ni(110) substrate [18-21]. The Pd atoms experience a strong compressive stress at the interface, due to the important mismatch between the Ni-Ni (2.49 Å) and the Pd-Pd (2.75 Å) bond distances. A part of the strain is relieved by the formation of vacancies separating rows of Pd atoms in the direction of the rows of the (110) surface [27]. However the Pd atoms remains in average under a compressive strain. However, in contrast with the previous examples, this compressed overlayers are more active than Pd(110) for the hydrogenation of the unsaturated C=C bonds of butadiene and olefins.

The chemisorption of ethene C_2H_4 on a strained Pd(111) surface has been studied from DFT calculations. A slab built of 4 Pd layers has been considered. The surface lattice vectors of the slab have been constrained in tensile or compressive mode compared to the calculated bulk equilibrium value (Pd-Pd = 2.8 Å). The atoms from the bottom layer of the slab have been kept fixed, while all the other atoms have been free to relax. The chemisorption of ethene was studied with a 2x2 super-cell corresponding to a coverage of 1/4 ML, which is close to the experimental coverage. A bridge site was considered for ethene, with each carbon atom interacting with a different Pd atom. This site was shown to be the most stable on Pd(111) [5]. The molecule has been completely optimized on the surface, and adsorbate induced relaxations of the substrate have been taken into account. The chemisorption energy Echem was calculated on each strained slab from
Echem = E (slab + molecule chemisorbed) - E(slab) - E(molecule in gas phase)

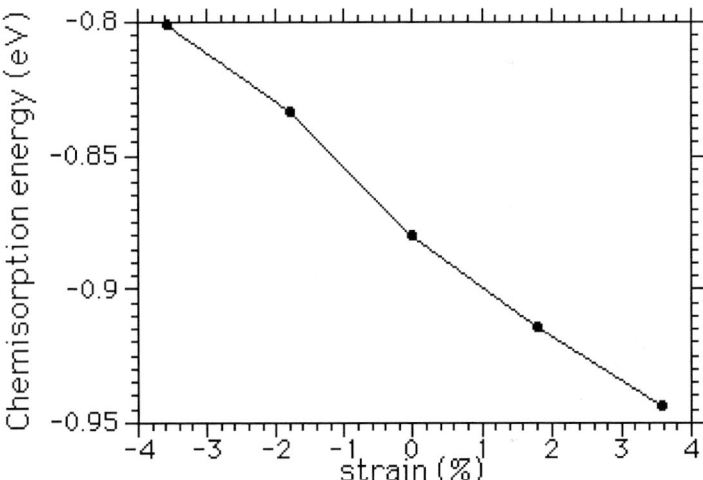

Figure 8: Ethene chemisorption energy (eV) on the four layer Pd slab as a function of the applied lateral strain. The strain is defined as the relative change in surface lattice constant $(a - a_{eq})/a_{eq}$.

The chemisorption energy of ethene on the strained Pd(111) slab is given in figure 8. There is a clear decrease of the absolute value of the adsorption energy for a compressive strain (reduced Pd-Pd distance) and an increase for tensile strain (elongated Pd-Pd). The effect is significant and almost linear in the considered range: it corresponds to -0.02 eV (or 2.4 % of chemisorption energy) for 1% strain. If the lattice is compressed, the molecule-molecule distance in the 2x2 array is reduced so that the effective molecular density on the surface is increased. This change in the molecule-molecule interaction has been tested on a fictitious layer of molecules with the geometry of the chemisorbed state. It results in a destabilisation of the molecule of 0.003 eV for a compression of 3.5 %. This is much less than the calculated change of chemisorption energy (0.08 eV) for the same distortion of the Pd lattice. The weakening of the chemisorption upon compressive strain is therefore not related to direct molecule-molecule interactions.

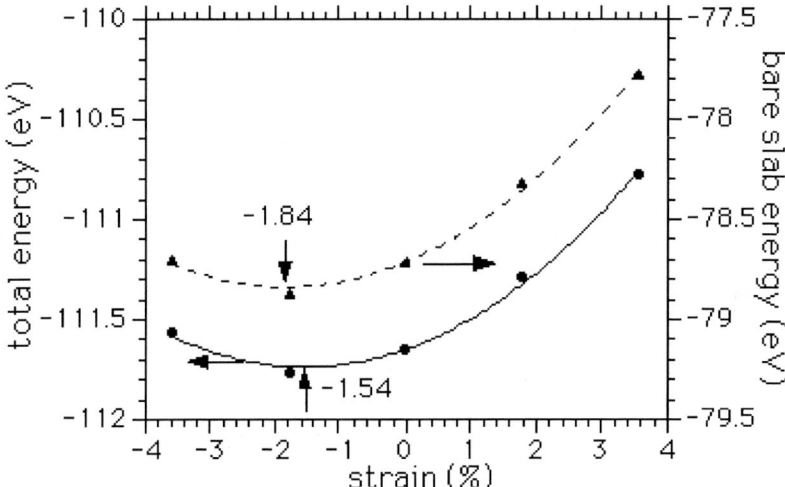

Figure 9: Total energy of the slab with the p(2x2) ethene layer (dots + full line, left) and of the bare slab (triangles + dotted line, right). The lines are parabolic fits and the coordinate for the minimum is indicated. The strain is defined as the relative change in surface lattice constant $(a - a_{eq})/a_{eq}$.

The total energy variations of the slab with the adsorbate and of the bare slab are illustrated on figure 9. For the discussion, the calculated values have been fitted to a parabola. Clearly, the surface lattice vectors for the energy minima are smaller than the Pd bulk lattice values. For a Pd(111) termination, the surface is submitted to a tensile stress which is a rather general phenomenon: the low coordination Pd atoms increase their interactions with the neighboring atoms. When ethylene is chemisorbed, bonds are formed between surface Pd atoms and adsorbates and part of this stress is released. As a consequence, the energy minimum for our 4 layer slab corresponds to a smaller lattice compression with the chemisorbed molecules. This different strain value at the energy optima for the bare and covered surfaces is directly related with the linear variation of the chemisorption energy with strain. If the curvatures of the energy parabola are supposed to be similar, then the slope of the chemisorption energy as a function of strain is directly proportional to the strain difference between the two optima. Therefore there is a clear relation between adsorbate induced surface stress and variation of chemisorption energy with strain: adsorbates which induce a large change in the surface stress will also show a large change in chemisorption energy with applied strain.

The geometry of the molecule is modified upon chemisorption. For the normal Pd lattice parameter, the C-C bond is elongated from 1.34 Å in the gas phase to 1.44 Å on the surface and the H atoms are tilted upward by 0.33 Å with respect to a planar geometry.

However, the molecule geometry is not modified in a significant way when a strain is applied to the surface. There is a strong adsorbate-induced relaxation of the substrate: the two Pd atoms which interact with ethene are displaced upward by 0.13 Å, while the other two Pd atoms in the unit cell move downward by 0.06 Å, compared to the bare surface position. However, again, there is no clear relation between changes in the relaxation and variation of chemisorption energy under applied strain.

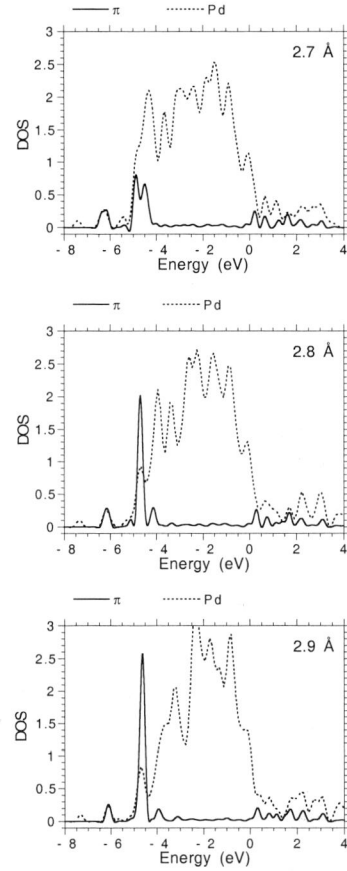

Figure 10: Density of States projected on the π orbital of ethene (full line) and on the first neighbor Pd atom (dotted line) for compressive (top), neutral (middle) and tensile (bottom) applied lateral strain in the slab. The associated Pd-Pd distance within the strained layer is indicated in the top right corner.

In order to better understand the change in chemisorption energy, partial density of states are given in figure 10. The molecular states which are the most involved in the chemisorption process are the frontier orbitals π and π^*. The π orbital is the in-phase combination of $2p_z$ atomic orbitals and is occupied by 2 electrons, while the π^* orbital is the antibonding partner and is vacant (z is the direction perpendicular to the surface). When the molecule is distorted on the surface, these levels get mixed with the sigma framework, but they keep a predominant $2p_z$ character [28].

The π^* orbital interacts with the d band and gets some small contributions below the Fermi level. As a result this previously vacant antibonding orbital is partially populated. This is the main driving force in the C-C elongation. Upon distortion of the molecule, the π^* orbital is lowered in energy, which increases its interaction with the d states. Straining the Pd slab has however no significant effect on this interaction. The amount of partial occupation is only changed by 5%. Moreover, if one wants to analyze this small difference, the expanded surface corresponds to a reduced mixing with the π^* orbital. This would result in a decrease of the chemisorption energy (in absolute value) which is just the opposite of the calculated effect.

The DOS projected on the occupied π orbital is given in figure 10 for three values of the surface lattice constant, together with the DOS projected a surface Pd atom interacting with the molecule. For the bulk Pd lattice constant case (fig10, middle), the π orbital mainly appears as a narrow peek positioned at the lower Pd d band edge. It slightly mixes with states above the Fermi level. The main variation with strain is however appearing near the lower Pd d band edge. When strain is applied one important change in the surface electronic structure is the d band width. When the lattice vector is shortened, as seen before, the d band gets wider and the π level fits better with the lower part of the d band. Therefore the mixing between π and d states is stronger, and the π level is more delocalized, appearing as a broader and less intense band. If the lattice is expanded, the d band gets narrower and the π ethene orbital is ejected out the d band. The interaction with the d band is decreased, and the π state is more localized . The π ethene orbital is occupied, and in the case of Pd the d states are almost fully occupied. Such an interaction between filled levels is unfavorable and is related with Pauli repulsion. In the band built from the π orbitals in fig10-top (between -5 and -4 eV) both bonding and antibonding contributions are populated. Therefore, the large interaction for compressed slabs reduces the chemisorption energy, while the release of this unfavorable mixing strengthens chemisorption when the slab is expanded. Influence of the interaction between occupied states of the surface and of the molecule have already been underlined for the chemisorption of unsaturated molecules on Pd and Pt surfaces [28, 29]. Compressive strain increases the Pauli repulsion term between the molecule and the surface, with only a small effect on the attractive part which mainly involves the π^* ethene orbital. Hence if ethene behaves similarly as CO, with a chemisorption energy increase for tensile strain, the reasons for the change in chemisorption energy are different.

3.3 Chemisorption and hydrogenation of ethene on pseudomorphic Pd overlayers

In a separate study, the chemisorption and reactivity of ethene has been described on a single layer of Pd atoms, in pseudomorphic deposit on Re(0001), Ru(0001), Pd(111) and Au(111) surfaces [30]. All surfaces present the same hexagonal symmetry, with modified interatomic distances: Re (2.76 Å) , Ru (2.71 Å), Pd (2.75 Å), Au (2.88 Å). The DFT computed binding energies of ethylene C_2H_4, ethyl C_2H_5 and atomic hydrogen are show in figure 11. The order of the substrates on the horizontal axis was chosen to correspond to an increasing binding energy (in absolute value). For Ru, Pd and Au, the trend is similar than in the previous section, a tensile strain being associated with a larger chemisorption energy. The Re substrate is not in agreement however, since it gives a weaker chemisorption, with a lattice parameter larger than that of Ru and Pd. This originates from an electronic effect of the Re surface, with a marked electropositive character of Re. The behavior is similar for hydrogen. Ethyl is the product of the first hydrogenation of ethylene. Its chemisorption energy follows again the same trend as a function of the substrate. However, it is important to note here that the magnitude of the chemisorption energy increase from Re to Au is smaller for ethyl than for ethylene. This point will be important for the reactivity.

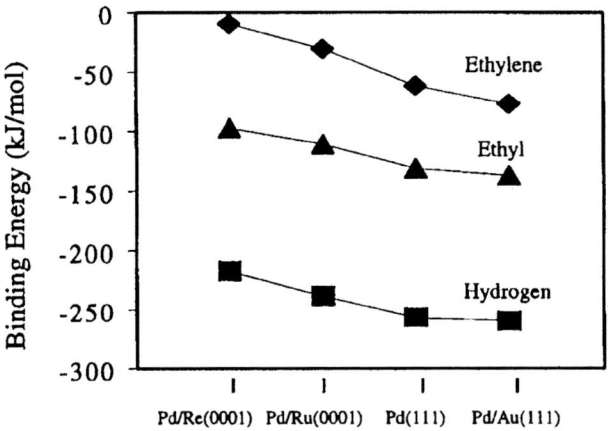

Figure 11: DFT-computed binding energies of ethene (or ethylene), ethyl and hydrogen on a monolayer of Pd deposited on Re(0001) (strain +0.4 ‰), Ru(0001) (strain −1.5 ‰), Pd(111) (strain +0 ‰) and Au(111) (strain +4.7 ‰). The strain is defined as the relative change in surface lattice constant between the Pd adlayer and the substrate. Reprinted from reference 30.

The different behavior of the Re substrate is elucidated if one analyses the d-band center for a Pd atom in the deposited monolayer, as shown in figure 12. The Re corresponds to a large shift downward of the d-band center. Indeed, Re being more electropositive, its electronic levels are higher in energy. The Pd electronic levels interact with the substrate Re levels, and as a consequence are shifted down in energy. This electronic effect is smaller for Ru, but it cumulates with the influence of the compressive strain. Finally the up-shift for the Au substrate is explained by the tensile strain. Hence, the chemisorption energy of ethene and ethyl again perfectly correlates with the d-band center energy. The only difference with previously is that this shift of the surface d-band center can originate both from strain and electronic effects in the case of a monolayer deposit.

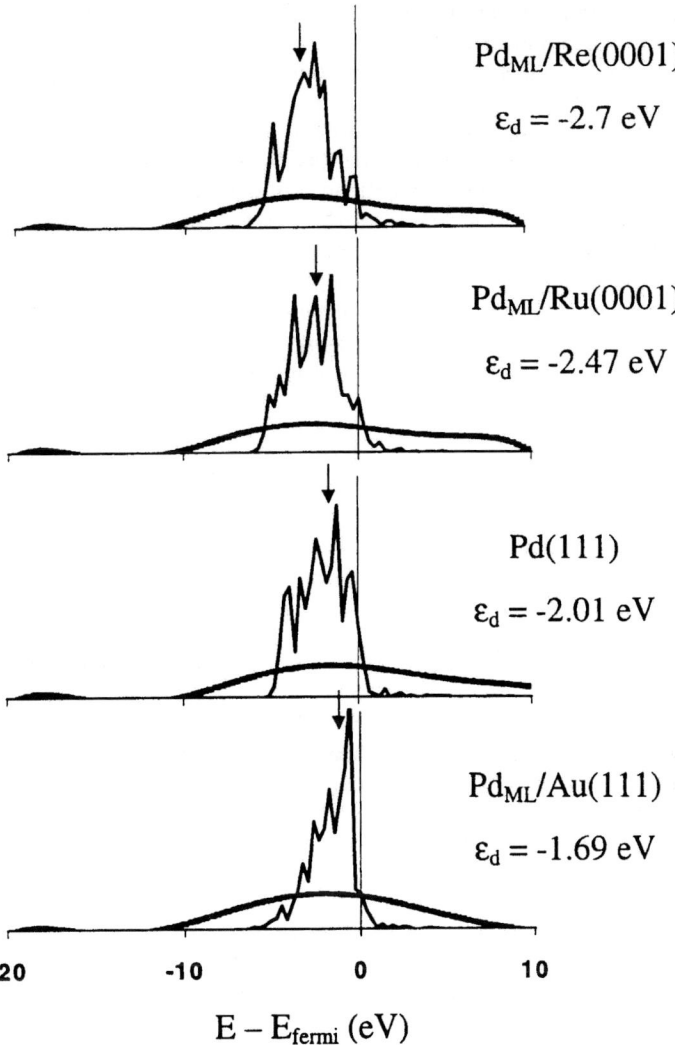

Figure 12: Density of states projected on the Pd surface atom for a deposit on Re(0001), Ru(0001) Pd(111) and Au(111) substrate. Both the flat sp and the sharp d bands are shown. The arrow indicates the center of the d band, and its energy ε_d is indicated relative to the Fermi level. Reprinted from reference 30.

The pathway for ethene hydrogenation on these deposits has been studied [30]. The barrier for the first hydrogenation of ethene to ethyl are respectively 78, 77, 82 and 104 KJ.mol^{-1} for the Pd monolayer on the Re(0001), Ru(0001), Pd(111) and Au(111) substrate. Hence, in contrast with the case of CO dissociation on Ru, the surface under tensile strain corresponds to the higher reaction barrier, and then to the less reactive situation. On the contrary, surfaces with a lower center of the Pd d band, either from an electronic effect or a compressive strain, are more active. This is in agreement with the observed increase of reactivity for Pd deposits on Ni surfaces, where the Pd is also under a compressive strain [18-21].

What is happening in these cases? From the shift to lower energy of the d-band, the chemisorption energy of ethene is weakened. This doesn't allow by itself to conclude on the reactivity, since the important point is the change in the reaction barrier. The energy of the transition state is also destabilized, but to a smaller extent. As a consequence, for ethene, this weaker chemisorption energy situation corresponds to an increased hydrogenation reactivity. Ethylene is an unsaturated molecule and the high lying π orbital overlaps in energy with the bottom of the metal d-band. The unsaturated character and the π orbital are lost in the case of the final state ethyl, and this is why the binding energy decrease is smaller in that case. The transition state resembles the final state, and is also less destabilized than ethene. Therefore, the reactivity is enhanced.

4. CONCLUSION

The previous examples show a clear relation between strain and chemical reactivity. Stress by itself is important because it generates strain, which in turn modifies the surface chemistry. The deformations are rather small, and so are the energy changes along the reaction pathway. However, these small energy changes can have important effect on the reaction rate, which is from the Arrhenuis law exponentially dependent on the activation barrier. Examples have be given in a large range of chemical processes, from organic chemistry to organo-metallic chemistry and finally to surface chemistry. For the metal surfaces, it is tempting to establish a correlation between the nature of the strain, compressive or tensile, and the reactivity change. Although in the case of CO, first studied, tensile strain was associated to an enhanced reactivity, such a relation is not general. Tensile strain is indeed associated to a stronger chemisorption energy of molecules, explained by the shift to a higher energy of the d band center. However, the key point is the change in the reaction barrier, ie the relative change of the transition state energy with respect to the initial state energy. Different changes in reactivity can hence be observed.

In the case of tensile strain, ruthenium is more active for CO dissociation since the transition state is stabilized in energy more than the chemisorption structure. The transition state is close in structure to the final state, where C and O are separated, which is strongly stabilized by the upward shift of the d band center. On the contrary, Pd under tensile strain would be less active for ethene hydrogenation. The transition state and the ethyl final state are indeed less stabilized than the chemisorbed state. In both cases, the transition state follows the trend of the final state. Therefore the key point for the reactivity is the compared implication of the metal d band in the initial and final state of the reaction. If this influence is stronger in the final state as for CO, tensile strain increases the reactivity. If it dominates in the chemisorption initial state as for ethene, then tensile strain reduces the reactivity.

REFERENCES

1. P.W. Atkins, Physical Chemistry, Oxford University Press, Oxford, 1994.
2. T.A. Albright, J.K. Burdett, M.-H. Whangbo, Orbital interactions in chemistry, John Wiley &sons Inc., 1985
3. K.J. Shea and J.-S. Kim, J. Am. Chem. Soc. 114 (1992) 4846
4. T.A. Albright, R. Hoffmann, J.C. Thibeault and D.L. Thorn, J. Am. Chem. Soc, 101 (1979) 3801
5. G.A. Somorjai, Introduction to Surface Chemistry and Catalysis; John Wiley &sons Inc., 1994
6. B. Hammer and J.K. Norskov, Advances in Catalysis, 45 (2000).
7. M. Gsell, P. Jokob, D. Menzel, Science 280 (1998) 717.
8. S.K. Kim, C. Petersen, F. Jona and P.M. Marcus, Phys. Rev. B 54 (1996) 2184.
9. H. Brune, H. Röder, C. Boragno and K. Kern, Phys. Rev. B 49 (1994) 2997.
10. H. Brune, M. Giovannini, K. Bromann and K. Kern, Nature 394 (1998) 451.
11. M. Sambi, G. Granozzi, Surf. Sci. 400 (1998) 239.
12. J.A. Rodriguez, D.W. Goodman, Science 257 (1992) 897.
13. J.A. Rodriguez, Heterogeneous Chemistry Reviews 3 (1996) 17.
14. E. Kampshoff, E. Hahn, K. Kern, Phys. Rev. Lett. 73 (1994) 704.
15. J. Heitzinger, S. C. Gebhard and B.E. Koel, Surf. Sci. 275 (1992) 209.
16. J.H. Larsen, I. Chorkendorff, Surface Science 405 (1998) 62.
17. M.O. Pedersen, S. Helveg, A. Ruban, I. Stensgard, E. Laegsgaard, J.K. Norskov, F. Besenbacher, Surface Science 426 (1999) 395.
18. P. Hermann, J.M. Guigner, B. Tardy, Y. Jugnet, D. Simon, J.C. Bertolini, J. Catal. 163 (1996) 169.
19. P. Hermann, B. Tardy, D. Simon, J.M. Guigner, B. Bigot, J.C. Bertolini, Surf. Sci 307-309 (1994) 422.
20. J.C. Bertolini, Surface Review and Letters 3 (1996) 1857.
21. J.C. Bertolini, P. Miegge, P. Hermann, J.L. Rousset, B. Tardy, Surf. Sci. 331-333 (1995) 651.
22. P. Hohenberg and W. Kohn, Phys. Rev. 136 (1964) B864
23. W. Kohn and L. Sham, Phys. Rev. 140 (1965) A1133
24. M.C. Payne, M.P. Teter, D.C. Allan, T.A. Arias and J.D. Joannopoulos, Rev. Mod. Phys. 64 (1992) 1045
25. G. Kresse and J. Furthmüller, computat. Mat. Sci. 6 (1996) 15.
26. M. Mavrikakis, B. Hammer and J. Norskov, Phys. Rev. Lett. 81 (1998) 2819.
27. J.S. Filhol, D. Simon, P. Sautet, Surf. Science. (2000) in press.
28. P. Sautet, J.-F. Paul, Catal. Lett. 9 (1991) 245.
29. J.-F. Paul, P. Sautet, J. Phys. Chem. 98 (1994) 10906.
30. V. Pallassana and M. Neurock, Journal of Catalysis 191 (2000) 301.

Strain measurement in ultra-thin films using RHEED and X-ray techniques

B. Gilles

Laboratoire de Thermodynamique et Physico-Chimie Métallurgiques, UMR-CNRS 5614, ENSEEG-INPG, BP 75, 38402 St Martin d'Hères, France

This article gives an overview of two of the main techniques which are commonly used for the determination of strain in thin and ultra-thin films, i.e. the *in-situ* Reflection High Energy Diffraction (RHEED) technique and the *ex-situ* x-ray Diffraction (XRD) technique. It intends to provide the main basis for undertaking these measurements. Different examples on ultra-thin films and multilayers grown by the Molecular Beam Epitaxy (MBE) technique are given.

1. INTRODUCTION

Strain in ultra-thin films or multilayers is often related to the structure and the physical properties of these materials. Therefore, it appears very important to develop experimental techniques which can provide reliable and accurate measurements of strain. Among the possible techniques, those which have a short response time and which may easily be handled at a reasonable cost are of course preferred. Because of these requirements, the most commonly techniques which are used nowadays are Transmission Electron Microscopy (TEM), X-ray Diffraction (XRD) and Reflection High Energy Diffraction (RHEED). This paper will focus on the two latter techniques, the former being developed in another paper. Experimental results obtained with these two techniques on metallic ultra-thin films and multilayers grown by Molecular Beam Epitaxy (MBE) will be given. These examples will highlight the possibilities and limits of the techniques.

2. STRAIN MEASUREMENT USING RHEED

2.1. Practical aspects

The MBE technique has been developed in the 70th for the growth of semi-conductors [1] and it has rapidly been extended to the growth of metals and oxides [2]. From the beginning, RHEED has appeared as a powerful tool for monitoring the structure of the deposited layers. Indeed, its easy use and capability to provide diffraction patterns in real time during the growth have made RHEED an indispensable tool even in MBE machines dedicated to production. The principle and geometry of RHEED have already been described in details [3-4]. An incoming e-beam of 10-50 kV strikes the surface at a near-grazing incidence (typically around 1°) and the diffraction pattern is projected on a fluorescent screen located at the specular reflected beam. Due to the high energy of the e-beam, the Ewald sphere may be roughly approximated with a plane. Within this simplified scheme and by using the Ewald construction, it may be easily shown that the diffraction pattern reproduces a section of the reciprocal space which is perpendicular to the surface and to the incoming beam [5]. In figure 1 we have reproduces the RHEED pattern taken on a Cu (001) layer deposited on a Si (001) substrate. The image has been acquired using a video-CCD camera and digitized with a frame-grabber card in a micro-computer. The azimuth, which is the direction of the incoming 40 keV beam, is along the [100] direction of the surface. In figure 2, the sample has been rotated by 45° around its normal and the azimuth is now along the [1$\bar{1}$0] direction of the surface. The (000) direct beam is located below the images. Because this beam is very intense, it is usually masked by a rotating shutter. The angle of incidence has been set just in-between the Bragg angles for the (002) and the (004) reflections, at the so-called anti-Bragg (003) position. Because the roughness of the surface is quite large (about 8-10 Å as deduced from STM measurements) the specularly reflected spot does not appear as an extra spot at the (003) position. However, a reinforcement of the intensity may be observed at this position. Also, it may be seen that the spots are becoming fainter and fainter as far as the Miller indexes increase. This is mainly because the Ewald sphere has a finite radius of curvature. The Ewald sphere, which intercepts the reciprocal lattice at exactly the (000) position, does not intercept

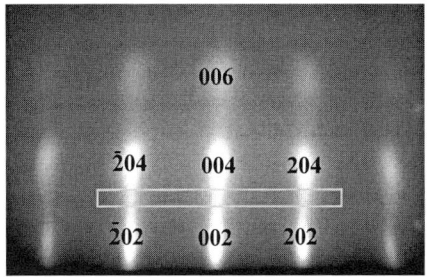
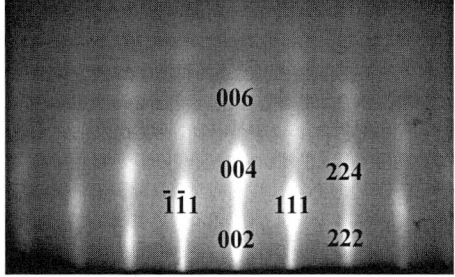

Figure 1. RHEED pattern on a Cu (001) layer with the azimuth along [010].

Figure 2. RHEED pattern on a Cu (001) layer with the azimuth along [1$\bar{1}$0].

the reciprocal points with high Miller indexes. Also, it may be seen on figures 1 and 2 that the sharp spots are connected with streaks which extend along the z-direction perpendicular to the surface. This streaks are due to the small thickness of the irradiated area due to the grazing incidence of the incoming e-beam. Therefore, in the reciprocal space, the 3D Bragg peaks are convoluted with rods. The RHEED pattern is produced by the intersection of the reciprocal space with the Ewald sphere. But both the reciprocal lattice rods and the Ewald sphere have finite thicknesses due respectively to lattice imperfections and to the electron energy spread and beam divergence. In figure 3, we have schematically depicted typical RHEED patterns for different kinds of surfaces. In figure 3.a the perfect crystal lattice with a flat surface may be idealized as a 2D object and the intersection of the Ewald sphere with the elongated thin rods produces a set of small spots located in the so-called Laüe zone of rods. Note that these spots do not coincide with the Bragg 3D positions. In figure 3.b the imperfections of the crystal lattice broaden the lattice rods and their intersection with the Ewald sphere extends along the z-direction. For rough surfaces, as in figure 3.c and 3.d, the intensity is confined near the 3D Bragg positions. During the growth of metals, the transition from a streaked pattern as in figure 3.b to a spotty pattern as in figures 3.c and 3.d is characteristic of the roughening of the growing surface. A nice RHEED pattern as figure 3.a is characteristic of a single crystal substrate of high quality but is not commonly observed on epitaxied layers. The feature of the

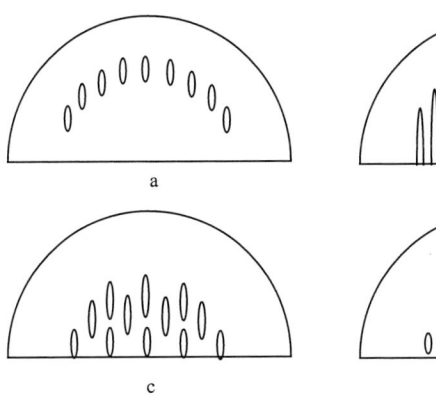
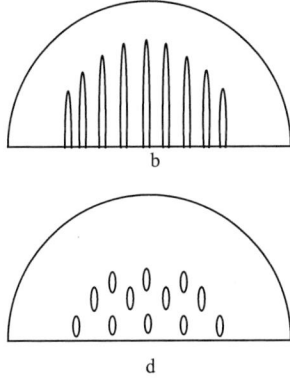

Figure 3. Schematics of the RHEED pattern for a) a flat surface of a perfect crystal b) a flat surface of a crystal having lattice imperfections c) a slightly rough surface d) a very rough surface.

RHEED patterns in figure 1 and 2 is somewhere in-between figure 3.b and figure 3.c. Of course, we have in figure 3 depicted only some of the RHEED patterns which are commonly observed. For instance, it is well known that faceted surfaces or crystals with mosaic structures produce characteristic RHEED patterns [3]. Also, it may be noted that for polycrystalline layers, RHEED may show well defined rings, provided that the surface is rough. But these aspects are beyond the scope of this paper which is mainly dedicated to the measurement of strain in epitaxied layers. In figure 4, we have reported the plot of the

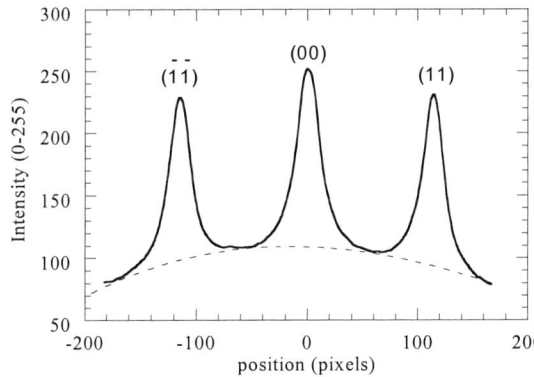

Figure 4. Horizontal plot of the intensity averaged over the vertical width of the rectangle in figure 2.a. The rods have been indexed using the 2D surface notation.

intensity along the horizontal direction defined by the rectangle in figure 1. Because the image in figure 1 has been digitized using a 8-bits converter, the intensity is defined in a integer scale between 0 and 255. Therefore, in order to improve the statistics, the intensity in figure 4 has been averaged over the vertical width of the rectangle of figure 1. Also, in figure 4 the ($\bar{2}$0L), (00L) and (20L) rods have been reported using the 2D surface notation [2,4], i.e. respectively

($\bar{1}\bar{1}$), (00) and (11). It has to be reminded that in the case of a FCC crystal, the 2D surface square mesh is rotated by 45° with respect to the 3D cubic mesh and that the 2D lattice vectors are smaller by a factor of $\sqrt{2}$ than the 3D lattice vectors. The dashed line in figure 4 is a polynomial fit used for the background subtraction. Indeed, the background contribution, when it is not flat, may change the position of the maximum of the peak. The variation of strain during growth is measured by taking the distance between the ($\bar{1}\bar{1}$) and (11) peaks. It is important to point out that the azimuth has to be carefully adjusted so that the intersection between the two rods and the Ewald sphere is symmetrical. When the azimuth is correctly set, it is easy to verify that the intensities of both peaks are nearly the same, as indeed it may be observed in figure 4.

2.2. Strain measurement for the growth of a Cu/Ni multilayer

In figure 5, we have reported the measurement of strain during the growth of a $Cu_{50}Ni_{50}$ alloy on a Cu (001) thick buffer layer grown on a Si(001) wafer. The deposition is carried out by co-depositing Cu atoms and Ni atoms using two different molecular beams. The starting RHEED pattern is indeed the image shown in figure 1 and the strain is calculated using the variation of the lateral distance between the two symmetrical peaks of figure 4. The lattice parameter of the Cu FCC lattice is 3.61 Å and the lattice parameter of the Ni FCC lattice is

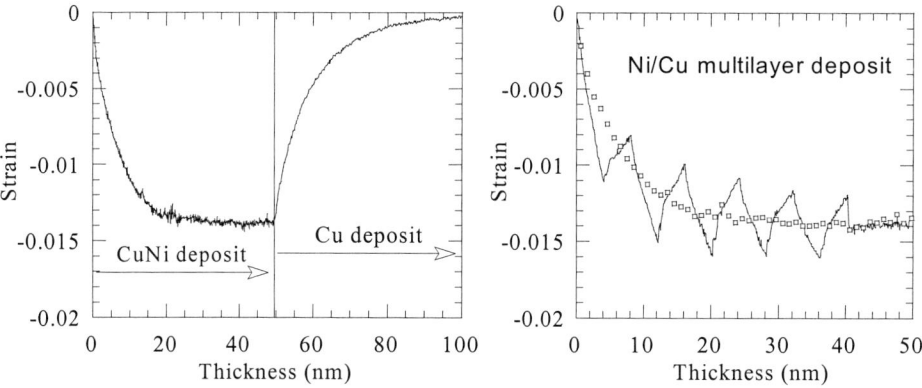

Figure 5. Strain measurement during the deposition on a Cu (001) surface of a CuNi alloy followed by the deposition of pure Cu.

Figure 6. Strain measurement during the deposition on a Cu (001) surface of a ($Ni_{4\ nm}$ / $Cu_{4\ nm}$)$_{x5}$ multilayer (line) compared with the deposition of the CuNi alloy (dots).

3.52 Å. Therefore, using a Vegard law, the epitaxial misfit between the CuNi alloy and the Cu surface is expected to be $\varepsilon_\Box = (a_{CuNi}-a_{Cu})/a_{Cu} = -0.013$ for the cube-on-cube epitaxial relationship. It may be seen in figure 5 that the in-plane lattice parameter decreases down to approximately this value in 20 nm. After 50 nm of the CuNi alloy have been deposited, the atomic flux of Ni atoms is masked with a shutter, so that only Cu atoms impinge the surface of the sample. The lattice parameter increases up to the original value of the starting Cu surface. In figure 6, we have reported the lattice strain measurement during the deposition of the CuNi alloy (dots) and the measurement in the case of the deposition of a Ni/Cu multilayer (full line). Indeed, the aim of this experiment was to compare the strain relaxation mechanism for a set of Ni/Cu multilayers with different periods with the strain relaxation for an alloy of the same composition. It may be seen in figure 6 that the lattice parameter oscillates about a mean value which follows the strain relaxation of the alloy. Therefore the strain relaxation rate in the whole multilayered stacking is similar to the relaxation rate in the alloy. On the contrary, in the case of multilayers of shorter periods, our preliminary results indicate that the relaxation rate is lower, suggesting that the misfit dislocations are blocked at the interfaces.

2.3. Strain relaxation for the growth of MgO on Fe (001)

MgO has a FCC structure with a lattice parameter of 4.21 Å and Fe has a BCC structure with a lattice parameter of 2.88 Å. The epitaxy of MgO / Fe (001) or Fe / MgO (001) follows the Bain epitaxial relationship with a lattice rotation of 45° as depicted in figure 7.

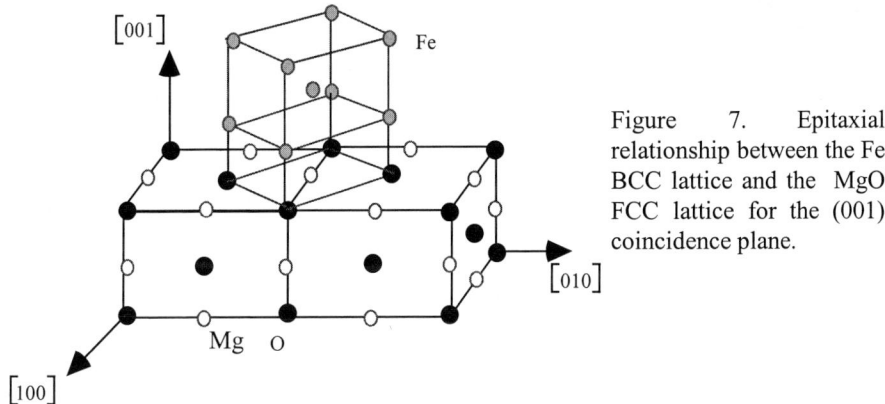

Figure 7. Epitaxial relationship between the Fe BCC lattice and the MgO FCC lattice for the (001) coincidence plane.

The epitaxial misfit is $\varepsilon_\square = (a_{MgO}-a_{Fe})/a_{Fe} = 0.038$. In figure 8, we have reported the strain relaxation measured at different growth temperatures. It may be seen that there is a critical thickness around 10 Å below which the MgO layer is perfectly coherent with the Fe lattice. Above this critical thickness, the strain relaxation depends on the growth temperature. It has been shown that the strain is relieved by perfect edge dislocations of Burgers vector 1/2<011> gliding on {101} inclined planes [6]. Because of the 4-fold symmetry of the (001) orientation, dislocations propagate to form two arrays of orthogonal dislocations : the first array is defined by dislocations gliding on the (101) or the ($\bar{1}$01) planes and the second array is defined by dislocations gliding on the (011) and (0$\bar{1}$1) planes [7]. Freund [8] has shown that the interaction between a moving dislocation and an orthogonal dislocation blocks the motion of this segment. This blocking mechanism may explain that the relaxation rate is lower than the relaxation rate predicted by an equilibrium calculation. In figure 9, we have reported the strain measurement at 300 K and two calculations using respectively an equilibrium model and Freund's mechanism [6].

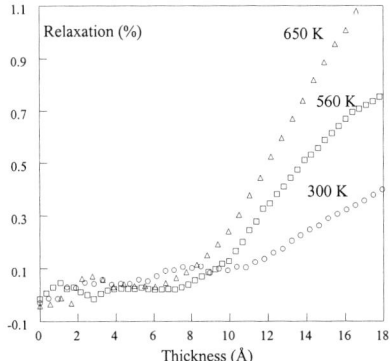 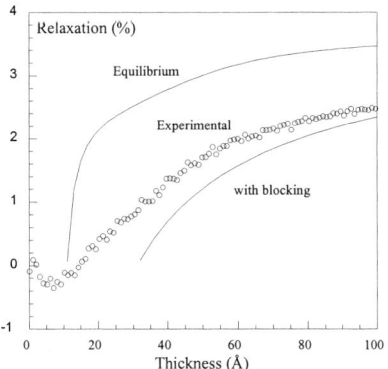

Figure 8. Strain relaxation during the growth of MgO on Fe (001) for different growth temperatures.

Figure 9. Experimental relaxation measured at 300 K during the growth of MgO on Fe (001) (dots). The full lines stand for two calculations using respectively an equilibrium model and the blocking mechanism.

In figure 10, we have reported the strain relaxation for the very first deposited monolayers of MgO. Also, we have reported in this figure the RHEED oscillations of the specular spot. The RHEED intensity oscillations are generally related to a layer-by-layer growth mode [9]. In the simplest picture, the variation of intensity results from interferences between the beams which are reflected from the step terraces. Therefore, the reflectivity of the surface oscillates during the 2D growth because of the sequential change of the step density (or mean terrace size). The RHEED oscillations are usually very strong for the specular reflected beam at very low incident angles but may also be observed at the Bragg spot or even in the diffuse background between rods. A basic introduction to the use in real time of RHEED oscillations related to growth effects may be found for example in [10]. In figure 10, it may be observed that the strain oscillations are in phase with the intensity oscillations. On the contrary, in the first experiment reporting such an effect in the case of the deposition of InGaAs on GaAs [11], it has been found that the strain oscillations are in anti-phase with the intensity oscillations. In that case, it has been argued that the elastic strain relaxation is expected to be higher at the edges of the atomic islands, as depicted in figure 11.a. Therefore, at the minima of the intensity oscillations, the step density is higher allowing an increase in strain relaxation whereas at the maxima this relaxation mechanism is not possible anymore because most of the atomic islands have coalesced. Figure 11.b depicts the opposite situation found in the case of the growth of MgO/Fe. Therefore, the strain oscillations observed in figure 10 may not be explained by the elastic strain relaxation at the island edges. Another explanation may be obtained by considering the ionic nature of the boundings. Indeed, Tasker and Duffy [12] have calculated that the ions located at the step edges of the atomic terraces on a vicinal MgO (1 0 11) surface are displaced toward the terrace. Their result, which is depicted schematically in figure 11.b is in agreement with the measurements reported in figure 10.

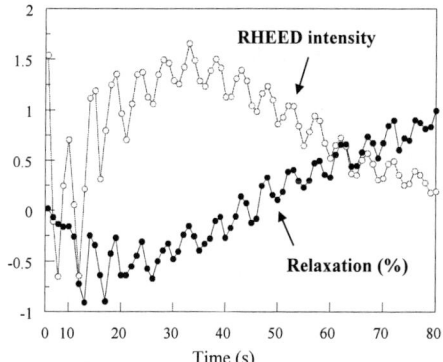

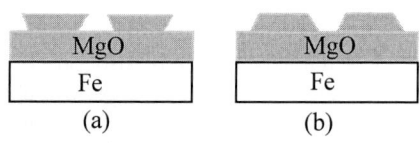

Figure 10. RHEED intensity oscillations (upper curve) and RHEED strain oscillations (lower curve) for the growth of MgO/Fe (001).

Figure 11. Schematics of the coherent atomic MgO islands with shapes expected from a) the elastic strain relaxation at the edges and b) the increased strain at the edges due to atomic boundings.

In table 1, we have reported the in-plane relaxation measured by RHEED and the in-plane relaxation measured *ex-situ* by the in-plane Grazing Incidence Diffraction (GID) technique which is explained in $3.3. As may be seen, there is a systematic difference between the two

Table 1
Relaxation measured by GID and RHEED.

Thickness (Å)	GID (%)	RHEED (%)
45	1.74	2.0
105	2.46	2.5

different measurements. Indeed, RHEED indicates a higher relaxation than X-rays and the difference is higher for the thinner layer. This may be due to the difference in the sensibility in depth for the two techniques: RHEED is very surface sensitive (a few atomic layers for the very low incident angle used for the above measurements) whereas the depth of penetration for the X-rays in this grazing-incidence geometry is about 10 nm according to formula 13. Although it is possible that a structural modification has occurred when the film has been put in the air, we believe that these differences are significant of an increase of the relaxation rate at the surface.

2.4. Perpendicular strain measurements using RHEED

So far in this paper, only the in-plane lattice parameter has been extracted from the RHEED images. However when considering for instance figure 1, it seems that there might be a straightforward way to measure the variation of the perpendicular lattice parameter in real time just by measuring the variation of the distance between the Bragg spots in the perpendicular direction. It has been done sometimes empirically but this method has often lead to wrong results for the following reasons which have been frequently neglected: 1) Due to the change in potential at the vacuum-solid interface, there is a strong refraction of the electron beams inside the crystal. As we will see below, the index of refraction for the e-beams is positive and therefore the angles inside the crystal are always greater than the angles in the vacuum. Because the Bragg relation is fulfilled *inside* the crystal, the position of the exit Bragg beams will depend on the corresponding exit angles. For instance, in figure 1 the (002) beam is shifted toward a lower angle than expected without refraction of the beam and this shift is larger than the shift of the (004) beam because the exit angle is lower. Of course, the scattering potential depends also on the material which is irradiated and therefore the refraction shifts of the Bragg peaks will change during an heteroepitaxy. 2) Kikuchi lines are often present on the diffraction patterns and the intersection between a Kikuchi line and a rod produces an increase of the intensity along the rod at this position. When this intersection is very close to a Bragg peak, this may cause a shift of the maximum intensity along the rod with respect to the exact Bragg position. 3) In figure 1, the intensity maxima along the z-direction do not exactly coincide with the expected Bragg 3D positions, because of the geometrical intersection between the Ewald sphere and the elongated Bragg spots. Indeed it may be seen in figure 1 that the Bragg spots are elongated in the perpendicular direction because of the convolution between the Bragg 3D peaks and the rods, as explained in $2.1. The extension of the rods in the perpendicular direction depends on the roughness of the surface, as described in figure 3 b,c,d. Therefore, the intensity maximum along the z-direction is shifted, depending on the position of the Bragg spots with respect to the Ewald sphere and also depending on the roughness of the surface. 4) Finally, the features of the RHEED pattern are often complicated due for example to secondary Bragg conditions or resonance conditions [13]. Indeed, it is well known since the early days of electron diffraction that in general the diffracted pattern cannot be described correctly in terms of single scattering models (first Born approximation or kinematical theory) but that multiple scattering process have to be taken into account (dynamical theory) [14]. At present three different methods are under development. The first follows the Block-wave approach which is now commonly used for X-rays and has been extended to the many-beam case for electron diffraction a long time ago [14,15]. The main problems for RHEED geometry are the surface matching of the waves and the identification of the excited Block waves in a semi-infinite crystal [16,17]. These problems have been recently discussed and limits have been found for this method [18,19,20]. The second method is the transfer of the fast numerical multislice method which is generally used in TEM image simulations to simulate RHEED surface reflection [21,22,23]. This method has recently been applied successfully to the case of GaAs(001)-2X4 surfaces [24] and the $Cu_3Au(111)$ surface [25]. Using the EMS package developed by Stadelmann [26] and the Doyle-Turner potentials [27], the authors have used an edge-patching method in which the crystal is sliced normal to the surface and to the incident beam. It has no be noted that the slice potentials do not have to be periodic along the incident beam or normal to the surface, thus allowing in principle to include structural defects, as for instance atomic steps, in a super-cell. However these calculations for complicated surfaces require enormous amounts of computer time. The third method which has been developed is a transfer of Low Energy Electron Diffraction (LEED) calculation methods to RHEED case, as pioneered by Maksym & Beeby [28] and independently by Ichimya [29]. In that case, the crystal is sliced in slabs parallel to the surface and the problem is treated as a series of 2D periodic layers. The rocking curve analysis of the specular (00) rod or along any (pq) rod is similar to the analysis of the LEED I-V curves. This method has been applied for example to the case of MgO (001) [30], Pt (110)-(2X1) [31,32], GaAs (001)-(2X4) [33] or Si(111)-(7X7) [34]. For all the examples given above, the information of the surface-normal individual atomic displacements of the surface atoms or the surface relaxation of the first atomic plane have been obtained through a refinement process in which the calculated whole pattern or the calculated rocking-curves have been compared to the experiment. No direct measurement of the perpendicular lattice spacing has been extracted, because of the complexity of the RHEED pattern which required a full dynamical simulation. However, as pointed out by Ichimya [35], it is possible to simplify the calculation by setting the azimuth out of alignment with a low index lattice plane to reduce the strong multiple scattering effect. This condition is usually defined as the two-beam condition (although Ichimya defines it as the one-beam condition) and the intensity of the specular beam is well-approximated considering only multiple scattering with the incident beam. This in turn is equivalent to the usual X-ray dynamical theory in the Bragg reflection geometry [36,37] provided that only one Bragg reciprocal position intersects the Ewald sphere. Since in that geometry the scattering vector is normal to the surface and multiple scattering with off-normal momentum transfer can be neglected, the specular beam intensity depends only on the surface-normal structure, i.e. the one-dimensional structure perpendicular to the surface [38]. In fact, this off-azimuth geometry has been used by Menadue [39] for a Si (111)-7X7 surface in one of the first rocking-curve measurements. Menadue found a rather good agreement between the intensities of the Bragg maxima with calculations derived from the 2-beam theory initially developed for X-rays [37]. Moreover, it has been found that the *positions* of the Bragg peaks are correctly given by a kinematical model, provided that refraction effects are taken into account. This last point is in turn a general result from the dynamical theory [37]. In a more recent paper, for the same Si (111)-7X7 surface Hanada [38] compared the rocking curve of the specular beam at the exact $[2\bar{1}\bar{1}]$ azimuth with the rocking curve measured with the azimuth rotated by 13° from the $[2\bar{1}\bar{1}]$ position. In the former case, multiple scattering occurs very clearly, giving rise to extra-peaks, whereas in the latter case the shape becomes considerably simpler with only the primary Bragg peaks of increasing orders. Similar observations have been carried out during the growth of Si/Si (111) [40].

In figure 12, we have reported a rocking curve measured on the Cu (001) surface grown on Si (001), corresponding to the RHEED patterns of figures 1-2. The azimuth has been tilted by several degrees (~ 10°) in order to prevent the (00) beam from overlapping the oblique Kikuchi lines (which are indeed very week because of the bad structural quality of the epitaxied Cu layer) and the specimen is nearly under the two-beam condition. For electron beams of high acceleration voltage, the wavelength is calculated using a relativistic expression [5] :

$$\lambda = h / \left\{ 2m_0 qV \left[V(1 + qV/(m_0 c^2)) \right] \right\}^{1/2}, \qquad (1)$$

where m_0 is the electron rest mass, q the magnitude of the electronic charge, c the speed of light and V the accelerating potential of the electron gun. This expression leads to the following numerical formula for the wavelength as a function of the accelerating voltage :

$$\lambda = 12.3 / \left[V(1 + 1.96 X 10^{-6} V) \right]^{1/2} \text{ Å} \qquad (2)$$
(with V expressed in volts).

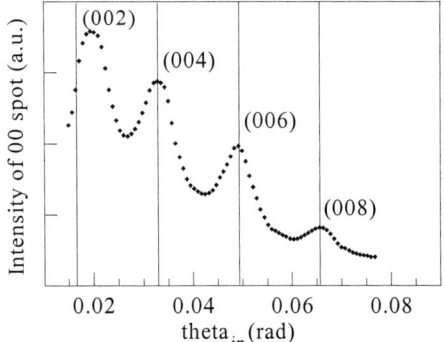

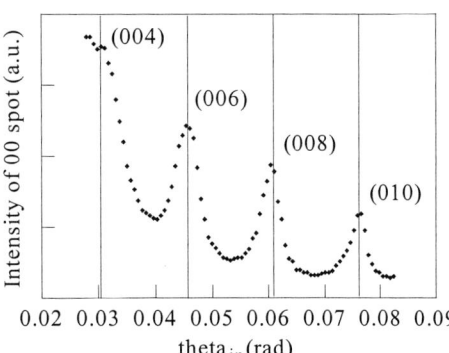

Figure 12. RHEED rocking curve on a Cu (001) thick film. Theta$_{in}$ is the angle inside the crystal. The vertical lines stand for the expected positions of the Bragg reflections for Cu bulk.

Figure 13. RHEED rocking curve on a Pd (001) thick film. Theta$_{in}$ is the angle inside the crystal. The vertical lines stand for the expected positions of the Bragg reflections for Pd bulk.

The refraction effects may be corrected by using the following relationship between the angle θ_{in} inside the solid and the angle θ_{out} outside [10]:
$$\cos \theta_{out} = (1 - V_0 / E)^{1/2} \cos \theta_{in}, \qquad (3)$$
where V_0 is the inner potential which is negative and represents the gain in potential energy that an electron of kinetic energy E undergoes when it enters the solid. In figure 12, we have reported the calculated angles inside the crystal, i.e. the angles calculated from the exit angles using (3) and an inner potential $V_0 = -10.5$ volts. This potential has been adjusted so that all the orders of the Bragg peaks, excepted the (002) peak, may satisfy the Bragg relation for the (00L) planes, with the wavelength calculated from (1), i.e.:
$$2 \frac{a_\perp}{L} \sin \theta_{in} = \lambda, \qquad (4)$$
where $a_\perp$ is the lattice parameter in the perpendicular direction. The value $a_\perp = 3.6$ Å corresponding to the bulk value has finally been obtained from this adjustment. In figure 12, we have reported the expected positions of the Bragg peaks for Cu bulk and it may be seen that a good agreement is obtained with the measurement. Indeed, it has been verified by X-rays that this 60 nm-thick Cu/Si (001) layer is fully relaxed and has the bulk cubic lattice cell with $a_\perp = a_{//} = a_{bulk} = 3.61$ Å. The (002) peak has not be included in this procedure because first the exit angle is too small and the measurement is very inaccurate in our experimental setup and also this peak is too much surface sensitive and may reflect surface relaxation of the topmost atomic layer [34]. In figure 13, we have reported the same measurement for a 60 nm-thick Pd/MgO (001) layer. It has been verified by RHEED in-plane measurements and ex-situ X-rays that the film is almost completely relaxed. The expected value of $a_\perp = 3.9$ Å has been

obtained, close to the expected value $a_{bulk} = 3.89$ Å. In this case the Pd surface is very flat giving rise to a very intense specular spot at very low angles. Therefore, the (002) and (004) Bragg reflections, which are very surface sensitive, have not been included in our determination of $a_\perp$.

In figures 14 and 15, we have reported rocking curve measurements on the (00) specular rod for a $(Au_{2\,nm})/[Ni_{2\,nm}/Au_{2\,nm}]_{\times 11}$ (111) multilayer grown on a Cu 60 nm-thick buffer layer on Si (001). In figure 14, the intensity has been measured on the Ni 11st layer of the periodic stacking and the extracted perpendicular parameter is $a_\perp = 3.55 +/-0.05$ Å. In figure 15, the intensity has been measured on the Au 12th layer, which is the last layer of the whole stacking, and the extracted perpendicular parameter is $a_\perp = 4.06 +/-0.05$ Å. The (111) peak has not been introduced in the determination of $a_\perp$ for the same reasons as explained above for the Cu and Pd films. Table 2 gives a comparison of the in-plane and out-of-plane lattice parameters deduced from RHEED measurements and the parameters derived from two different X-ray techniques. "X-ray I" refers to GID results ($3.3) combined with the asymmetrical Bragg ($3.4) and the simulation profiles ($3.5) whereas "X-ray II" refers to the $Sin^2\psi$ analysis ($3.6). As may be seen, the different measurements are in good agreement.

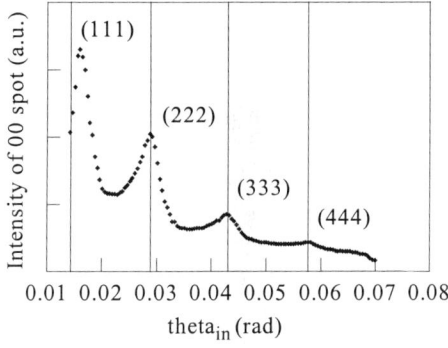
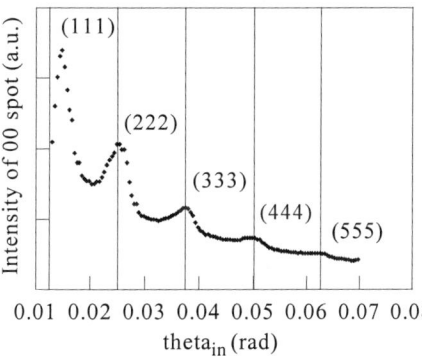

Figure 14. RHEED rocking curve on a Ni (111) ultra-thin film. Theta$_{in}$ is the angle inside the crystal. The vertical lines stand for the positions of the Bragg calculated reflections.

Figure 15. RHEED rocking curve on a Au (111) ultra-thin film. Theta$_{in}$ is the angle inside the crystal. The vertical lines stand for the positions of the Bragg calculated reflections.

Indeed, the Au layers are found to be in a compressive strain ($a_{//} < a_\perp$) and the Ni layers on the contrary are under tensile strain ($a_{//} > a_\perp$). This is expected because the FCC lattice parameter of Au ($a_{bulk} = 4.08$ Å) is much greater than that of FCC Ni ($a_{bulk} = 3.52$ Å). However, it may be noted that both the in-plane and the out-of-plane parameters of the Ni layer are greater than that of bulk Ni. This has been interpreted as the effect of the interdiffusion of Au atoms in the growing Ni layer, due to the surfactant behavior of the Au atoms during the MBE growth. This alloying effect changes the stress-free parameter of the Ni nominal layer (i.e. the cubic unstrained lattice parameter, once corrected for the biaxial measured strain), which slightly increases due to a Vegard law. Auger analysis has put in evidence this alloying mechanism of the Ni nominal layers. When considering now the differences between RHEED measurements

Table 2
Lattice parameters (Å) obtained with RHEED and X-ray measurements I and II.

	Au ⊥	Au //	Ni ⊥	Ni //
RHEED	4.06	4.03	3.55	3.58
X-ray I	4.10	4.03	3.59	3.61
X-ray II	4.12	4.04	3.60	3.61

and X-ray measurements, it appears that the RHEED values are slightly shifted toward the bulk relaxed values, although the accuracy about the RHEED values is rather poor. The question which is raised concerns the homogeneity of the lattice parameter in the Au or the Ni layers. Indeed, X-rays are sensitive to the mean lattice parameter of the layers, averaged over the whole stacking which is irradiated by the X-ray beam, whereas RHEED is sensitive to the topmost atomic layers of the last deposited Au or Ni film. Concerning the sensitivity of RHEED in depth, it is difficult to calculate an accurate value, because this will depend on the extinction depth, which is well defined in the dynamical theory for a perfect crystal but is very difficult to estimate in the case of a crystal with defects and a rough surface with a high density of atomic steps. It has to be reminded that the extinction depth is different from the depth of penetration which is usually used in the kinematical theory. The depth of penetration is much higher than the extinction depth because the calculation ignores the strong interaction between the diffracted beam and the incident beam [37]. In any case, the sensitivity in depth will depend on the angles, and it is expected that the low Bragg indexes are more surface sensitive than the high indexes. We have deduced an estimation of the surface sensitivity of RHEED by performing rocking-scans similar to the intensity curves shown in figure 14 and 15 but for thinner Ni (respectively Au) films (i.e. ~ 1 nm). Indeed, we found in that case secondary maxima on the sides of the Bragg reflections which could be identified to the peak positions of the underlying Au (respectively Ni) layers. These secondary maxima were more intense for the high reflection angles, as expected. Therefore, as far as rather rough surfaces are considered (a RMS roughness of 10-12 Å has been deduced from STM measurements), it appears that the sensitivity of RHEED in depth is of the order of 1 to 2 nm. Of course, the contribution of the topmost atomic layers to the intensity profile will be dominant. For all these reasons, it is difficult to know whereas the difference between the lattice parameter measured ex-situ by x-rays and in-situ by RHEED is significant of any increase of the relaxation process at the topmost atomic layers or is below the accuracy of the measurements.

As a conclusion, RHEED measurement of the variation of the in-plane lattice parameter has been proved to be an efficient way of measuring the relaxation process in thin films and is largely used nowadays in MBE experiments. The perpendicular relaxation measurement is a complicated problem which has not been addressed so far in the literature for the growth of epitaxied thin films. We have shown that the rocking curve measurement technique is a possible direction which may be developed, provided that dynamical multiple reflections may be neglected as explained above. However, for films consisting of only a few atomic layers, a simulation of the intensity profile is required to extract information in depth. Also, for thicker epitaxied films, the calculation of the intensity is needed to extract any possible relaxation profile in the depth of the film. Because of the poor structural quality of most of the epitaxied metallic films, calculation methods developed so far for perfect single crystals will not be adequate. Therefore a theoretical approach for "heavily perturbed" crystals has to be developed.

3. STRAIN MEASUREMENTS USING X-RAYS

3.1. Introduction

X-ray diffraction has been used for almost a century to study the structural properties of crystalline materials on an atomic scale. The reason why it has been so widely used is that in many cases the interaction of X-rays with matter is weak so that a single scattering approximation is often sufficient to quantitatively reproduce the experiments. Surface sensitivity may be achieved by simply decreasing the angle of incidence of the incoming beam. The depth of penetration of the beam, which may be defined as the depth at which the intensity is reduced by a factor of 1/e, is reduced because the incoming beam propagates in the material very close to the surface. Therefore, the penetration of the incoming beam into the bulk material is limited. Scattering from the bulk is then highly reduced and the surface sensitivity is increased. A typical glancing angle for such an experiment is ~1°. Marra and Eisenberger used this principle in their GID pioneering work [41,42]. If now the incident angle is set below or equal to the critical angle for total external reflection, the incoming beam inside the material becomes evanescent, as described by the Fresnel optical equations [43,44]. The evanescent beam is strongly damped in the bulk and therefore the surface sensitivity is highly enhanced. Furthermore, part of the beam is specularly reflected from the surface.

The first part of this section will give the principles for the calculation of transmission and refraction of the beams at grazing angles. Then the second part will be dedicated to the in-plane Bragg diffraction using the GID geometry, which has become now since the work of Marra and Eisenberger cited above a very efficient and straightforward way of measuring in-plane relaxation. The third part will concern asymmetrical Bragg diffraction for the measurement of the diffraction by atomic planes inclined with respect to the surface. The fourth part will be de dedicated to the symmetrical diffraction geometry and to the intensity simulation for periodic multilayers. Finally, in the last section, we will briefly give the principle of the $Sin^2\psi$ analysis.

3.2. Refraction and total external reflection at glancing angles

The index of refraction n for X-rays is complex and the real part is slightly smaller than 1. The real and imaginary correction factors δ and β are of the order of 10^{-6} and can be determined from the following equations [44]:

$$n = 1 - \delta - i\beta , \quad (5)$$

where $\delta = \dfrac{r_e \lambda^2}{2\pi} \sum_i (Z_i + \Delta f_i') N_i$ and $\beta = \dfrac{r_e \lambda^2}{2\pi} \sum_i \Delta f_i'' N_i .$ \quad (6)

In this equation, $r_e = 2.82 \times 10^{-15}$ m is the classical electron radius, λ is the wavelength, Z_i is the atomic number of atoms i, N_i is the number of atoms i per volume unit, $\Delta f'$ and $\Delta f''$ are the dispersion corrections [45]. The imaginary part β may be related to the absorption coefficient μ by:

$$\beta = \frac{\lambda \mu}{4\pi}. \qquad (7)$$

If the angle of incidence is less than a critical angle defined by $\cos \alpha_c = n$, which for small angles leads to $\alpha_c = \sqrt{2\delta}$, then total external reflection occurs. The range of the values for α_c is typically 0.2° - 0.6° for X-ray wavelengths around 1.5 Å. We will consider the situation depicted in figure 16, where an incoming beam of wave vector $k = \frac{2\pi}{\lambda}$ is reflected upon the interface between air and a material of index n:

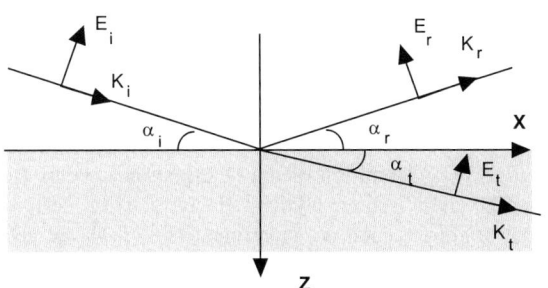

Figure 16. Reflection and transmission of a linearly polarized plane wave upon an interface between air and a material.

The incident plane wave is linearly polarized with an electric field E_i and has a wave vector K_i. It impinges the surface with an angle of incidence α_i. The reflected beam has an electric field E_r, a wave vector K_r and makes the angle α_r with the surface. The transmitted beam has an electric vector E_t, a wave vector K_t and makes the angle α_t with the surface. It can be shown by using the Fresnel formulae [46] that the angles are related by:

$$\cos \alpha_t = \frac{\cos \alpha_i}{n}, \qquad (8)$$

which for small angles may be approximated by:

$$\alpha_t^2 = \alpha_i^2 - \alpha_c^2. \qquad (9)$$

If $\alpha_i < \alpha_c$ then α_t becomes imaginary: this means that the z-component of the wave vector K_t which is $K_t^z = k\, n \sin(\alpha_t)$ has an imaginary component. Therefore there is an exponential damping of the transmitted wave. The depth of penetration $\tau(\alpha)$ of the evanescent wave is [46]:

$$\tau(\alpha_i) = \frac{\lambda}{4\pi \mathrm{Im}\left(\sqrt{\alpha_i^2 - 2\delta - 2i\beta} \right)}. \qquad (10)$$

For the two perpendicular polarizations of the electric field (parallel and perpendicular to the diffracted plane X0Z), it can be shown by applying the boundary conditions of continuity at the interface together with the Maxwell equations that the amplitude of the reflected and refracted waves are for small angles [46]:

$$E_r = R\, E_i \quad E_t = T\, E_i \;, \tag{11}$$

$$\text{where } R = \frac{\alpha_i - \alpha_t}{\alpha_i + \alpha_t} \quad T = \frac{2\alpha_i}{\alpha_i + \alpha_t}. \tag{12}$$

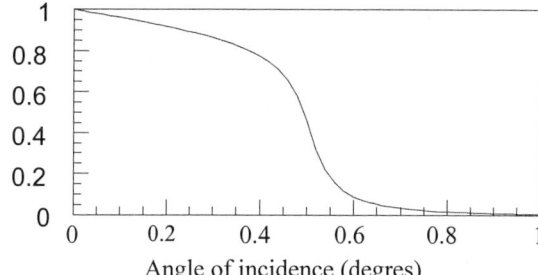

Figure 17. Reflectivity curve calculated for nickel.

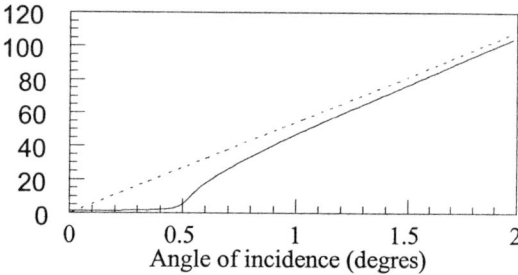

Figure 18. Penetration depth (in nm) of the X-rays (λ=1.54 Å) in nickel. The full line is calculated with the refraction and the dashed line without.

Figure 17 shows the reflectivity $|R(\alpha_i)|^2$ calculated for nickel. The wavelength of the X-ray beam is 1.54 Å and the critical angle for nickel is $\alpha_c = 0.5°$. Below α_c the incoming beam is totally reflected: the intensity would be exactly equal to 1 if there were no absorption (β=0). Figure 18 shows the penetration depth calculated in nickel (full line). The dashed line is the calculation of the penetration depth when ignoring the total reflection phenomenon, by using the "classical" attenuation factor $\sin(\alpha_i)/\mu$. The transmission coefficient $|T(\alpha_i)|^2$ has also been plotted for nickel in figure 19. It may be seen that at $\alpha_i = \alpha_c$ the value is larger than 1.

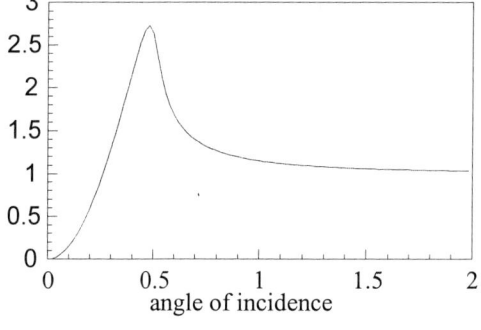

Figure 19. Plot of the transmission coefficient $|T(\alpha_i)|^2$ for nickel. The wavelength of the X-ray beam is 1.54 Å.

Neglecting absorption, the transmission coefficient T would be equal to 2 at this angle: this may be understood as a standing wave phenomenon. Because the reflection coefficient R is equal to 1 at the critical angle, the incoming beam and the reflected beam are in phase and the total external electric field which is $E_i + E_r = 2\,E_i$ is therefore enhanced.

3.3 In-plane Bragg diffraction using grazing incidence

The GID geometry is the only geometry which allows the diffraction by planes perpendicular to the surface with a high signal-to-noise ratio [41]. Both the angle of incidence

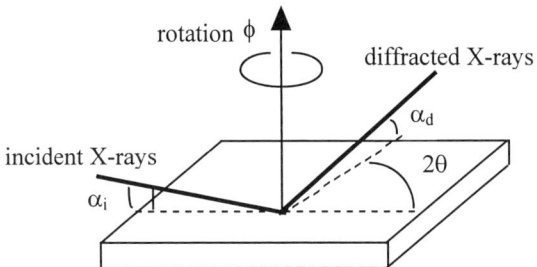

Figure 20. Schematic of the in-plane Bragg measurement combining a grazing incidence and a grazing emergence.

and the angle of emergence are grazing angles with $\alpha_i \cong \alpha_d \cong \alpha_c$. Note that α_d may not necessarily be equal to α_i. The problem of solving the intensity variation of an in-plane Bragg diffraction for different exit angles is not easy to handle. Following the early work by Becker et al. [47] we can use the well-known reciprocity theorem from optics which states that [43]: "a point source at P_0 will produce at P the same effect as a point source of equal intensity placed at P will produce at P_0". Therefore the intensity of the outgoing wave measured in vacuum for an exit angle α_d is identical to the intensity of an incoming wave measured at the surface for an incident angle $\alpha_i = \alpha_d$. Becker et al. [47] have experimentally demonstrated this theorem by measuring the fluorescence yield in two equivalent configurations: the geometry which combines a grazing incidence and a normal emergence gives the same variation of the fluorescence yield when the incident angle α_i is varied than the inverse geometry which combines a normal incidence and a grazing emergence when α_d is varied. This approach has been also applied to the calculation of the diffraction [48-50] under GID. The main results are:

* The diffracted wave inside the crystal is an evanescent wave.

* The scattering depth is:

$$\Lambda = \frac{\lambda}{2\,\text{Im}(q'_z)}, \qquad (13)$$

where q'_z is the z-component of the momentum transfer inside the crystal:

$$q'_z = \frac{2\pi}{\lambda}\left[\sqrt{\alpha_i^2 - 2\delta - 2i\beta} + \sqrt{\alpha_d^2 - 2\delta - 2i\beta}\right]. \qquad (14)$$

* The diffracted intensity is:

$$I(\alpha_i, \alpha_d) = |T(\alpha_i)|^2 |T(\alpha_d)|^2 \frac{(\lambda/a_0)^2}{1 + \left[(2\lambda/a_0)\sin(\mathrm{Re}(q_z')a_0/2)\right]^2}, \quad (15)$$

for an homogeneous crystal with a lattice constant a_0 along z. $T(\alpha)$ is the transmission factor at the interface for a beam at angle α, as defined in (12).

Figure 21 is an example of a ϕ-scan. The Bragg condition $2 d \sin(\theta) = \lambda$ is verified in the plane, as depicted in figure 20. Both the incident angle and the emergent angle are close to the critical angle, which correspond to the maximum intensity according to formula 15. A ϕ-scan is a rotation of the sample around the surface normal. The specimen is the same Au/Ni multilayer which has been measured in-situ using RHEED in figures 14 and 15 of $2.4. The gold and nickel layers are polycrystalline with a (111) growth axis. The Cu substrate is a Si (001) single crystal and the epitaxial relationship between the crystallographic structures is:

Au(111)//Ni(111) and Au[$0\bar{2}2$]//Ni[$0\bar{2}2$] for the multilayer

Au(111)//Cu(001) and Au[$0\bar{2}2$]//Cu[200] for the first Au layer on Cu.

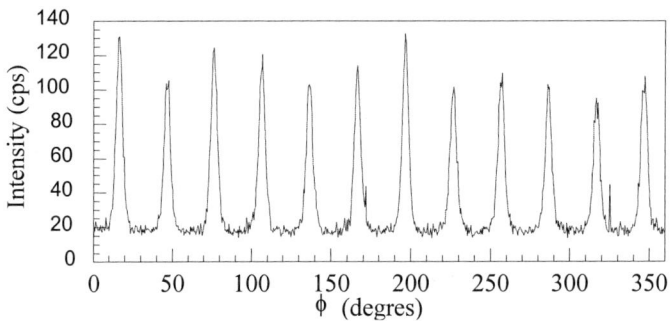

Figure 21. GID ϕ-scan for a the Au (022) reflection of a Au/Ni (111) multilayer.

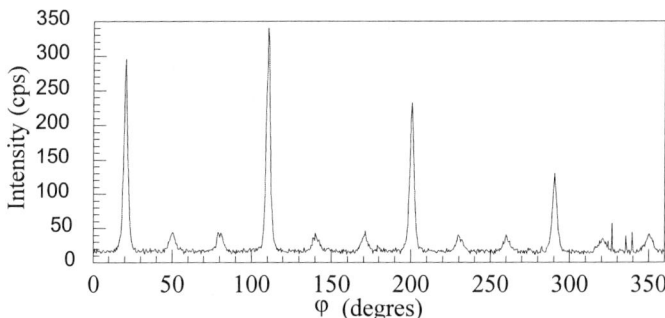

Figure 22. GID ϕ-scan for a the Ni (022) reflection of a Au/Ni (111) multilayer. The 4 main peaks are the Cu (220) substrate reflections.

Because the gold grains have 3 equivalent <022> in-plane directions and because the Cu substrate has 4 equivalent <200> in-plane directions, there is a 12-fold symmetry in the pole figure, corresponding to the 4 possible in-plane orientations of the Au(111) grains on Cu(001). In figure 21 the in-plane Bragg angle selects the (022) planes of Au. When rotating the specimen around the surface normal, 12 equivalent peaks are measured. In figure 22, the same φ-scan has been performed but the in-plane Bragg angle has been increased to select the (022) planes of Ni. There are also 12 peaks at the same positions than the 12 Au (022) peaks of figure 21 but 4 equidistant peaks are more intense than the others. They correspond to the Cu (220) peaks of the substrate, which cannot be resolved from the Ni (022) peaks. The analysis of these two figures has lead to the epitaxial relationship described above.

Figure 23 is a GID φ–2θ scan on a Au/Ni (111) multilayer, the period of which being 56 Å. This measurement is the in-plane equivalent to a conventional perpendicular XRD θ–2θ scan. The two main peaks are respectively the Au (022) reflection and the Ni (022) reflection. Note that only 8 scans out of the 12 equivalent possible φ–2θ scans will avoid the Cu (022) contribution (see figure 22). From the positions of these peaks the in-plane strained lattice parameters of the Au layers and the Ni layers may be determined and compared to the bulk values of the (022) reflections from Au and Ni in the bulk (respectively at 32.3° and 38.3°). This method has lead to the in-plane parameters which have been referred as "X-ray I" results in table 2 of $2.4 for a multilayer of 40 Å-thick period. As already explained, the Au layers are in a compressive strain whereas the Ni layers are in a tensile strain because of the partial coherency of the crystallographic structures at the interfaces. However, the stress-free parameter of the Ni layers has been found to be higher than the bulk value. The alloying of the nominal Ni layers with Au atoms which segregate at the growing Ni surface (surfactant atoms) may explain this increase of the lattice parameter. Using a Vegard law, a composition of $Ni_{80}Au_{20}$ may be deduced from this stress analysis. Also, from a quantitative analysis of the *integrated intensities* of the two peaks in figure 23 and following the analytical description in formula 15, it has been found that the composition of the so-called nominal Ni layer was indeed around $Ni_{80}Au_{20}$.

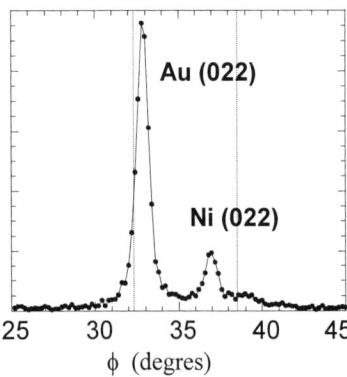

Figure 23. GID φ–2θ scan on a Au/Ni (111) multilayer. The peak at 33° is a Au (022) peak and the peak near 37° is a Ni (022) peak. The vertical lines indicate the position of the corresponding expected bulk values. The broad underlying contribution comes from the overlapping with a Au (311) peak at 13° out of the surface.

3.4 Asymmetric Bragg diffraction: reciprocal maps

In Figure 24, we have depicted an asymmetric Bragg diffraction geometry using a 2-circle diffractometer. The total momentum transfer q_t may coincide with a Bragg node, thus providing the in-plane lattice parameter $a_{//}$ through the in-plane component q_x and the out-of-plane component $a_\perp$ through the perpendicular component q_z. Figure 26 is an example of such a measurement, showing two reciprocal maps around the (113) Bragg position for two $Au_{1-x}Ni_x$ (001) alloys epitaxied on two different buffer layers. The samples have been oriented so that the (113) reciprocal position is in the so-called plane of diffraction, which is defined by the direction of the incident beam and the direction of the diffracted beam. In figure 25.a, the composition of the alloy is $Au_{50}Ni_{50}$ and the substrate is a Pt thick buffer layer. It may be

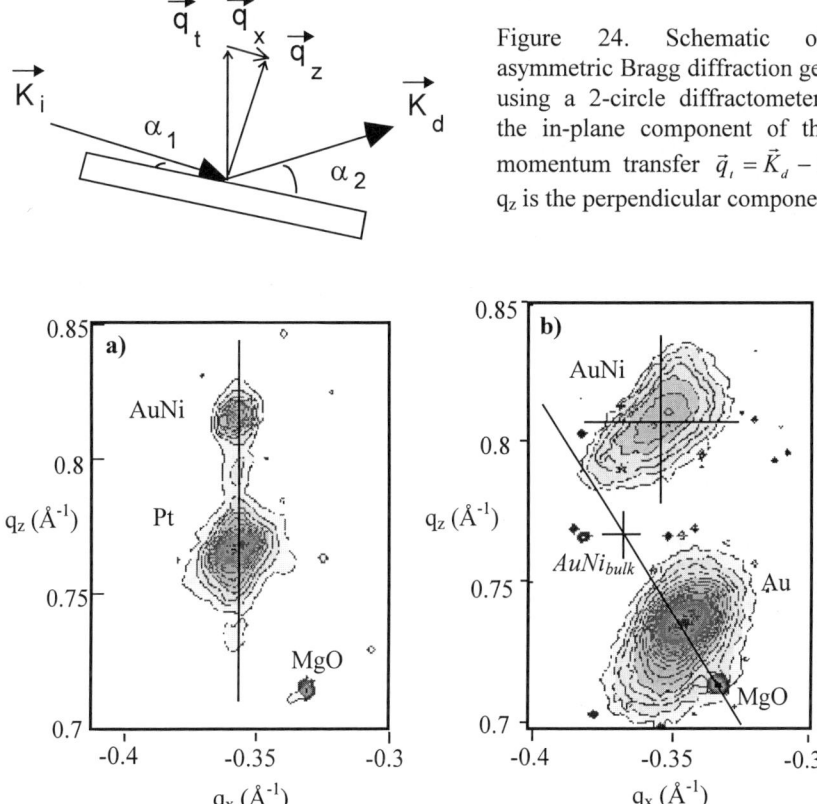

Figure 24. Schematic of the asymmetric Bragg diffraction geometry using a 2-circle diffractometer. q_x is the in-plane component of the total momentum transfer $\vec{q}_t = \vec{K}_d - \vec{K}_i$ and q_z is the perpendicular component.

Figure 25. Reciprocal map around the (113) Bragg position for an $Au_{1-x}Ni_x$ alloy epitaxied on two different buffer layers deposited on a MgO (001) substrate. In a) the alloy of composition $Au_{50}Ni_{50}$ is fully coherent on the buffer layer Pt (001) and in b) the alloy of composition $Au_{60}Ni_{40}$ is partially coherent on the buffer layer Au (001). The in-plane residual strain is $\varepsilon_r = 2.8\%$ in a) and $\varepsilon_r = 3.0\%$ in b). The position of the (113) Bragg spot for the bulk $Au_{60}Ni_{40}$ alloy is indicated by the cross in b). The straight line in b) join the Au (113), the MgO (113) and the origin.

observed that the $Au_{50}Ni_{50}$ layer is fully coherent with the Pt (001) buffer layer, because both in-plane lattice parameters are the same, whereas in figure 25.b the $Au_{60}Ni_{40}$ alloy is partially relaxed with respect to the Au (001) buffer layer, because the in-plane lattice parameter is slightly different from that of the Au layer. However, the epitaxial misfit is not fully relaxed, because both the in-plane and the out-of-plane lattice parameters are different from the bulk expected value, indicated by the cross in figure 26.b.

In figure 26, we have depicted an extremely asymmetric Bragg diffraction geometry, using a 4-circle diffractometer. The angle of incidence α_i is close to the critical angle α_c in order to be very surface sensitive. Note that the angle of emergence α_d may not be necessarily low.

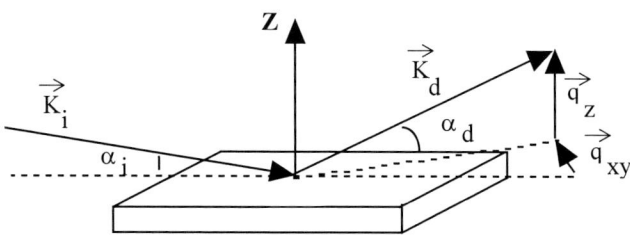

Figure 26. Schematic of the extremely asymmetric Bragg diffraction with a four-circle diffractometer using a grazing incidence to be more surface sensitive.

This geometry has first been used in the dynamical theoretical framework [51] and further extended to the kinematical case [52,53]. In §3.2 it has been found that at grazing angle of incidence the incident beam becomes evanescent inside the material, with a penetration $\tau(\alpha)$ and a transmission factor $T(\alpha)$. Therefore, if the material is a crystal in Bragg position, the amplitude F(h,k,l) of this Bragg reflection will be [52,53]:

$$F(hkl) = \int_0^\infty \left\{ \iint_s \rho(xyz) \exp[-2i\pi(hx+ky)] \, dxdy \right\} T(\alpha) \exp\left[-\frac{z}{2\tau(\alpha)}\right] \exp[-2i\pi lz] \, dz, \quad (16)$$

where $\rho(x,y,z)$ is the electron density in the crystal, s is the surface illuminated by the X-ray beam and z is the axis perpendicular to the surface. The intensity and the form of the Bragg reflection at a grazing incident angle will be the intensity and the form of the Bragg reflection in perpendicular incidence multiplied in convolution by the function [53]:

$$P(l) = \frac{T^2(\alpha)}{[\tau(\alpha)]^{-2} + (2\pi l)^2}, \quad (17)$$

the half-maximum of which being:

$$l_0 = \frac{1}{2\pi\tau(\alpha)}. \quad (18)$$

If $\alpha < \alpha_c$ then the half maximum l_0 increases. The Bragg intensity in the reciprocal space is spread over the l-direction, along rods perpendicular to the surface. If we ignore the form and the structure factor of the unit cell, the intensity variation of the Bragg reflection with the incident angle α will be [53]:

$$I(\alpha) \propto T^2(\alpha) \tau(\alpha). \quad (19)$$

In figure 27, two reciprocal map measured on two Au/Ni (111) periodic multilayers are shown. These multilayers have already been described in $2.4 and $3.3. The angle of incidence is close to the critical angle (~ 0.5°) in order to lower the contribution coming from the underlying Cu buffer layer and the Si substrate. The Au layers and the Ni layers have the same individual thicknesses and only the period is different between the two measurements. In figure 27.a the period of the stacking is 28 Å and the (022) contributions of the Au layers and the Ni layers are well separated whereas only a single peak is measured in figure 27.b, which corresponds to a periodic stacking of 9 Å. Indeed, in that case the in-plane lattice parameters of the Au and Ni layers are very close (the epitaxy is almost coherent) and the two contributions cannot be separated. Note that in figure 27.a the Ni (022) peak overlaps the Au (311) peak and therefore an accurate measurement of the position of the Ni (022) peak is difficult.

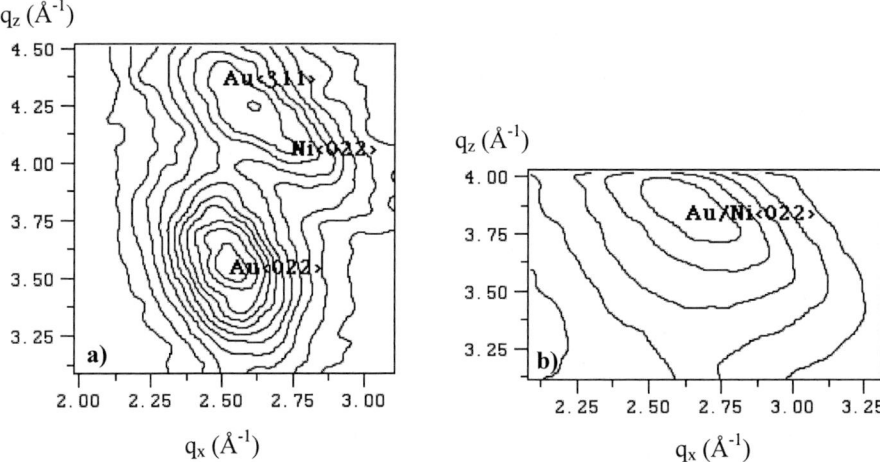

Figure 27. Reciprocal maps for two Au/Ni (111) periodic multilayers around the (022) Bragg positions which are inclined by 32° with respect to the (111) growth axis. In a) the period of the stacking is 28 Å and the Ni (022) and the Au (022) are well separated, whereas in b) the period is only 9 Å and a single broad peak is measured.

3.5 Symmetrical Bragg measurement for multilayers

A great deal of work has been devoted to the simulation of the diffraction pattern from periodic superlattices [54]. In the framework of the kinematical theory, one of the earliest pioneering works has used a Fourier analysis in order to extract directly a chemical profile from the integrated intensities of the satellites coming from the superlattice [55]. Indeed, it may be reminded that multilayers have first been elaborated in order to study diffusion couples for metals [56]. However, a simple Fourier analysis is not adequate for disordered stacking showing for instance fluctuations of the period during the growth or cumulative roughening of the interfaces. Recently, a very efficient model including a statistical description of such kind of disorder has been proposed by E. Fullerton [54]. In this model, the entire diffraction profile is described by an analytical formulation based on the calculation of the structure factor F of a periodic stacking of N bilayers of materials A and B in the symmetrical θ–2θ geometry :

$$F = \sum_{i=1}^{N} e^{iqx_i}\left[F_{A_i} + e^{iqt_{A_i}} F_{B_i}\right], \qquad (20)$$

where x_i is the distance from the substrate to the bilayer i:

$$x_i = \sum_{j=1}^{i-1} t_{A_j} + t_{B_j}, \qquad (21)$$

with t_{A_i} (t_{B_i}) being the thickness of material A (B) in the bilayer i, and F_{Ai} (F_{Bi}) being the structure factor of layer i of material A (B):

$$F_{A_i} = \sum_{j=1}^{n_i} f^{j}_{A(B)} e^{iq\sum_{k=1}^{j} d_k}, \qquad (22)$$

in which f^j is the atomic diffusion factor [45] of the atomic plane j in layer i having n_i diffracting atomic planes and d_k is the interplane distance between the atomic planes k and k-1. In formula 20, q is the momentum transfer for the wavelength λ and is defined by:

$$q = \frac{4\pi Sin(\theta)}{\lambda}. \qquad (23)$$

As a matter of fact, fluctuations of the thickness of the deposited layers along the stacking are often encountered, because of the fluctuation in time of the deposition rate of the materials. Also, there might be lateral discrete fluctuations of the actual deposited numbers of atomic planes inside a layer. For instance, atomic steps are expected to produce such discrete fluctuations, as depicted in figure 28:

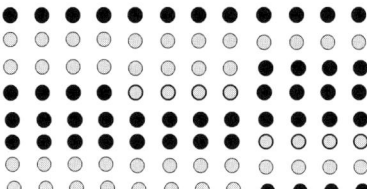

Figure 28. Schematic of the MBE deposition of a multilayer with two different materials. The actual number of deposited atomic planes shows discrete fluctuation in the growth direction as well as in the direction parallel to the interfaces.

Therefore, the expressions F_{Ai} (F_{Bi}), t_{Ai} (t_{Bi}) and x_i in equation 20 are not constant and have to be averaged along the stacking and in each individual layer. When calculating the intensity, the *total averaged intensity* has to be calculated, i.e.:

$$I(q) = <FF^*>, \qquad (24)$$

and not only the *specular part*:

$$I(q) = <F><F^*>. \qquad (25)$$

Fullerton used the Hendrix-Teller formula [57] for the calculation of the theoretical profile corresponding to equation 24 including the effect of discrete fluctuation without neighboring correlation. Indeed, the layer thicknesses are supposed to vary in a random fashion throughout the layering: for example, the value of x_i in equation 20 is independent of the actual value of x_{i-1}. Therefore, in the calculation of 24, the averaged quantities $<F_A>$, $<F_B>$, $<F_A F_A^*>$ and $<F_B F_B^*>$ are calculated. It is usually assumed that the number of atomic planes in each layer A and B is always an integer value with a Gaussian probability distribution around an averaged value which may be non-integer [54]. Also, it is easy to include the relaxation of the interplane distance at the interfaces by including in equation 22 a variation of the parameters d_k close to the interfaces with respect to the value in the core of the layer. The effect of interdiffusion is also taken into account by changing the values of the atomic diffusion factors f^j in equation 22. Finally, a Simplex or Marquard algorithm is used to refine the calculated profile with respect to the measured X-ray profile. A least-square fitting procedure may be used which minimizes the χ^2 parameter, defined by:

$$\chi^2 = \sum_{i=1}^{N_{pt}} \left(\log_{10}\left[I_c(i)\right] - \log_{10}\left[I_{exp}(i)\right] \right) \quad (25)$$

where N_{pt} is the number of measured points, I_c is the calculation using formula 22 and I_{exp} is the experimental value. Details of the calculation may be found in [54] and the modification including chemical interdiffusion profiles may be found in [58].

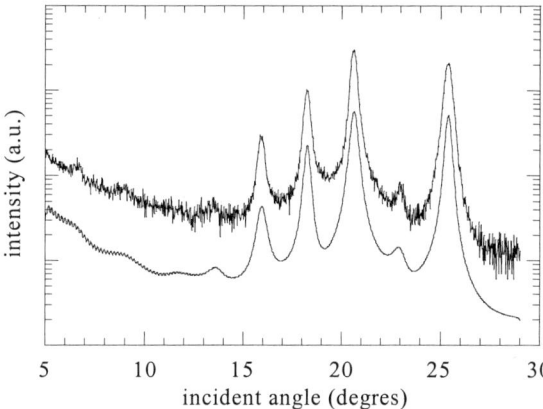

Figure 29. θ–2θ scan in a log-scale for a $(Au_{10 Å} / Ni_{10 Å})_{x10}$ multilayer. The growth direction is (111) and all the peaks are the satellites coming from the multilayer except the peak near 25 degrees which is the Cu (002) peak from the buffer layer. The upper curve is the measurement and the lower curve (shifted for clarity) is the result from the profile refinement.

Figure 29 is an example of a θ–2θ scan on a $(Au_{10 Å} / Ni_{10 Å})_{x10}$ (111) multilayer. The structure of this kind of multilayer deposited on Cu/Si (001) has already been described in $2.4 (figures 14 and 15), $3.3 (figures 21, 22 and 23) and $3.4 (figure 28). The values d_k of the (111) interatomic distances have been refined in equation 22, as well as the values of the atomic diffusion factors f^j which are related to the interdiffusion profile. The result is reported in figure 30. As may be seen, the variation of the interatomic distance between the (111) planes is not the square variation expected from a simple model without relaxation. After the refinement process, only 3 atomic planes have been found to be pure Au planes whereas 6 atomic planes have been found to be actually Ni-rich alloys. Because the same thickness in Au and Ni has been deposited, this suggest that Au atoms have interdiffused into the growing Ni layers, as already explained. Also, it may be seen that the second deposited Au plane (indexed as -3 in figure 30) is highly strained in compression as expected with a perpendicular interplane distance increased due to a Poisson effect. On the contrary the variation of the interplane distance of the Ni-rich planes is not consistent with strain in tension. Only the two last deposited planes (indexed as 3 and 4 in figure 30), which have been found to be nearly pure Ni planes, show an interplane distance lower that the bulk value, as expected from a

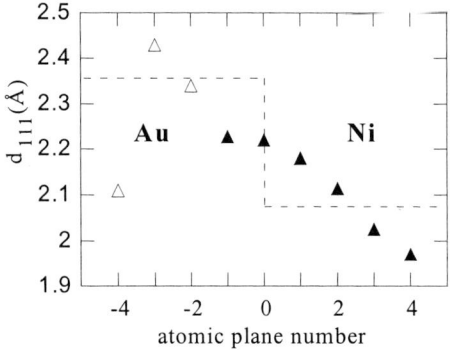

Figure 30. Interatomic distances of the (111) planes for a Au/Ni bilayer deduced from the simulation in figure 29. The horizontal axis is the position of the atomic plane along the stacking with the growth direction toward the right (the origin is arbitrary set at the center of the bilayer). The open triangles stand for the Au atomic planes and the filled triangles stand for the Ni-rich atomic planes. The dotted line indicates the expected interatomic distance for Au and Ni in the bulk.

Poisson effect for strain in tension. The other Ni planes have been alloyed with Au atoms and the variation of the lattice parameter follows a Vegard law with less and less Au atoms diluted in the growing atomic layers as far as the thickness is increased. It is important to point out that the information on the strain is indirectly obtained through a simulation of the intensity profile. Therefore, the reliability of the result may not be compared to the straightforward measurement in $3.3 or in $3.4. However, information on interdiffusion is also obtained through the refined values of the mean atomic diffusion factors f^j of the atomic planes, which are adjusted between the values of pure nickel and pure gold using a Vegard law. Also, the simulation provides not only the averaged lattice parameter values but the entire lattice parameter profile along the period, as depicted in figure 30.

3.6 The $\text{Sin}^2\Psi$ method

The $\text{Sin}^2\Psi$ analysis is based on the formulation of the strain $\varepsilon_{hkl} = (d_{hkl}-d_{khl}^0)/d_{khl}^0$, where d_{khl}^0 is the interplane distance with no strain, which is measured for a (hkl) reflection inclined by an angle Ψ with respect to the normal to the surface of the sample. For a biaxial stress (i.e.

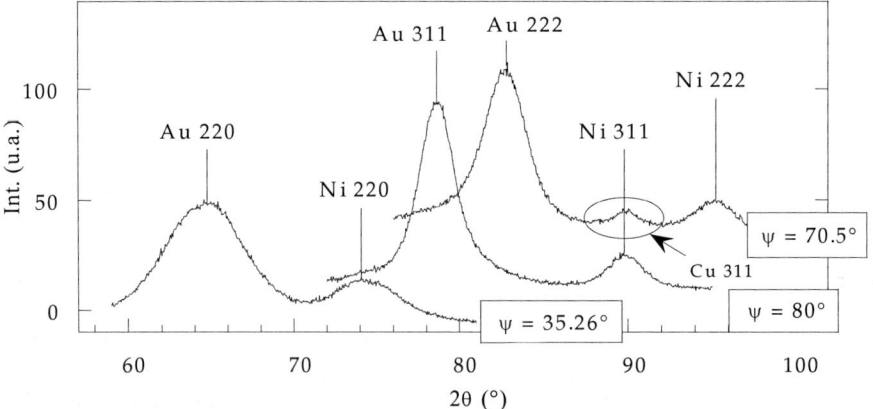

Figure 31. Intensity profiles of θ–2θ scans for different ψ values. The period of the multilayer is 35 Å (from [60]).

only the $\varepsilon_{xx} = \varepsilon_{yy}$ and ε_{zz} components of the strain tensor do not vanish), it easy to show that ε_{hkl} depends only on the angle ψ and may be expressed by [59]:

$$\varepsilon_{hkl} = \varepsilon_\psi = (\varepsilon_{xx} - \varepsilon_{zz})\sin^2\psi + \varepsilon_{zz}.$$
(25)

Therefore, a plot of ε_ψ versus $\sin^2\Psi$ for a number of reflections (hkl) may provide directly the values of the in-plane and out-of-plane components of the strain. In figure 31, we have reported three different θ–2θ scans corresponding to three different Ψ values measured on a Au/Ni (111) multilayer, the period of which being 35 Å. In figure 32, we have reported the measured d_{hkl} values versus $\sin^2\Psi$ for two Au/Ni (111) multilayers having same period but different thickness ratios between the Au and Ni layers.

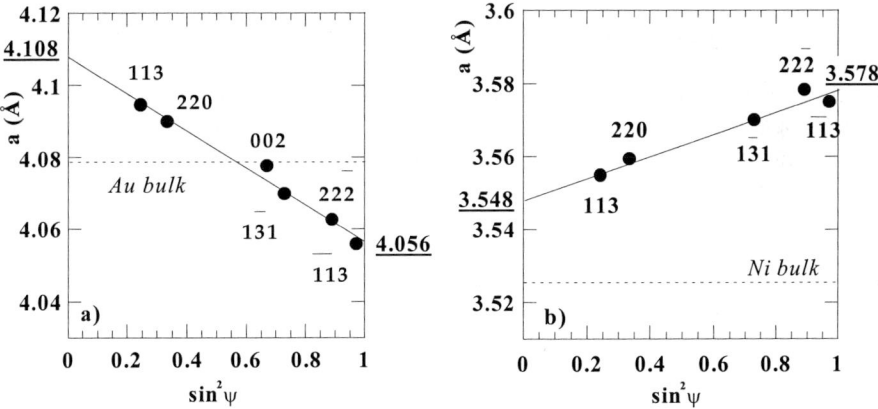

Figure 32. Lattice parameters for a) the Au layer in a $(Au_{32 Å}/Ni_{8 Å})_{x21}$ multilayer and b) the Ni layers in $(Au_{8 Å}/Ni_{32 Å})_{x21}$ multilayer (from [60])..

As may be seen, all the measured interplane distances may be connected with straight lines, thus providing the in-plane lattice parameter ($\sin^2\psi=1$) and the out-of-plane lattice parameter ($\sin^2\psi=0$) for the individual Au or Ni layers. For the $(Au_{20 Å}/Ni_{20 Å})_{x21}$ multilayer, the measurements of the lattice parameters using the $\sin^2\psi$ method have been reported in table 2 as "X-ray II" values and a good agreement has been found with the RHEED results.

The $\sin^2\psi$ method has the advantage to provide a strain analysis based on a large number of different (hkl) reflections. Therefore the reliability of the result is increased compared to a strain analysis based only on a few reflections. Also, the method is still efficient for complicated stress states, including for example shear components. However, great care as to be taken when different reflections overlap. For example the Ni (220) reflection in figure 31 for the scan at $\psi=35.21°$ is very close in the reciprocal space to the Au (311) Bragg position, as may be observed on the intensity map in figure 27.a. Therefore, in a θ–2θ scan which corresponds to an oblique trajectory passing through the Au (220) and Ni (220) positions in figure 27.a, the intensity maximum of the Ni (220) reflection may be shifted because of the overlap with the nearby Au (311) peak. In figure 32, the nice straight lines which connect all the measured d_{hkl} values are an indication that the final results are exempt from such artifacts. Finally, it should be stressed that the $\sin^2\psi$ method requires a 4-circle diffractometer equipped with a Chi-circle and that sometimes a synchrotron source is necessary to be able to measure several reflection from an ultra-thin layer with disorder [59].

4. Conclusion

RHEED is a very efficient tool for measuring in-situ the in-plane strain variation during the epitaxial growth. One of the main advantages of the RHEED technique is also the possibility to combine the *strain* measurement with another kind of measurement, as for instance the curvature measurement which provides directly the *stress* variation [61]. However, we have shown that the measurement of the perpendicular component of the lattice parameter is not straightforward, though rocking-curves measurements may sometimes provide reasonable measurements. Also, the lattice parameter which is determined from the RHEED pattern is averaged over a few atomic planes, the exact number of which being difficult to be accurately calculated or measured. Therefore the problem of the relaxation profile within a few atomic layers, as the profile deduced from the X-ray simulation in figure 30, has not been addressed so far by the RHEED technique. A simulation of the RHEED diffracted intensity is required and it has been done sometimes for perfect single crystals termination using dynamical theories. But so far diffraction theories including disorder have not been developed for RHEED.

Different X-ray techniques may be used for strain measurements. The GID technique is fast and easy to use. It provides a straightforward and accurate measurement of the in-plane lattice parameter and may be combined with a $\theta-2\theta$ measurement to determine the out-of-plane lattice parameter. For simple cubic epitaxial relationships, these measurements are often enough to describe the biaxial strain of a single epilayer. GID however requires a specific diffractometer with an accurate control of the incident angle. We have shown that the asymmetrical Bragg measurement, using a standard $\omega-2\theta$ diffractometer, may be an alternative to GID. Reciprocal intensity maps may indeed provide both the in-plane and the out-of-plane components of the lattice parameters of thin films with a very good accuracy, provided that the alignment of the diffractometer and the sample are done with great care. Finally, the $Sin^2\psi$ method may be used in the case of complicated stress states, or for measuring the strain in each individual components of a periodic multilayer. In the case of a multilayer with a very short period, the individual in-plane lattice parameters of the layers are expected to very close and it is not possible anymore to distinguish between the individual Bragg components (see for instance figure 27.b). Therefore the simulation of the intensity profile is the only way to extract the variation of the strain inside a period, through a refinement process. We shall stress that this method is also the only method which may provide not only the averaged lattice parameter for both individual layers but also the chemical information and the strain profile along a period of the multilayer, as depicted in figure 30.

ACKNOWLEDGMENT

A. Marty, Pr. O. Thomas and S. Labat are acknowledged for their contribution to part of the work described.

REFERENCES

[1] A.Y. Cho and J.R. Arthur, Progr. Solid State Chem. 10 (1975) 157.
[2] S.T. Purcell, B. Heinrich and A.S. Arrott, Phys. Rev. B 35 (1987) 6458.
[3] B.A. Joyce, J.H. Neave, P.J. Dobson and P.K. Larsen, Phys. Rev. B 29 (1988) 2176.
[4] A.Y. Cho, The technology and physics of molecular beam epitaxy, edited by E.H.C. Parker, Plenum, 1985.
[5] J.E. Mahan, K.M. Geib, G.Y. Robinson and R.G. Long, J. Vac. Sci. Technol. A 8 (1990) 3692.
[6] J.L. Vassent, M. Dynna, A. Marty, B. Gilles, G. Patrat, J. Appl. Phys. 80 (1996) 5727.
[7] P. Politi, G. Grenet, A. Marty, A. Ponchet, J. Villain, Physics Reports 324 (2000) 271.
[8] L.B. Freund, J. Appl. Phys. 68 (1990) 2073.
[9] C.S. Lent, P.I. Cohen, Surf. Sci. 139 (1984) 121.
[10] P.J. Dobson, B.A. Joyce, J.H. Neave, J. Zhang, J. of Cryst. Growth 81 (1987) 1.
[11] J. Massies, N. Grandjean, Phys. Rev. Lett. 71 (1993) 1411.
[12] P.W. Tasker, D.M. Duffy, Surf. Sci. 137 (1977) 309.
[13] P.K. Larsen, P.J. Dobson, J.H. Neave, B.A. Joyce, B. Bolger, J. Zhang, Surf. Sci. 169 (1986) 176.
[14] H.A. Bethe, Ann. Phys. (leipzig) 87 (1928) 55.
[15] K. Shinohara, Inst. Phys. Chem. Res. (Tokyo) 18 (1932) 223.
[16] R. Colella, Acta Cryst. A28 (1972) 11.
[17] A.R. Moon, Z. Naturforsch. A27 (1972) 11.
[18] P.G. Self, M.A. O'Keefe, P.R. Busek, A.E.C. Spargo, Utramicroscopy 11 (1998) 35.
[19] Y. Ma, L.D. Marks, Acta Cryst. A46 (1990) 11.
[20] Y. Ma, L.D. Marks, Acta Cryst. A47 (1991) 707.
[21] J.M. Cowley, A.F. Moodie, Acta Cryst. 10 (1957) 609.
[22] J.M. Cowley, A.F. Moodie, Acta Cryst. 12 (1959) 353.
[23] L.M. Peng, J.M. Cowley, Acta Cryst. A43 (1986) 545.
[24] Y. Ma, S. Lordi, P.K. Larsen, J.A. Eades, Surf. Sci. 289 (1993) 47.
[25] Y. Ma, S. Lordi, C.P. Flynn, J.A. Eades, Surf. Sci. 302 (1994) 241.
[26] P.A. Stadelmann, Ultramicroscopy 21 (1987) 131.
[27] P.A. Doyle, P.S. Turner, Acta Cryst. A24 (1968) 390.
[28] P.A. Maksym and J.L. Beeby, Surf. Sci. 110 (1981) 423.
[29] A. Ichimiya, J. Appl. Phys. Jpn. 22 (1983) 176.
[30] P.A. Maksym, Surf. Sci. 149 (1985) 157.
[32] U. Korte, G. Meyer-Ehmsen, SurF. Sci. 271 (1992) 616.
[32] U. Korte, G. Meyer-Ehmsen, SurF. Sci. 227 (1992) 109.
[33] J.M. McCoy, U. Korte, P.A. Maksym, G. Meyer-Ehmsen, Surf. Sci. 261 (1992) 29.
[34] T. Hanada, S. Ino, H. Daimon, Surf. SCi. 313 (1994) 143.
[35] A. Ichimiya, Surf. Sci. 192 (1987) L893.
[36] M.V. Laüe, Ergeb. Exakt. Naturw. 10 (1931) 33.
[37] B.W. Batterman, H. Cole, Rev. Mod. Phys. 36 (1964) 684.
[38] T. Hanada, S. Ino, H. Daimon, Surf. SCi. 313 (1994) 143.

[39] J.F. Menadue, Acta Cryst. A28 (1972) 1.
[40] Y. Shigeta, Y. Fukaya, H. Mitsui, K. Nakamura, Surf. Sci. 4902 (1997) 313.
[41] W.C. Marra, P. Eisenberger, A.Y. Cho, J. Appl. Phys. 50 (1979) 6927.
[42] P. Eisenberger, W.C. Marra, Phys. Rev. Lett. 46, (1981) 1081.
[43] M. Born, E. Wolf, Principles of Optics, 5 ed., Permamon, New York, 1986.
[44] R.W. James, The Optical principles of the Diffraction of X-Rays, Ox Bow, Connecticut, 1982.
[45] International Tables of Crystallography, Vol. IV, Eds. J.A. Ibers and W.C. Hamilton, The International Union of Crystallography (1974).
[46] L.G. Paratt, Phys. Rev. 95 (1954) 359.
[47] R.S. Becker, J.A. Golevechenko and J.R. Patel, Phs. Rev. Lett. 50 (1983) 153.
[48] R. Feidenhans'l, Surf. Sci. Rep. 10(3) (1989) 105.
[49] S. Dietrich and H. Wagner, Phys. Rev. Lett. 51 (1983) 1469.
[50] H. Dosh, Phys. Rev. B35 (1987) 2137. See also H. Dosch, B.W. Batterman, D.C. Wack, Phys. Rev. Lett. 56(11) (1986) 1144 and U. Pietsch, H. Metzger, S. Rugel, B. Jenichen, I.K. Robinson, J. Appl. Phys. 74(4) (1993) 2381.
[51] S. Kishino and K. Kohra, Jap. J. Appl. Phys. 10(5) (1971) 551.
[52] G.H.Vineyard : Phys. Rev. B 26 (1982) 4146.
[53] M. Brunel and F. de Bergevin : Act. Cryst. A 42 (1986) 299.
[54] E.E. Fullerton, I.K. Schuller, H. Vandetstraeten et Y. Bruynseraede, Phys. Rev. B 45(16) (1992) 9292 and references therein.
[55] R.M. Fleming, D.B. McWhan, A.C. Gossard, W. Wiegmann, R.A. Logan, J. Appl. Phys. 51 (1980) 557.
[56] P.D. Dernier, D.E. Moncton, D.B. McWhan, A.C. Gossard, W. Wiegmann, Bull. Am. Phys. Soc. 22 (1977) 293.
[57] S. Hendricks, E. Teller, J. Chem. Phys. 10 (1942) 147. See also J. Kakinoki, Y. Komura, J. Phys. Soc. Jpn. 7 (1952) 30.
[58] B. Gilles, A. Marty, Mat. Res. Soc. Symp. Proc. 356 (1995) 379.
[59] B.M. Clemens B.M., J.A. Bain, MRS Bulletin 17 (1992) 713.
[60] S. Labat, P. Gergaud, O. Thomas, B. Gilles, A. Marty, J. Appl. Phys. 87 (1999) 1172.
[61] S. Labat, P. Gergaud, O. Thomas, B. Gilles et A. Marty, Appl. Phys. Lett. 75 (2000) 914.

Measurement of displacement and strain by high-resolution transmission electron microscopy

M. J. Hÿtch

Centre d'Etude de Chimie Métallurgique-CNRS, 15 rue G. Urbain, 94407 Vitry-sur-Seine, France

Abstract : The measurement of strain using high-resolution electron microscopy (HREM) is presented. In particular, the geometric phase technique for measuring the distortion of lattice fringes in the image is described in detail. The method is based on the calculation of the "local" Fourier components of the HREM image by filtering in Fourier space. The accuracy and inherent limitations of strain mapping will be discussed, such as image distortions, the projection problem and thin film effects. Examples are given where displacements are measured to an accuracy of 0.003 nm at nanometre resolution, strain variations to within 1%, and rotations to within 0.2°. Applications to the study of compositional variations in strained $In_xGa_{1-x}As$ islands grown on GaAs, relaxation of $La_{0.67}Sr_{0.33}MnO_3/SrTiO_3$ heterostructures, and misfit dislocation networks in thin epitaxially grown layers will be presented.

1. INTRODUCTION

High-resolution electron microscopy (HREM) differs from conventional electron microscopy in that images are formed of the atomic lattice. Lattice fringes are created by the interference of diffracted beams with the transmitted beam, though in general, the number of lattice fringes is limited to two or three. Practically all the information contained in an HREM image can be obtained by analysing these few components to the image intensity. It is not the aim of this course to explain how the images are formed, it is sufficient for the present purposes to know that the lattice fringes seen in the image are closely related to the atomic planes in the specimen. How closely will be discussed later.

Figure 1 shows a typical image with its decomposition into different lattice fringes. In this case the image contrast is dominated by the fringes corresponding to the {111} atomic planes; the other periodicities are only weakly present in the image. The crossing of the different fringes produces the dot-like contrast corresponding to the atomic columns viewed in projection. The principle behind strain mapping in HREM is the measurement of these atomic column positions from the image and relating this to a displacement field in the specimen.

The measurement is schematically presented in Figure 2. An undistorted reference lattice is chosen and the displacements u_n calculated with respect to it. This is a very important step as strain in elastic theory is defined with respect to the undeformed initial state. There is no way of knowing this from electron microscopy. The nearest that can be obtained is to use a region of crystal which is undistorted, far from defects and interfaces. A second point is that the local chemistry is not always known. Strain is defined by the deformation of the undeformed unit cell. If the composition varies the lattice spacing will vary irrespective of any strain present. These considerations are independent of the way the displacements are measured, and must be born in mind throughout the ensuing desciption.

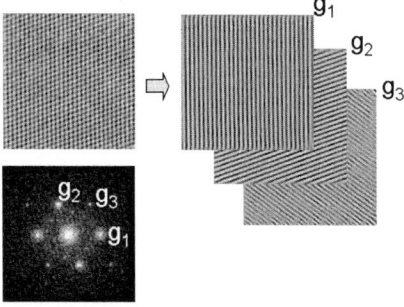 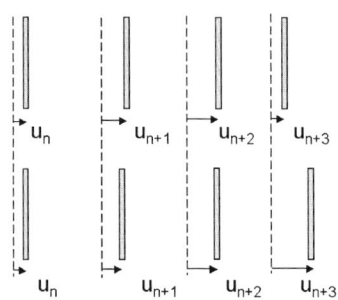

Figure 1. Lattice fringe images: (a) typical high-resolution image of the atomic lattice in [011] projection of aluminium; (b) Fourier transform of image intensity showing that the image is dominated by 2 periodicities, g1 and g2; (c) individual lattice images corresponding to the planes (111), (-111) and (200) respectively.

Figure 2. Principle of the measurement of displacements. The dotted lines represent the reference lattice of constant spacing. The local displacements are given by u_n. A change in the overall spacing produces a linear displacement field (bottom).

There are two ways of carrying out the measurements. One is to localise the maxima of intensity in the image, the so-called peak-finding method [1-3]. A number of crosses are shown to this effect on Figure 3a. In the following, we will assume that the images have already been digitised and are formed of pixels. A detail of the technique is that the intensity distribution of the peaks can be fitted using suitable functions to order to find the maximum to sub-pixel accuracy, as indicated in Figure 3b. Each peak is catalogued and compared to a reference grid.

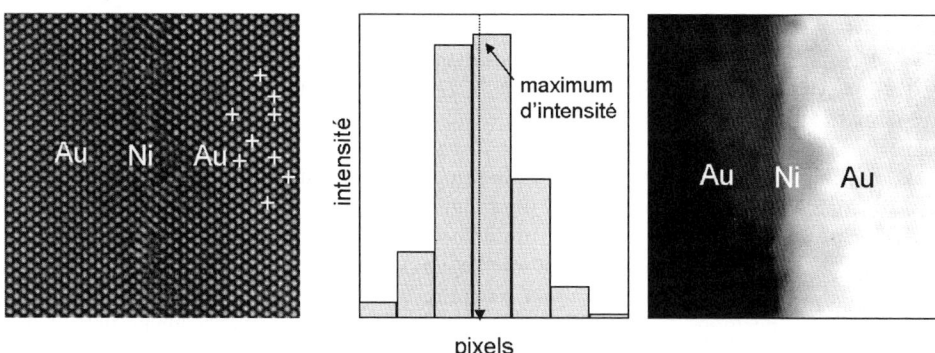

Figure 3. Methods of measurement: (a) high resolution image of a thin Ni layer sandwiched between two Au layers, Ni layer nominally 2 atomic layers wide lying vertically; (b) measurement of a peak position, due to image pixelisation the position of the maximum is estimated by a fitting procedure; (c) geometric phase image showing continuous displacement field with respect to the Au lattice.

Here, the first layer of Au can be chosen as the reference lattice. The lattice fringes spacing in the Ni layer are much smaller than that in the Au. Not all of this is due to strain, the majority is simply that Ni has a smaller lattice parameter (14% smaller to be more precise). All of the subsequent measures should be interpreted in this light: deformation is measured with respect to a given reference lattice.

The geometric phase method is another way of measuring displacements and this is the method we shall concentrate on here [4-5]. The image shown in Figure 3a is transformed into that of Figure 3c and gives an image of the continuous displacement of the lattice fringes. An explanation of the method follows.

2. GEOMETRIC PHASE METHOD

As mentioned previously, the high resolution image is a sum of individual lattice fringes. The geometric phase method analyses each of these lattice fringe images separately. The intensity variation, $B_g(\mathbf{r})$, for a perfect set lattice fringes (see Figure 4a) can be written :

$$B_g(\mathbf{r}) = 2A_g \cos \{2\pi \mathbf{g}.\mathbf{r} + P_g\} \tag{1}$$

where $\mathbf{g}$ is the periodicity of the fringes and $\mathbf{r}$ the position in the image. The amplitude of the fringes is given by A_g and the phase, P_g, gives the lateral position of the fringes with respect to the origin. A change in the phase of π would translate the fringes by exactly half a lattice spacing. More generally, if the displacement is $\mathbf{u}$, then:

$$\mathbf{r} \rightarrow \mathbf{r} - \mathbf{u} \tag{2}$$

and

$$B_g(\mathbf{r}) = 2A_g \cos \{2\pi \mathbf{g}.\mathbf{r} - 2\pi \mathbf{g}.\mathbf{u} + P_g\}. \tag{3}$$

The additional phase shift introduced by the translation is therefore given by:

$$P_g = -2\pi \mathbf{g}.\mathbf{u}. \tag{4}$$

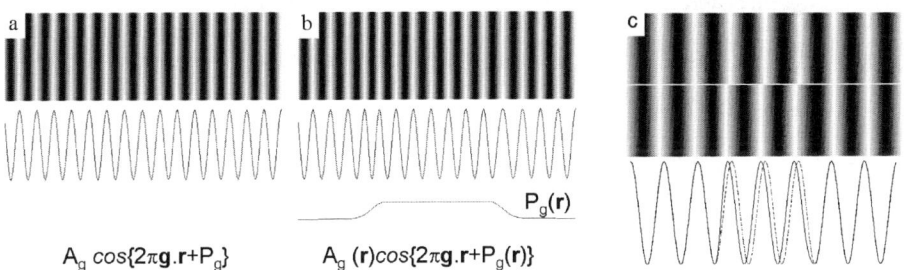

Figure 4. Lattice fringes analysis: (a) perfect fringes, intensity profile and mathematical description; (b) modulated fringes, profile, and phase; (c) reference lattice, modulated fringes, profiles (reference dotted line).

The originality of the phase method is that the phase (and more generally, the amplitude) is allowed to be a function of position in the image:

$$B_g(\mathbf{r}) = 2A_g(\mathbf{r}) \cos \{2\pi \mathbf{g}.\mathbf{r} + P_g(\mathbf{r})\} \tag{5}$$

where $A_g(\mathbf{r})$ and $P_g(\mathbf{r})$ correspond to the "local" amplitude and phase [4]. This allows variations in structure to be described. It is assumed that in the presence of a variable displacement field $\mathbf{u}(\mathbf{r})$, Equation (4) still holds and that:

$$P_g(\mathbf{r}) = -2\pi \mathbf{g}.\mathbf{u}(\mathbf{r}) \tag{6}$$

as illustrated in Figure 4b. The fringes in the centre of the image are displaced slightly to the left which produces a positive variation in the phase. A summary is shown in Figure 4c.

The phase is always measured with respect to a reference lattice and choosing a different reference, changes the resulting phase. If the local lattice parameter differs then:

$$\mathbf{g} \rightarrow \mathbf{g} + \Delta\mathbf{g} \tag{7}$$

where $\Delta\mathbf{g}$ is the change in the reciprocal lattice vector. Following an argument similar to that given for the displacement field, the phase shift is found to be:

$$P_g(\mathbf{r}) = 2\pi\Delta\mathbf{g}.\mathbf{r} \tag{8}$$

which can be seen to be a linearly increasing phase ramp. This can be understood referring back to Figure 2. A linearly varying displacement field indicates a difference in the local lattice parameter with respect to the reference. Taking the gradient of Equation (8) gives:

$$\nabla P_g(\mathbf{r}) = 2\pi\Delta\mathbf{g}(\mathbf{r}) \tag{9}$$

showing that the gradient of the phase gives the local reciprocal lattice vector. These results are only strictly true for small distortions.

2.1 Vectorial displacement fields

According to Equation (6), the phase only gives the information concerning the component of the displacement field in the direction of $\mathbf{g}$. This is only natural, as displacements can only be measured perpendicularly to a set of fringes. It is however possible, by combining the information from two non-colinear lattice fringes, to calculate the 2-dimension displacement field [6]. Given the measurement of the phase for two reciprocal lattice vectors $\mathbf{g}_1$ and $\mathbf{g}_2$:

$$P_{g1}(\mathbf{r}) = -2\pi\, \mathbf{g}_1.\mathbf{u}(\mathbf{r}) \text{ and } P_{g2}(\mathbf{r}) = -2\pi\mathbf{g}_2.\mathbf{u}(\mathbf{r}) \tag{10}$$

it is possible to show that [5]:

$$\mathbf{u}(\mathbf{r}) = \frac{-1}{2\pi}[P_{g1}(\mathbf{r})\mathbf{a}_1 + P_{g2}(\mathbf{r})\mathbf{a}_2] \tag{11}$$

where $\mathbf{a}_1$ and $\mathbf{a}_2$ are the basis vectors for the real space lattice corresponding to the reciprocal lattice defined by $\mathbf{g}_1$ and $\mathbf{g}_2$. The relationship is illustrated in Figure 5. This is a pleasing result and allows us to pass from one coordinate system to another very simply.

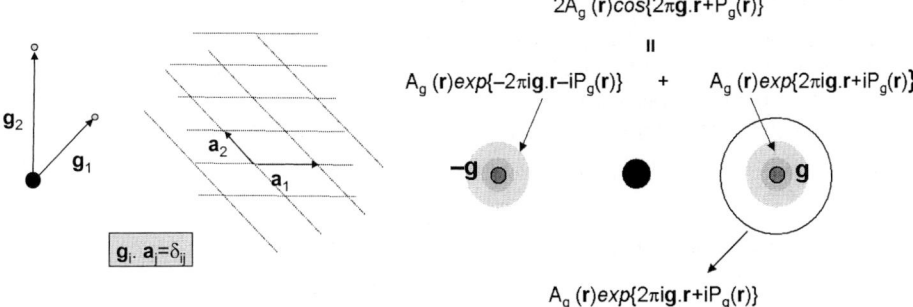

Figure 5. From on space to another: a relationship between basis vectors of a lattice defined in real and reciprocal space.

Figure 6. Method of calculating phase images. A Fourier transform decomposes a cosine into two "side-bands". Selecting one produces a complex image from which the phase is easy to calculate.

2.2 Method of calculating phase images

The method of calculating phase images is illustrated in Figure 6. The Fourier transform of a cosine function contains two "side-bands" at **g** and −**g**. The information concerning the variations is contained in the diffuse intensity around the principle periodicities. Application of a mask followed by a inverse Fourier transform produces a complex image containing the desired information. The phase is then obtained by calculating the arctangent of the imaginary part divided by the real part, point by point in the image. (Note that it is not possible to carry out this operation on the original cosine function because of the zeros.) The method is identical to the way the phase is calculated from a hologram, hence the reference to side-bands. The phase calculated in this way is called the "geometric" phase to distinguish it from the phase of the electron wave function which is obtained via electron holography. An example will illustrate the full procedure. All image analysis has been carried out using routines specially written for the software package Digital Micrograph [7].

Figure 7a shows an image of a ferroelectric-ferroelastic domain wall [5]. Due to the slight difference between the 'a' and 'c' parameter of 4%, domains are formed at angles of 3.7°. The Fourier transform of the whole area shows the mirror plane of the domain structure. The image was Fourier transformed and the two reciprocal lattice vectors selected one after the other. The mask for the (100) lattice fringes is shown on the complex Fourier transform (Figure 7c and 7d). Conventionally, a complex image is represented by two images, one containing the real part pixel by pixel (Figure 7c) and the other, the imaginary part (Figure 7d). Rather than a hard round mask (1 inside and 0 outside) a Gaussian mask was in fact used. The size and shape of the mask are parts of the details of the method: a smaller mask allows greater noise reduction but at the expense of spatial resolution. The phase is calculated after the inverse Fourier transform.

Given the two reciprocal lattice vectors used the displacement field is given by :

$$\mathbf{u}(\mathbf{r}) = \frac{-1}{2\pi} \{P_{100}(\mathbf{r})[1\ 0\ -1] + P_{101}(\mathbf{r})\ [0\ 0\ 1]\} \quad (12)$$

where the square brackets indicates vectors in real space.

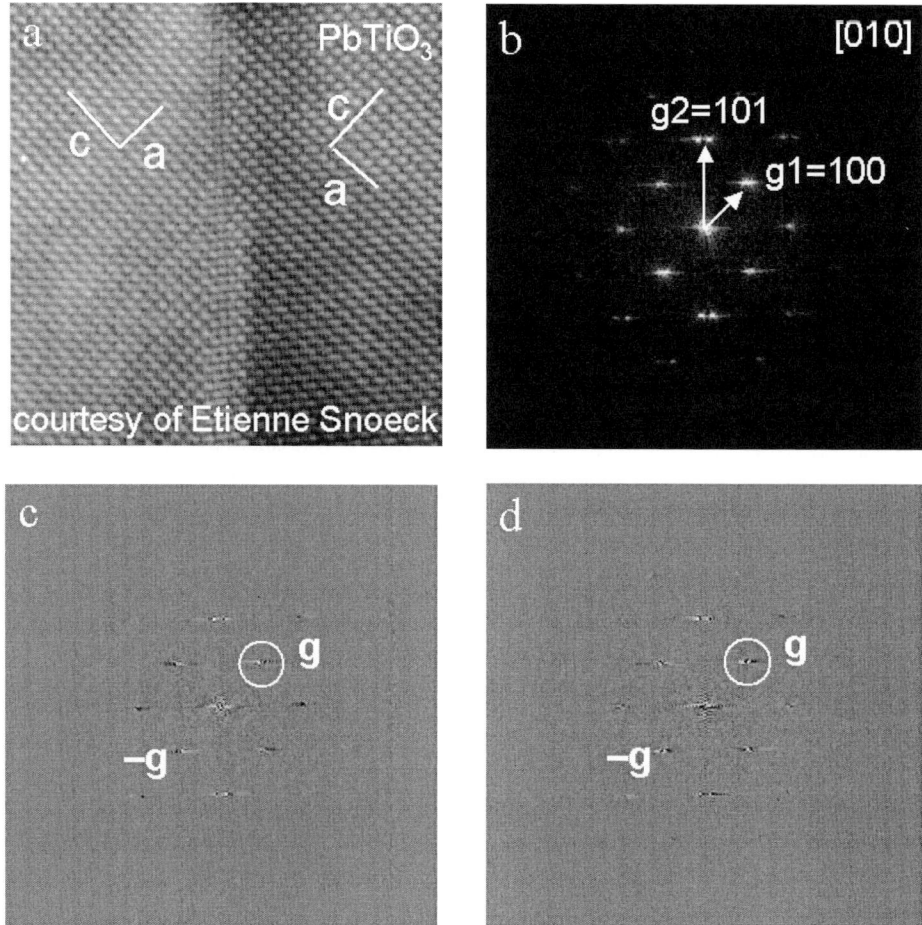

Figure 7. Calculation of phase images: (a) high resolution image of a ferroelectric-ferroelastic domain wall in PbTiO$_3$, notice the slight rotation of the lattice and the inversion of the unit cell vectors; (b) power spectrum of the image (modulus squared of the Fourier transform) showing **g** vectors used for phase analysis; (c) real part of the Fourier transform showing mask used for the (100) fringe analysis; (d) imaginary part (grey indicates zero).

Equation (12) as images is presented in Figure 8. The phase image for the (100) and (101) fringes are shown as a vector on the right hand side of the equation. This is a kind of image algebra where the equation acts at each position of the image, i.e. pixel by pixel. The reference has been taken on the left hand side of the interface so the phase is constant here. On the other side a net gradient is seen in the phase. This is because the local reciprocal lattice vector differs from the reference lattice. The black lines are due to the normalisation of the phase between ±π: each time the phase deceases below -π it is renormalized to +π. The positions of these discontinuities have no physical significance, however, they have the same spacing as moiré fringes due to the superposition of the reference lattice and the local fringes.

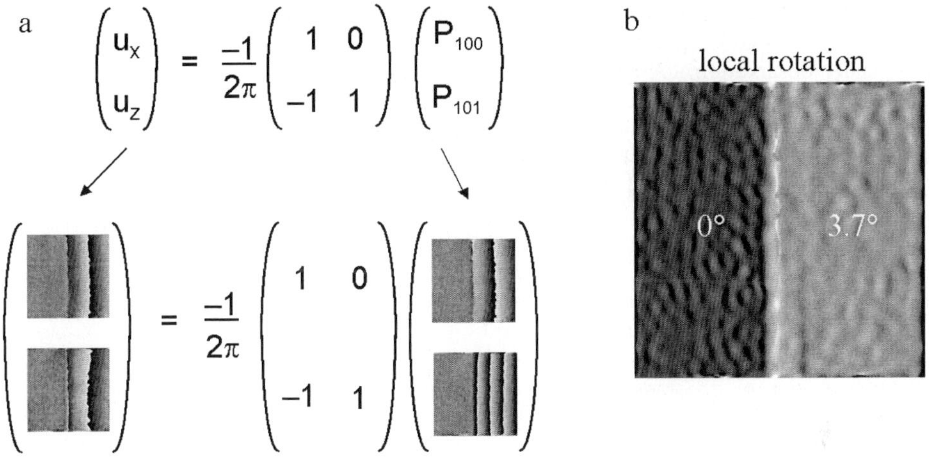

$$\begin{pmatrix} u_x \\ u_z \end{pmatrix} = \frac{-1}{2\pi} \begin{pmatrix} 1 & 0 \\ -1 & 1 \end{pmatrix} \begin{pmatrix} P_{100} \\ P_{101} \end{pmatrix}$$

$$\begin{pmatrix} \text{image} \end{pmatrix} = \frac{-1}{2\pi} \begin{pmatrix} 1 & 0 \\ -1 & 1 \end{pmatrix} \begin{pmatrix} \text{image} \end{pmatrix}$$

Figure 8. Vector displacement field calculation: (a) Equation (12) written in matrix form (top) and with corresponding images (bottom); (b) rigid rotation of the lattice calculated from the deformation matrix (black = 0°, white = 4°, average 3.7°).

2.3 Deformation and strain

Once the vectorial displacement field has been obtained, the deformation matrix can be immediately calculated by differentiation. We can define a matrix:

$$\mathbf{e} = \begin{pmatrix} e_{xx} & e_{xy} \\ e_{yx} & e_{yy} \end{pmatrix} = \begin{pmatrix} \frac{\partial u_x}{\partial x} & \frac{\partial u_x}{\partial y} \\ \frac{\partial u_y}{\partial x} & \frac{\partial u_y}{\partial y} \end{pmatrix} \tag{14}$$

which gives the 2-dimensional lattice distortion in its most general form. This matrix can, in turn, be separated into a symmetric part ε and an anti-symmetric part ω:

$$\epsilon = \tfrac{1}{2}\{\mathbf{e} + \mathbf{e}^T\} \quad \text{and} \quad \omega = \tfrac{1}{2}\{\mathbf{e} - \mathbf{e}^T\} \tag{15}$$

where T denotes the transpose of the matrix. The strain is given by ε and the local rigid rotation by ω, for small distortions. The local lattice rotation for the example given is shown in Figure 8b. For the full deformation matrix refer to [5].

3. FRINGES POSITIONS AND ATOMIC PLANES

Before presenting some applications, the central assumption behind the method needs to be examined. For the results to have any meaning, the displacement of the fringes in an image need to correspond to the displacement of the atomic planes in the specimen. There are a number of factors affecting the accuracy of this statement.

3.1 Dynamical scattering, lens aberrations and the projection problem

As for any imaging system, the electron microscope provides a distorted image of the object under observation. A schematic illustration of the imaging process can be seen in Figure 9. In a transmission electron microscope, the electron arrives as a near plane wave on the specimen. It then interacts with the specimen and exits with a modulated wave function. For our case, the specimen will be crystalline and oriented to a zone axis. In the limiting case of a thin specimen, the resulting wave function will have a phase variation proportional to the projected potential provided by the atomic nuclei (and shielded by the surrounding electrons).

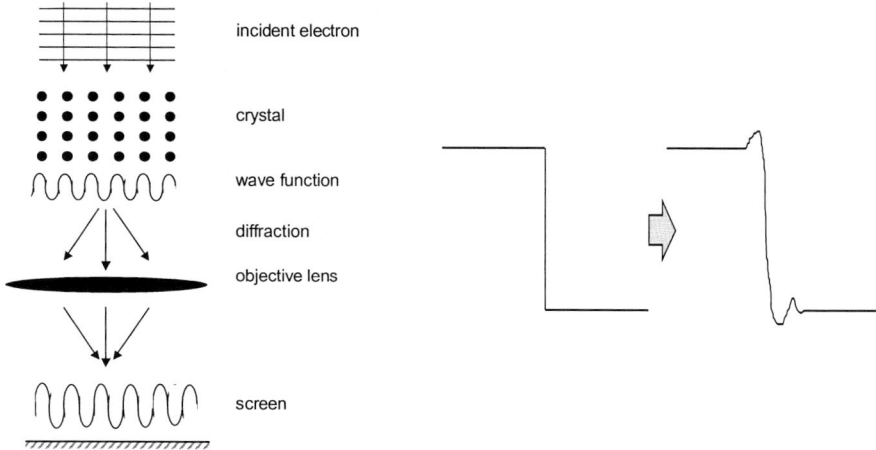

Figure 9. Lens effects: (a) schematic representation of the microscope, from the incident electron beam to the screen, each part of the imaging process introduces distortions; (b) a sharp interface becomes diffuse and ripples appear due to the action of the objective lens.

Displacements of atomic columns will therefore feed directly into the exit wave function.

For thicker specimens, dynamical scattering takes place and the relation between the structure and the exit wave function is not so simple. However, if we imagine moving a crystal bodily to the left, say, the wave function must follow. In the limit of a slowly varying displacement field, the lattice displacements should be faithfully respected in the wave function. This breaks down for rapid variations (typically less than 1 nm) but in this regime the lens effects are far more important.

Where dynamical scattering poses the most problems is for specimens with significant variations in thickness. For non-centrosymmetric crystals, for example, the positions of lattice fringes depend on the thickness of the specimen. Interpreting shifts of the fringes would be hazardous in this case.

3.2 Experimental verification of the method

In order to test the method experimentally, we need to measure a displacement field for a known specimen. There are many arguments which can be brought forward to justify interpreting lattice fringe displacements in terms of displacements of the atomic planes in the sample. Nevertheless, there is nothing better than showing the method works for a specimen with a known displacement field on the atomic level. Finding such a specimen is not, however, trivial. Figure 10a shows a rare example: an edge dislocation seen end-on. The sample is silicon (a = 0.543 nm) viewed along the [110] direction on a JEOL 200CX microscope operating at 200kV with the dislocation at the step in the grain boundary in the centre of the image. The analysis will not hindered by the presence of the grain boundary

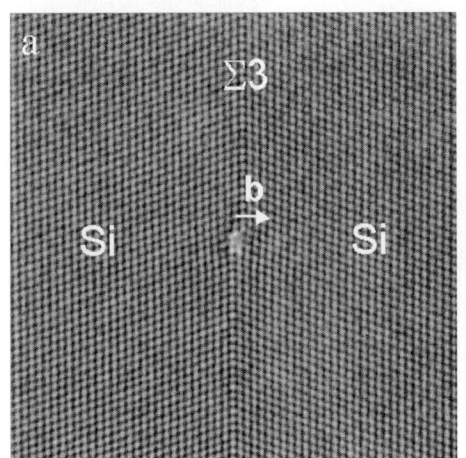 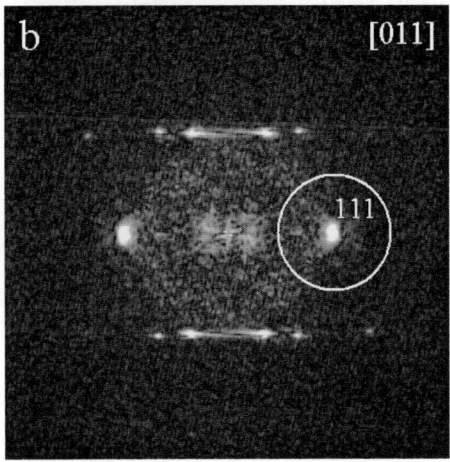

Figure 10: Edge dislocation in silicon: (a) high-resolution image in [011] orientation, dislocation core seen end-on at grain boundary step, **b**=1/3<111>; (b) power spectrum of image.

because the (111) lattice planes are common to both grains (see Figure 10b).

The reason why this image has been chosen, rather than one of an isolated dislocation, is due to the exceptional quality of the image (courtesy of Jean-Luc Putaux). This can be attributed to a number of experimental factors. Firstly, the presence of the grain boundary helps the alignment of both the crystal and the microscope itself. Any asymmetry is immediately visible as the Σ3 boundary should present a perfect mirror. Secondly, the dislocation is pinned to the boundary and cannot move during the observations. Finally, and most importantly, the dislocation is held in a vertical position by the step and the boundary. Isolated dislocations tend to curve at the free surfaces of the specimen.

The resulting phase is shown in Figure 11a. The first thing to notice is the point discontinuity in the phase at the position of the dislocation core. The "black" line is just due to the phase normalisation; its position moves when a constant phase is added. The wiggles are due to noise. The point discontinuity is, however, the physically significant feature and can be understood from the definition of the Burgers vector. The integral of the displacement field, **u(r)**, around a closed loop, L, containing a dislocation gives the Burgers vector:

$$\mathbf{b} = \oint_L \nabla \mathbf{u}.\mathbf{dl} \tag{13}$$

Translating this in terms of the phase using to Equation (6), we find the following result:

$$2\pi n = \oint_L \nabla P_g.\mathbf{dl} \tag{14}$$

if **g** is in the direction of **b** and n is the number of extra planes. A discontinuity of 2π is found on circumnavigating the dislocation core in the example, indicating a single extra plane. The position of the black line where the phase changes abruptly by 2π is due the

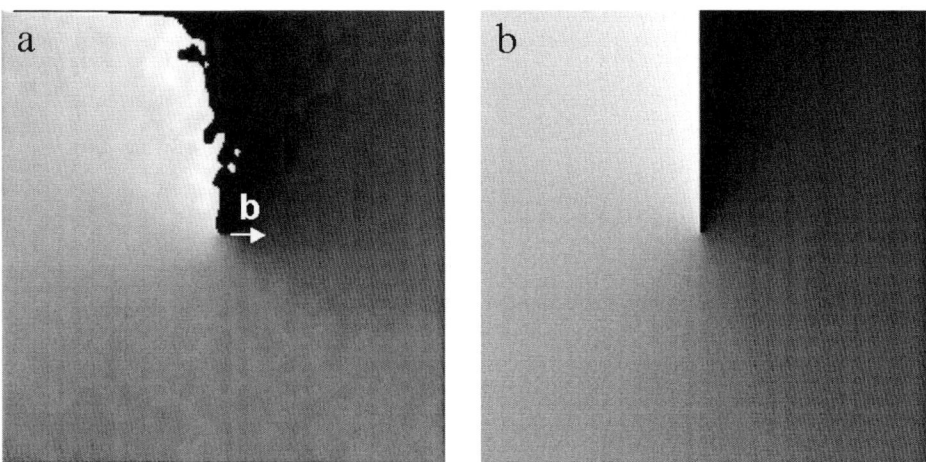

Figure 11. Displacement field: (a) experimental (111) phase image, point discontinuity at dislocation core; (b) predicted (111) phase from linear elastic theory.

normalisation of the phase : the position moves when a constant phase is added.

According to isotropic elastic theory the component of the displacement field parallel to the Burgers vector, u_b, is given by:

$$u_b = \frac{2\pi}{d}\left(\theta + \frac{\sin 2\theta}{4(1-v)}\right) \quad (15)$$

where θ is the angle subtended to the core, and v the Poisson's constant [8]. According to Equation (6):

$$P_{111}(\mathbf{r}) = -2\pi \mathbf{g}_{111}.\mathbf{u}(\mathbf{r}) = -\frac{2\pi}{d}u_b \quad (16)$$

and substituting into Equation (15) we find that:

$$P_{111}(\mathbf{r}) = -\theta - \frac{\sin 2\theta}{4(1-v)}. \quad (17)$$

Given that v is typically of the order of 0.3, the first term dominates the displacement field, and this is what is seen in the result (Figure 11). The phase increases monotonically around the dislocation core. There is a very good qualitative agreement between the experimental phase and that predicted by theory. To test this result quantitatively, we shall subtract the first term from the experimental result to see if the second term is detectable. Figure 12 shows the result.

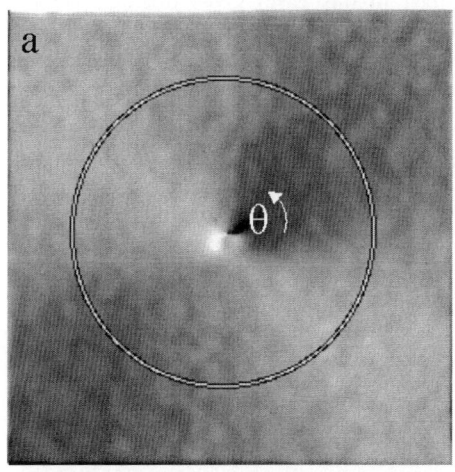
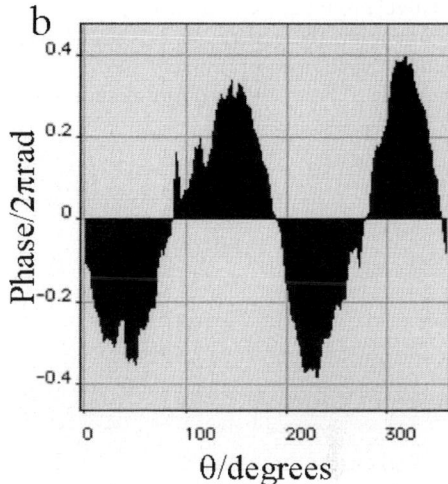

Figure 12: Quantitative comparison: (a) experimental term in sin2θ; (b) variation of the phase around the circle marked in (a), $P_g/2\pi$ gives the displacement as a fraction of the fringe spacing, according to Equation (16), where $d_{111} = 0.31$ nm.

There is good qualitative agreement with a variation due to a sin 2θ function. It can be seen though that at the dislocation core, the phase differs strongly from this model. It is expected that the theory should break down in this region given that the model is non-atomistic and for small. However, if the phase is plotted at a reasonable distance (in this case 6 nm) from the core (Figure 12b), the agreement with theory is very good. The average deviation is ±1% of the planar spacing or ±0.003 nm (theoretical curve not shown). This was evaluated using a value of ν = 0.30 in the isotropic model. In fact, silicon is highly anisotropic but an almost identical result is obtained when the full anisotropic calculation is carried through. Indeed, the sine wave can be seen to be slightly leaning to the right, which is reproduced in the anisotropic calculation. The theoretical limit for the accuracy is given by the signal to noise ratio as a fraction of the planar spacing.

4. APPLICATIONS OF STRAIN MAPPING

4.1 Strained semiconductor islands

The first application comes from a study $In_xGa_{1-x}As$ (x is nominally 0.35) islands grown by molecular beam epitaxy (MBE) on a GaAs substrate (Figure 13) [9]. In this [110] projection only three principal image periodicities are responsible for the image contrast: the $(1\bar{1}1)$ and $(\bar{1}11)$ and the $(2\bar{2}0)$ lattice fringes. It is interesting to note that visually there is very little difference between the island and the substrate. This illustrates one of the characteristics of high-resolution electron microscopy that chemical information does not necessary appear in a direct way. For the defocus value chosen, there is little difference in contrast. However, for other defocus values the island and the substrate could differ in various ways, none of them being directly interpretable in terms of a change in the indium concentration. Whilst the relative brightness of the features (dots of contrast in this case) gives difficulty usable information, the separation of these features does provide interpretable information. The lattice fringe spacing is very closely related to the actual atomic planar spacing in the specimen. From this we can deduce reliable information concerning the chemistry and/or

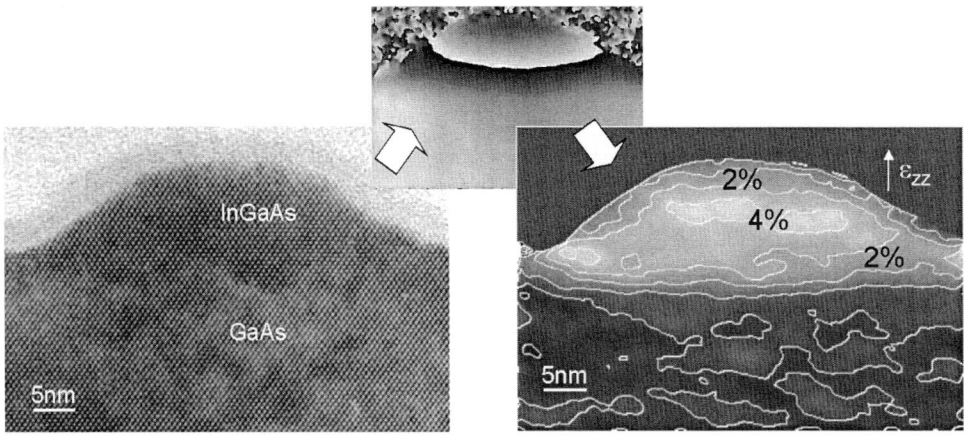

Figure 13. High-resolution image of $In_{0.35}Ga_{0.65}As$ island growth on a GaAs substrate by molecular beam epitaxy (MBE). Image is composed of lattice fringes corresponding to the atomic planes $(1\bar{1}1)$ and $(\bar{1}11)$ and the $(2\bar{2}0)$. The phase of the (002) lattice planes can be calculated from the $(1\bar{1}1)$ and $(\bar{1}11)$ phases (see text for details) and this in turn gives the deformation in the z-direction with respect to the substrate (contours shown every 1%).

strain.

Figure 13 shows the phase image calculated for the (002) lattice planes. The reference has been defined in a region of the substrate far from the island where the strains can be considered to be small. The (002) phase was obtained by adding the phases calculated from the $(1\bar{1}1)$ and $(\bar{1}11)$ lattice fringes according to the relation:

$$P_{g1+g2}(\mathbf{r}) = P_{g1}(\mathbf{r}) + P_{g2}(\mathbf{r}). \tag{13}$$

This can be shown directly from Equation (10) and is an alternative to using the general Equation (12) for simple cases. The (002) phase directly gives the component of the displacement in the z direction from which the deformation ε_{zz} can be calculated, as shown on Figure 13.

The average deformation in the substrate is zero, as this is where the reference was chosen. The lattice spacing increases within the island reaching a maximum of 4.5% at the centre before decreasing towards the surface. By comparison, the mismatch between bulk $In_{0.35}Ga_{0.65}As$ and GaAs is 2.7%.

Deformation, strain and chemistry

This illustrates the important fact that we measure deformation, which can be due to both a change in chemistry and the strain. Let us assume that the composition is uniform in the island. For the hypothetical unstrained state the measured deformation should be 0% in the substrate and 2.7% in the island. To calculate the strain we therefore need to subtract these values from the deformation map. We immediately come across the problem of defining the position of the interface between the two materials. It could be where the deformation increases rapidly. In that case, the $In_{0.35}Ga_{0.65}As$ is in compression close to the interface and the surface but in extension of 1.8% in the centre of the island. However, displacing the interface would indicate that a region of the substrate would be in extension. It is obvious that the results need to be compared with models of the elastic strain field.

If on the other hand, the chemistry is allowed to vary, some of the deformation could be explained by In segregation. Often in interpreting deformation, Vergard's law is assumed:

$$a(In_xGa_{1-x}As) = x\,a(InAs) + (1-x)\,a(GaAs) \tag{13}$$

where the lattice parameter 'a' is linearly related to the composition. The deformation of 4.5% at the centre of the island could therefore be attributed to an increase of the indium concentration from 35% to 58%. Again, the only way to understand the results are by carrying out some modelling.

Modelling, 2-d projections and the thin film effect

Figure 14 shows the results from finite element simulations of the island system [9]. The models where carried out assuming the indium concentration to be constant in the island with no chemical diffusion. For this system, the displacement field is 3-dimensional and varies in the viewing direction. Ideally, image simulations should therefore be carried out using the full 3-D structure. However, the variation in the viewing direction is small and we can approximate the results by averaging the displacement field over the [110] direction. Results for the 2-dimensional deformation maps are shown in Figure 14b and 14c.

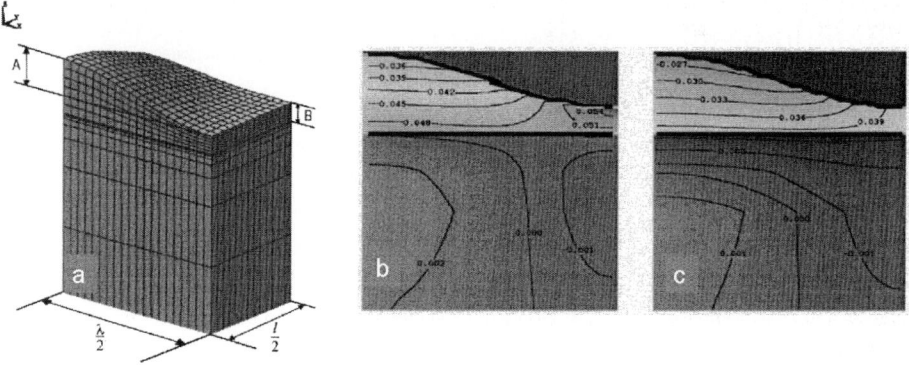

Figure 14. Finite elements calculation: (a) mesh used for the modelling showing one quadrant of the island structure; (b) projected deformation for a complete island; (c) projected deformation for a sectioned island. Note the lower deformation but similar form for the thin film.

A second point is that in the thin film specimen preparation for the microscope observations, there is a high probability that a section has been taken of the island. Models were therefore carried out assuming this to be the case (Figure 14c). Comparing the deformation values in Figures 14b and 14c, we find that for the thin film, the resulting deformations are lower. This is typically the result for thin films: the free surfaces allow for easy relaxation of the strains.

It can be seen that it will be difficult to compare the experimental results with the simulated cases as the exact geometry of the specimen is not known. However, in all of the cases studied in simulations, the deformation increases steadily and monotonically towards the surface. Neither did a detailed study of the possible artefacts due to the imaging conditions show a variation from this model [10]. On the other hand, the experimentally measured deformation had a maximum at the centre of the island. We can therefore conclude that the specimen examined had a segregation of indium to the centre of the island.

Ex-situ measurements

It must also be remembered that the vast majority of electron microscopy is carried out ex-situ. This means that the specimen could evolve with repect to the initial structure. In the example studied, it might be suggested that oxidation of the specimen surface explains the depletion of the indium (see the amorphous layer on Figure 13 on top of the island). Great care therefore needs to be taken when carrying out the experiments. However, for this case, the evidence for indium segregation is strong and has been supported theoretically [11].

4.2 Strained oxide layers

Our second example comes from a study of a $La_{0.67}Sr_{0.33}MnO_3/SrTiO_3/La_{0.67}Sr_{0.33}MnO_3$ heterostructure grown by pulsed laser deposition on a substrate of $SrTiO_3$ [12].

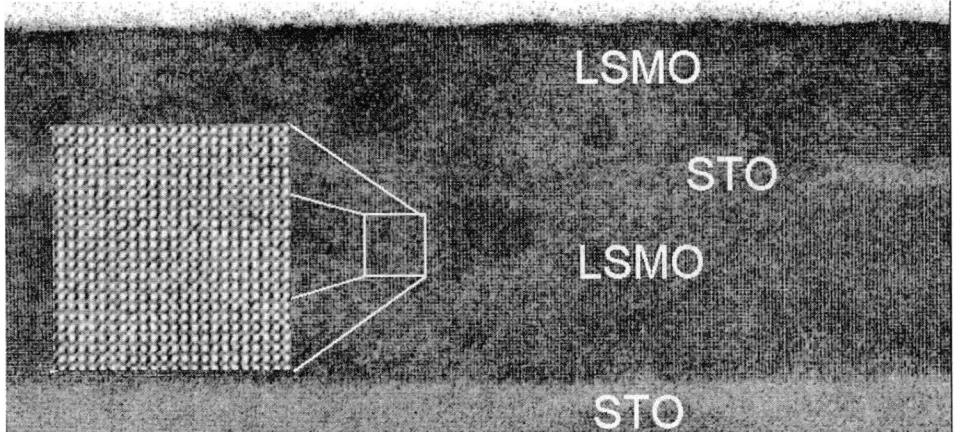

Figure 15: $La_{0.67}Sr_{0.33}MnO_3/SrTiO_3/La_{0.67}Sr_{0.33}MnO_3$ heterostructure grown on a $SrTiO_3$ (STO) substrate, total layer thickness of 40 nm. Zoom shows high-resolution detail, present over whole area.

The conduction electrons in $La_{0.67}Sr_{0.33}MnO_3$ (LSMO) are 100% spin polarised so conduction is theoretically zero across two domains having opposite magnetisations. This means that in the bulk material the electrical resistance depends strongly on the applied magnetic field: the so-called giant magneto-resistance effect. In this case the thin layer of $SrTiO_3$ creates a tunnel junction between the "electrodes" of $La_{0.67}Sr_{0.33}MnO_3$ thus creating a device which has a theoretically infinite magneto-resistance. In practice, the performances are less than the theoretical value and depend on the quality of the grown layers, their structure and the structure of the interfaces. Here we are particularly interested by the state of relaxation of the grown thin film. A comparison of the bulk structures of STO and LSMO is shown in Figure 16.

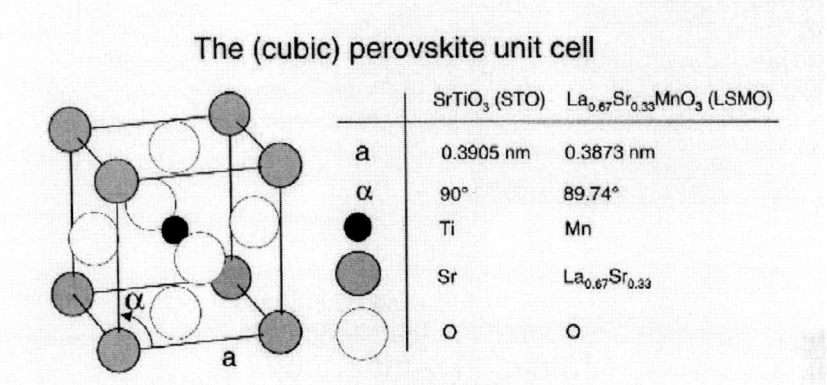

Figure 16: Bulk structures of $La_{0.67}Sr_{0.33}MnO_3$ (rhombohedral) and $SrTiO_3$ (cubic).

The results for the lattice planes parallel to the (001) planes in the $SrTiO_3$ are shown in Figure 17. The reference lattice was chosen as the average value of **g** in the deposited layer and not the value in the substrate. We can see that the phase has a zig-zag variation in the [010] direction parallel to the interface. A gradient in the phase indicates a difference in the local reciprocal lattice from that of the reference according to Equation (9). In this case the gradient is perpendicular to the reference lattice vector, indicating a rotation of the lattice. The vectors Δ**g** on Figure 17 have been exaggerated: the variations are in fact very slight and can be calculated as being 0.26°±0.03° about the average. The results for the (010) lattice fringes show that they remain constant across the image and continuous between the layer and the substrate. The conclusion is that a twinned structure has been formed as presented in Figure 17c. This twinning is due to the relaxation of the LSMO into its bulk rombohedral structure.

This relaxation can be monitored from the substrate to the surface of the layers as shown by the line traces in Figure 17b. The variation in the phase can be seen to accentuate gradually towards the surface and indeed, the values of the rotation are closest to the values expected for twinning in the bulk structure. It is interesting to not the continuity of the twinned structure across the thin layer of STO indicating the strains are transmitted across the layer.

The resulting phase images for the two sets of {111} fringes is shown in Figure 19, with the reference lattice taken in the GaAs and the direction of **g** indicated by the arrows. As for the silicon example, the positions of the dislocation cores are given by the point discontinuities in the phase (also arrowed). In the GaSb layer, the phase has a constant gradient due to the difference in reciprocal lattice vector Δ**g** (not to scale).

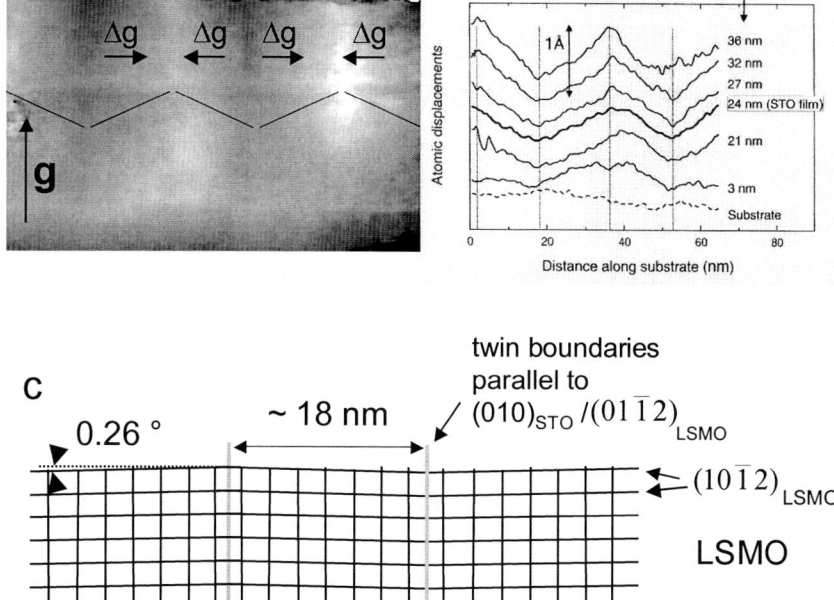

Figure 17: Twin boundary structure in strained layer: (a) measured (010) phase, reference lattice chosen as average LSMO lattice, phase gradients indicate local reciprocal lattice vector variations; (b) phase variation as a function of distance from the substrate showing relaxation towards the surface; (c) schematic model of twin structure deduced from the slight rotations of the lattice.

4.3 Relaxation of epitaxially grown GaAs/GaSb

The final example comes from the growth of GaSb on (100)GaAs by molecular beam epitaxy (MBE) [6,13]. Figure 18a shows a HREM image of the GaAs/GaSb interface taken in [110] orientation. The bulk lattice mismatch is 8% between GaSb (a = 0.609 nm) and GaAs (a = 0.565 nm). The difference in lattice parameter between the two layers is clearly seen in the power spectrum of the image (Figure 18b).

If we combine the results to obtain the deformation in x and y we have the Figure 19. It is interesting to note that the interface defined by the deformation in y lies exactly on the line of dislocations. The deformation in x (perpendicular to the interface) is slightly on the side of the GaAs. It will be necessary to carry out elastic simulations to interpret these results. At the bottom of Figure 19 are the profiles of the deformation to show how the interface thickness can be defined using this method of measurement [6].

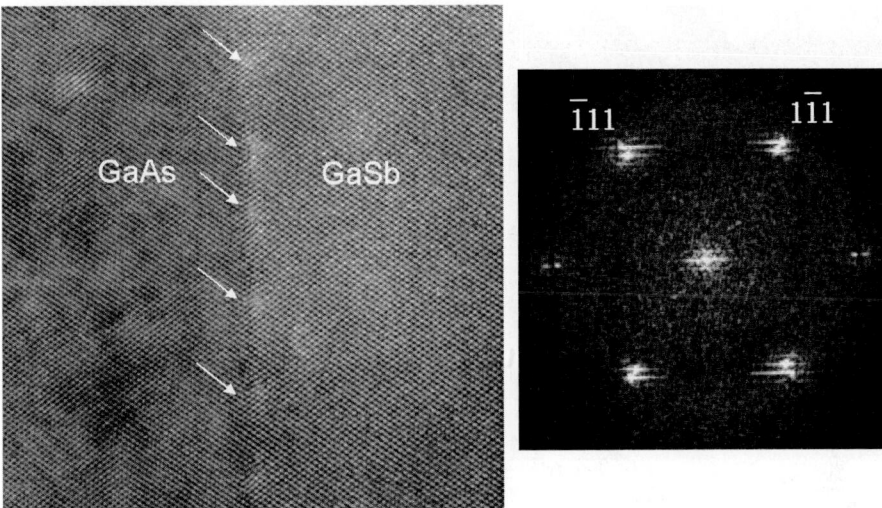

Figure 18: GaAs/GaSb interface observed along [110]: (a) HREM image with arrows indicating dislocation cores; (b) power spectrum of the image, splitting due to 8% mismatch.

5. CONCLUSIONS

High-resolution electron microscopy is a powerful technique for measuring strain in thin layers. The accuracies of the measurements of displacement can be as high as 0.003 nm at nanometre resolution, strain variations to within 1%, and rotations to within 0.2°. As we have seen however, there are a number of factors to bear in mind:
- the projection problem;
- dynamical scattering effects;
- objective lens distortions;
- thin film relaxation;
- ex-situ measurements.

The objective lens distortions have recently been analysed in detail and theory shows that for the measurement of rotations or for the analysis of centro-symmetric crystals the effects are minimal [14]. Imaging conditions are also likely to be greatly improved when aberration corrected microscopes become fully available. The remaining challenge in terms of the theory is to well characterise the effect of dynamical scattering within the crystal and to study the effect of misoriented or bent atomic columns.

From a practical point of view, the rate limiting step is not so much the taking of the images nor indeed their analysis but the preparation of good quality samples for the microscope observations.

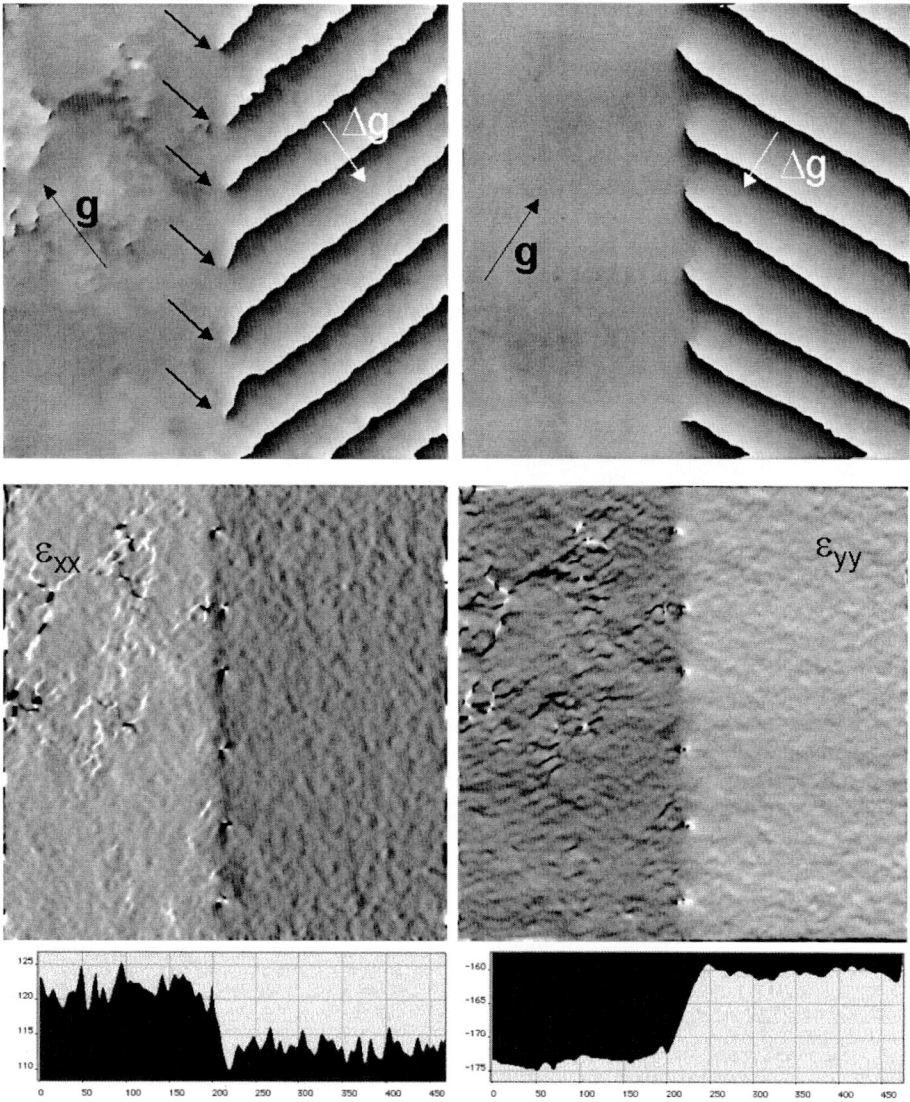

Figure 19: Deformation measurements: (a) phase image for the $(1\bar{1}1)$ lattice fringes; (b) $(\bar{1}11)$ phase image; (c) deformation map ε_{xx} (profile across interface also shown); (d) deformation map ε_{yy} (profile across interface also shown).

6. REFERENCES

1. R. Bierwolf, H. Hohenstein, F. Philipp, O. Brandt, G. E. Crook and K. Ploog, Ultramicroscopy 49 (1993) 273.
2. P. Bayle, T. Deutsch, B. Gilles, F. Lançon, A. Marty, J. Thibault, Ultramicroscopy 56 (1994) 94.
3. P. H. Jouneau, A. Tardot, B. Feuillet, H. Mariette and J. Cibert, J. Applied Physics 75 (1994) 7310.
4. M. J. Hÿtch, Microscopy Microanalysis Microstructure 8 (1997) 41.
5. M. J. Hÿtch, E. Snoeck and R. Kilaas, Ultramicroscopy 74 (1998) 131.
6. Digital Micrograph version 2.5, Gatan Inc., phase routines available at http://ncem.lbl.gov.
7. E. Snoeck, B. Warot, H. Ardhuin, A. Rocher, M. J. Casanove, R. Kilaas and M. J. Hÿtch, Thin Solid Films 319 (1998) 157.
8. J.P. Hirth and J. Lothe, Theory of dislocations, McGraw-Hill, New York, 1968.
9. S. Kret, T. Benabbas, C. Delamarre, Y. Androussi, A. Dubon, J.-Y. Laval and A. Lefebvre, Journal of Applied Physics 86 (1999) 1988.
10. T. Plamann, M. J. Hÿtch, S. Kret, J.-Y. Laval and C. Delamarre, Inst. Phys. Conf. Ser. 164 (Inst. of Phys., London, 1999) p. 23.
11. J. Tersoff, Physical Review Letters, 81 (1998) 3183.
12. R. Lyonnet, J.-L. Maurice, M. J. Hÿtch, D. Michel, J.-P. Contour, Applied Surface Science 162-163 (2000) 245.
13. A. Rocher and E. Snoeck, Materials Science & Engineering B, 67 (1999) 62.
14. M. J. Hÿtch and T. Plamann, Ultramicroscopy (2001) 1 *in press*.

Stress measurements of atomic layers and at surfaces

D. Sander [a] *,

[a]Max-Planck-Institut für Mikrostrukturphysik,
Weinberg 2, D-06120 Halle, Germany

The application of the cantilever bending technique to measure mechanical stress in atomic layers from the curvature of a thin substrate is described. The effects of substrate clamping and elastic anisotropy on the proper evaluation of the curvature is discussed. Examples for stress measurements during growth of epitaxial Fe and Ni monolayers are presented. It is shown that curvature measurements are well suited to detect stress on 100 μm thin substrates with sub-monolayer sensitivity. Stress measurements detect structural transitions in monolayers and identify film stress as an importing driving force for structural transitions. For sub-monolayer coverage, changes of surface stress due to the adsorption process can be the dominant source of stress as compared to lattice misfit arguments. The application of the cantilever bending technique to measure magnetoelastic coupling constants from stress measurements during magnetization processes is briefly described.

1. Introduction

Almost all electronic devices rely on the proper functioning of often quite complex film and multilayer structures that are deposited on a substrate. The deposition of a film on a substrate is generally connected with the build-up of mechanical stress and consequently the elastic energy of the system increases accordingly. This stress can be caused by the thermal treatment during film preparation due to different thermal expansion coefficients. Epitaxial misfit between film and substrate will also induce stress. Consequently, most multilayered structures are in a metastable state and are thriving to reduce their total energy, which is often dominated by the elastic energy. Possible mechanisms to reduce the elastic energy are changes of the film morphology, e.g. the formation of islands instead of layers, the introduction and growth of misfit dislocations, the formation of cracks, the phase separations of alloys and other mechanisms. Such transitions are often driven by film stress, and they have a detrimental impact on the device performance. Therefore, experimental studies of film stress have a long tradition and many reviews are devoted to this topic [1–6].

However, it is important to realize that current electronic and magneto-electronic devices work with film structures where the single layer thickness is as thin as a few dozen atomic layers [7]. Therefore the venerable stress measurement of thin films, where *thin*

---

*presently at C.R.M.C.2-C.N.R.S., Luminy, case 913, F-13288 Marseille Cedex 9, France stood for a thickness in the micrometer range, has to be boosted in sensitivity by at least three orders of magnitude to investigate stress in (sub)nanometer (nm) *ultrathin* films.

This article describes a stress measurement technique that gives sub-atomic layer sensitivity. This high sensitivity allows to investigate the correlation between atomic rearrangements in a monolayer and the resulting stress. Thus, in contrast to the former work on micrometer thick films, the presented stress measurements work in a regime where film and substrate are well defined and characterized on an atomic scale which makes a contact between experiment and first principles calculations on stress in atomic layers and at surfaces feasible [8–13].

The interest in stress measurements on atomic layers is not limited to its impact on nm thin devices but is also based on many open questions regarding the relevant physical processes that determine stress. Is lattice mismatch between film and substrate of any relevance? Stress measurements suggest, that epitaxial Ni films as thin as two atomic layers induces a tensile stress in a Cu(100) substrate, that is quantitatively described by the lattice mismatch between Ni and Cu [14,15]. However, in the first layer of pseudomorphic Fe on W stress opposite in sign as expected from lattice mismatch is measured [16]. The classical stress-strain relation fails and there is not yet a physical model established to describe the different stress behavior. Therefore, stress measurements on well defined atomic layers contribute to a better understanding of fundamental aspects that govern stress in monolayers and at surfaces.

In the following section the basic idea of a stress measurement based on the curvature analysis is presented. The effect of elastic anisotropy and substrate clamping on the data analysis is discussed in section 2.1. Examples for experimental set-ups to measure stress with sub-monolayer sensitivity are presented in section 2.2. Examples of stress measurements in Fe and Ni monolayers are presented in section 3. The application of the curvature technique to measure the magnetostrictive properties of ferromagnetic monolayers follows in section 4.

2. Experimental determination of stress changes

The most powerful method to measure film stress during growth is to exploit the bending action of a surface stress imbalance between front- and backside of a thin substrate. A curvature results, see Fig. 1, from which the stress is calculated. The relevant equations for the curvature analysis are presented, and the effect of clamping and elastic anisotropy is discussed in section 2.1.

Surfaces in which the substrate is bent with radii of curvature that are of the same sign, are called *synclastic* surfaces. If the radii of curvature are of opposite sign, like a saddle, one speaks about an *anticlastic* surface [17]. Both curvature states are observed in experiments, as discussed below.

It is important to note, that stress changes on both surfaces contribute to the resulting curvature. For a measurement of film stress during growth it is usually sufficient to prevent a direct exposure of the backside of the substrate to the molecular beam of the film material by orienting the substrate with its front surface facing the evaporation source. A resulting curvature can than be ascribed to the forces acting at the front surface of the substrate.

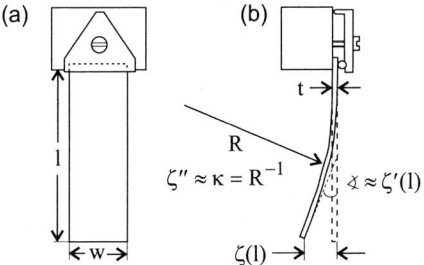

Figure 1. Principle of the stress measurement. (a) A thin substrate of length l and width w is clamped at its top end, the bottom end is free. (b) The side view shows how a stress imbalance between front and back surface leads to a deflection ζ, a change of slope ζ', and a change of curvature $\kappa = R^{-1} \approx \zeta''$ of the substrate with thickness t.

However, if gas phase epitaxy is involved, or when the effect of gaseous adsorbates on the surface stress is to be investigated, special care has to be devoted to a defined exposure of only one surface. This can be achieved by using a gas doser, that creates a substantially larger partial pressure of the adsorbate above the front surface as compared to the back surface. This works fine as long as the adsorbate uptake at the front surface does not slow down, as it is often the case with adsorption studies due to the coverage dependence of the adsorbate sticking coefficient. The much smaller partial pressure at the back side will then lead finally to an appreciable adsorbate coverage also on the back-side, reversing the initial curvature back to zero. This behavior can be exploited to verify the proper preparation of both surfaces. For Si and W substrates, the crystal is prepared under ultra-high vacuum (UHV) conditions by a quick heating to 1200 K or 2600 K, respectively. This procedure leads to equivalent surface qualities on both sides, and adsorption on the back side will reduce the effect of adsorption on the front side, as verified experimentally [18,16]. When ion sputtering is involved in the sample preparation, the back side usually does not get cleaned at all and can be assumed to remain inert for most adsorbates. For bending measurements in a liquid, a a protective varnish can be permanently applied to one surface. This has been done to perform bending measurements in an electrochemical cell [19].

The mechanics of crystal bending due to film stress has been reviewed in great detail [19, 20]. We are interested in the effect of ultrathin films (nm) on reasonably thick substrates (100 μm), and the ratio between film and substrate thickness is of the order of 10^{-5}. This negligible ratio allows numerous simplification in the treatment of the elasticity problem, as described in [21].

We proceed with a simplified treatment of the relation between bending moment and substrate curvature, which is illustrated by Fig. 2 [21]. In a first step, we neglect the important effect of elastic anisotropy and consider the effect of biaxial bending moments

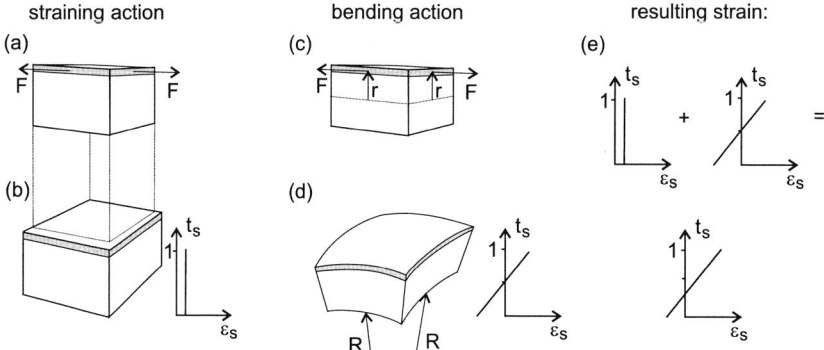

Figure 2. Superposition of the straining and the bending action of forces acting in the surface region. (a) The forces F induce a constant strain in the substrate, shown in (b). (c) The forces F induce a bending moment, as they are acting in a distance r of the neutral plane, a biaxial curvature with radius if curvature R results, see (d). The bending induces a strain that varies within the sample thickness. (e) The superposition of the straining and the bending action reveals that the neutral plane is shifted away from the middle plane of the substrate.

M_x and M_y, acting along the substrate width w and substrate length l, respectively [17]:

$$\frac{M_x}{w} = \frac{Y_S t_S^3}{12(1-\nu_S^2)}\left(\frac{1}{R_x} + \nu_S \frac{1}{R_y}\right) \qquad \frac{M_y}{l} = \frac{Y_S t_S^3}{12(1-\nu_S^2)}\left(\frac{1}{R_y} + \nu_S \frac{1}{R_x}\right). \tag{1}$$

The subscript S denotes substrate properties, the substrate length is oriented along the x-axis, the width along the y-axis. The bending moment M_x induces a curvature along the length, M_y induces a curvature along the width. The bending moments M are due to a film stress τ. Performing the separation of bending and straining action of the film stresses, the bending moments are simply given by $M_x = \tau_x w t_F t_S/2$ and $M_y = \tau_y l t_F t_S/2$, and the expressions for the biaxial film stress as a function of the radii of curvature follow as:

$$\tau_x t_F = \frac{Y_S t_S^2}{6(1-\nu_S^2)}\left(\frac{1}{R_x} + \nu_S \frac{1}{R_y}\right) \qquad \tau_y t_F = \frac{Y_S t_S^2}{6(1-\nu_S^2)}\left(\frac{1}{R_y} + \nu_S \frac{1}{R_x}\right) \tag{2}$$

Equation 2 gives the in-plane stress components τ_x and τ_y as a function of the experimentally determined radii of curvatures R_x and R_y measured along *two* directions on a substrate with isotropic elastic properties. Tungsten happens to be elastically isotropic and eq. 2 has been used to derive the biaxial stress in Fe monolayers on a W(110) substrate, as will be discussed in Section 3.

For isotropic stress on an elastically isotropic substrate the so-called Stoney equation $\tau = Y_S t_S^2/(6(1-\nu_S)Rt_F)$ follows, which states that the curvature $1/R$ is proportional to the product of film stress and film thickness, $\tau t_F \sim 1/R$.

The substrate curvature during film growth changes proportionally to τt_F. A constant stress τ induces a constant slope of the plot curvature vs thickness. The stress τ is then given by the slope of the curve.

At a given point of the curvature measurement, the curvature is proportional to the force per unit width acting on the sample. This is exactly what one defines as surface stress. Changes of surface stress due to adsorption can be directly calculated from the curvature signal.

Note, that this curvature analysis relies on a knowledge of the elastic constants of the substrate, and *not* of those of the film material. This is an important aspect, as elastic properties of monolayers cannot be assumed to correspond to the respective bulk properties a priori. However, for situation where film strain and epitaxial orientation are well defined, the validity of stress-strain relations that are based on bulk properties can be checked. The stress induced by Ni and Co monolayers on Cu(100) indicates, that already a few monolayer thin films exhibit bulk-like elastic properties [15].

The bending stiffness of the substrate is determined by its geometry and surface orientation, as will be shown in the following section 2.1. Therefore, the calculation of stress from curvature can be performed with an high accuracy in the percent range. The largest source of error are a possible inhomogeneity of sample thickness t_f and experimental errors in the determination of the film thickness t_f, which is usually known from quartz oscillator measurements for thicker films, and/or from scanning tunneling microscopy (STM) for monolayers.

Two techniques have been proven useful to check experimentally the Young modulus Y of the substrate against the calculated value for consistency [15]. The loading of a clamped substrate with known weights leads to a sample deflection, from which the quantity $Y_S t_S^3$ can be extracted [22]. Secondly, the clamped sample performs flexural vibrations that can be excited externally to determine the frequencies f_i of the first vibrational modes. The frequencies are given by $f_i = \frac{t_s \beta_i}{4\pi l^2}\sqrt{\frac{Y_S}{3\varrho}}$, with $\beta_i = 3.516, 22.035, 61.698$ for the first three modes. The frequency of the first mode is for a typical crystal (W-crystal, l:14 mm, w:3 mm, t_S: 0.1 mm) of the order of a few hundred Hz, leading to a few kHz for the third mode. These experimental test of the substrate rigidity reveal consistently a experimentally determined rigidity that is a few percent smaller than the calculated value. This slight deviation is ascribed to the influence of the not rigorously defined clamping of the sample.

2.1. Sample clamping and elastic anisotropy

In most experimental situations the sample is clamped to a manipulator. Then, the idealized conception of a thin substrate sheet, that is free to bend along the sample width and along the sample length is not justified near the sample clamping. Dahmen, Lehwald and Ibach performed finite element method (FEM) calculations to elucidate this issue [23]. The result of their work reveals that the larger the length-to-width ratio of the sample, the less important is the influence of the clamping on the bending.

To make full use of their calculations, one has to realize that there are three experimental approaches to calculate the curvature $\kappa = R^{-1}$ from a measurement. For simplicity, we concentrate on the free end of the sample, at a distance l away from the clamping. We assume a constant curvature along the sample length, then the approximate mathematical expression for curvature as the second derivative of a function can be integrated to derive the deflection $\zeta(l) = l^2/(2R)$, the slope $\zeta'(l) = l/R$ and the curvature $\zeta''(l) = 1/R$ at the sample end. All three measurements, deflection, slope, and curvature could be used to calculate R.

These experimental methods are affected to different degrees by the effect of clamping. The largest effect is expected for the deflection measurement, which is realized in a capacitive technique described in section 2.2. Here, the deflection builds up over the total length l, and a hindered curvature near the clamping is expected to have the strongest impact on the calculated curvature κ_ζ. In an optical deflection experiment, a laser beam is reflected from the substrate to a position sensitive detector, as described in the following section, and the curvature is calculated from the slope κ'_ζ. The best approach is to measure the change of slope, i.e. the curvature directly, κ''_ζ. This can be realized with a two beam optical technique described below. The curvature information is extracted from a localized region, far away from the clamping, and the effect of clamping is expected to be the smallest. This qualitative reasoning is supported by the FEM calculation presented in Fig. 3.

The results of Fig. 3 are given in terms of the dimensionality D for a system under isotropic biaxial stress. A dimensionality of $D = 2$ indicates perfectly free two dimensional bending, $D = 1$ indicates mere one-dimensional bending with no curvature along the sample width. These extremes can be realized for a given sample width by very long or by very short samples, respectively. Even pretty large length-to-width ratios of five lead to a significant deviation of the dimensionality of 0.1 from its ideal value $D = 2$ for a deflection measurement, whereas already a length-to-width ratio of two produces no appreciable deviation for curvature measurements, see Fig. 3(c).

The dimensionality D as extracted from Fig. 3 can be plugged in the following generalized Stoney equation to calculate the stress from one of the above mentioned curvature measurements [23]

$$\tau t_F = \frac{Y t^2}{6(1-\nu)(1+(2-D))\nu\kappa}. \tag{3}$$

In addition to the impact of clamping, the elastic anisotropy needs to be considered for a meaningful data analysis.

Elastic anisotropy is described by the tensor transformation of the elastic compliances s_{ijkl} [24,25]. Examples for various transformations are presented in [21,23]. The most

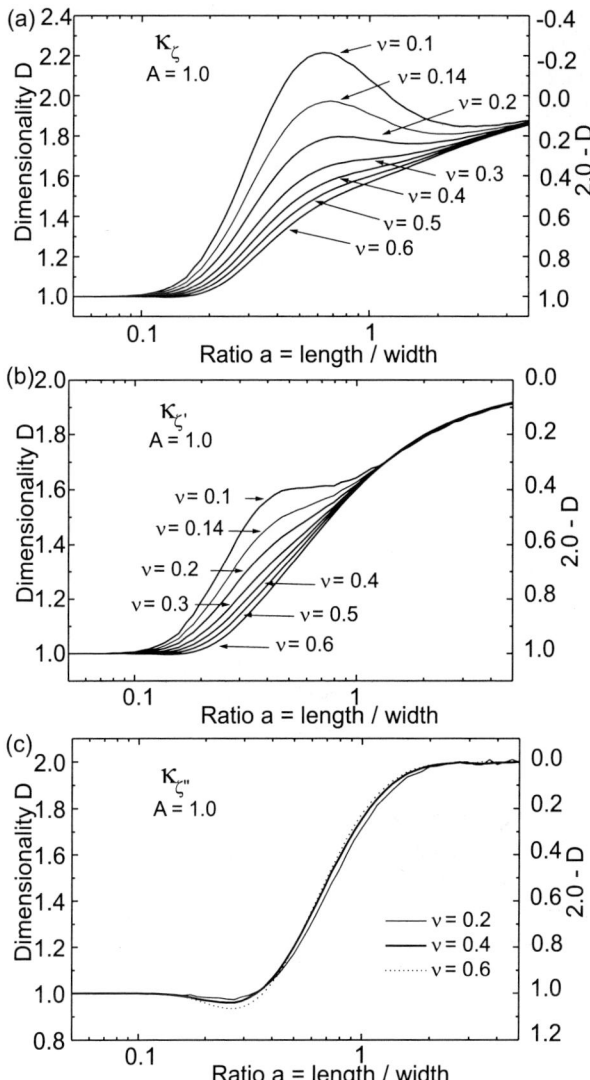

Figure 3. Dependence of the dimensionality D of the bending on the length-to-width ratio a of an elastically isotropic sample, $A = 1$, and different Poisson ratios ν [23]. (a) The curvature κ_ζ calculated from the sample deflection influenced most by the clamping as compared to the curvature κ'_ζ determined from the sample slope. (c) A direct measurement of the curvature κ''_ζ is influenced the least by the clamping as indicated by $D \approx 2$ already for a moderate value of $a = 2$.

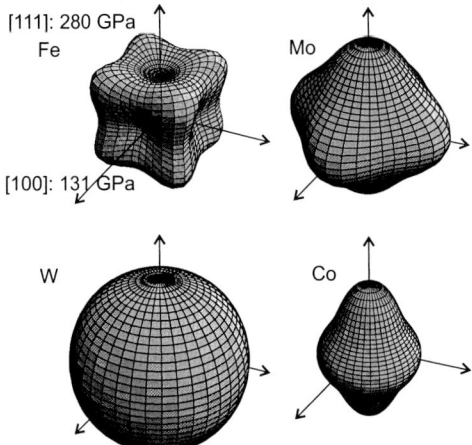

Figure 4. Anisotropy of the Young modulus Y for different elements. These polar plots give the magnitude of Y as the length of the radius vector. Note the pronounced anisotropy for Fe, which is soft along the cubic axes [100], 131 GPa, and more than twice as hard along the body diagonal [111] with 280 GPa. Mo shows the opposite dependence of Y on these directions, whereas W is elastically almost isotropic. Hexagonal Co is stiff along the c-axis, and the value of Y depends only on the angle with respect to this axis.

important result is the directional dependence of the Young modulus $Y' = 1/s'_{11}$ and of the Poisson ratio $\nu' = -s'_{12}/s'_{11}$. Primes indicate values along a direction that might differ from the cubic axes. These directions are given by the directional cosines l_i and m_i. One direction suffices to characterize the directional dependence of Y. The Poisson ratio, describing the contribution of a stress in an orthogonal direction to a stress along another direction, requires two direction cosines to describe its anisotropy. The relevant transformations are:

$$\text{cubic}: \quad 1/Y' = s'_{11} = s_{11} - 2(s_{11} - s_{12} - \frac{1}{2}s_{44})(l_1^2 l_2^2 + l_1^2 l_3^2 + l_2^2 l_3^2) \tag{4}$$

$$\text{cubic}: \quad \nu' = -\frac{s'_{12}}{s'_{11}} = -\frac{s_{12} + (s_{11} - s_{12} - \frac{1}{2}s_{44})(l_1^2 m_1^2 + l_2^2 m_2^2 + l_3^2 m_3^2)}{s_{11} - 2(s_{11} - s_{12} - \frac{1}{2}s_{44})(l_1^2 l_2^2 + l_1^2 l_3^2 + l_2^2 l_3^2)} \tag{5}$$

The important result of eq. 4 is that the amount of elastic anisotropy for cubic elements depends on the directional factor $(l_1^2 l_2^2 + l_1^2 l_3^2 + l_2^2 l_3^2)$ and on the magnitude of the anisotropy term $(s_{11} - s_{12} - \frac{1}{2}s_{44})$. The larger the anisotropy term is, the larger is the correction to s_{11} and a pronounced elastic anisotropy results. Figure 4 illustrates the different degrees of

Table 1
Elastic stiffness constants c_{ij} (GPa) and elastic compliance constants s_{ij} (TPa)$^{-1}$ from [26, 27]. Young's modulus Y(GPa) and Poisson's ratio ν for cubic and hexagonal elements. Y and ν are calculated from equations (4) and (5) for cubic elements for directions parallel to the crystal axes and for hcp Co for directions within the basal plane.

Element	c_{11}	c_{12}	c_{44}	c_{13}	c_{33}	s_{11}	s_{12}	s_{44}	s_{13}	s_{33}	Y	ν
hcp Co	307	165	75.5	103	358	4.73	-2.31	13.2	-0.69	3.19	211	0.49
fcc Co	242	160	128			8.81	-3.51	7.83			114	0.40
bcc Fe	229	134	115			7.64	-2.81	8.71			131	0.37
fcc Ni	249	152	118			7.53	-2.86	8.49			133	0.38
Si	165	63	79.1			7.73	-2.15	12.7			129	0.28
MgO	293	92	155			4.01	-0.96	6.48			249	0.24
bcc Mo	465	163	109			2.63	-0.68	9.20			380	0.26
bcc W	517	203	157			2.49	-0.7	6.35			402	0.28

anisotropy for various elements with a three dimensional representation of the directional dependence of Y.

Polar plots of the elastic properties of Fe in different planes are presented in Fig. 5. The pronounced anisotropy is reflected in the non-circular shape of the plots for the (100) and (110) planes. However, the biaxial modulus $Y/(1-\nu)$, that enters the stress calculation from a $D=2$ curvature measurement, is isotropic within the (100) plane. A reduced dimensionality $D=1$ leads to $Y/(1-\nu^2)$, and this expression is *not* isotropic. It is fourfold symmetric with a large value of 218 GPa along 45° $\equiv$ [110] and a smaller value of 151 GPa along [001].

Table 1 summarizes the elastic constants for selected elements [21].

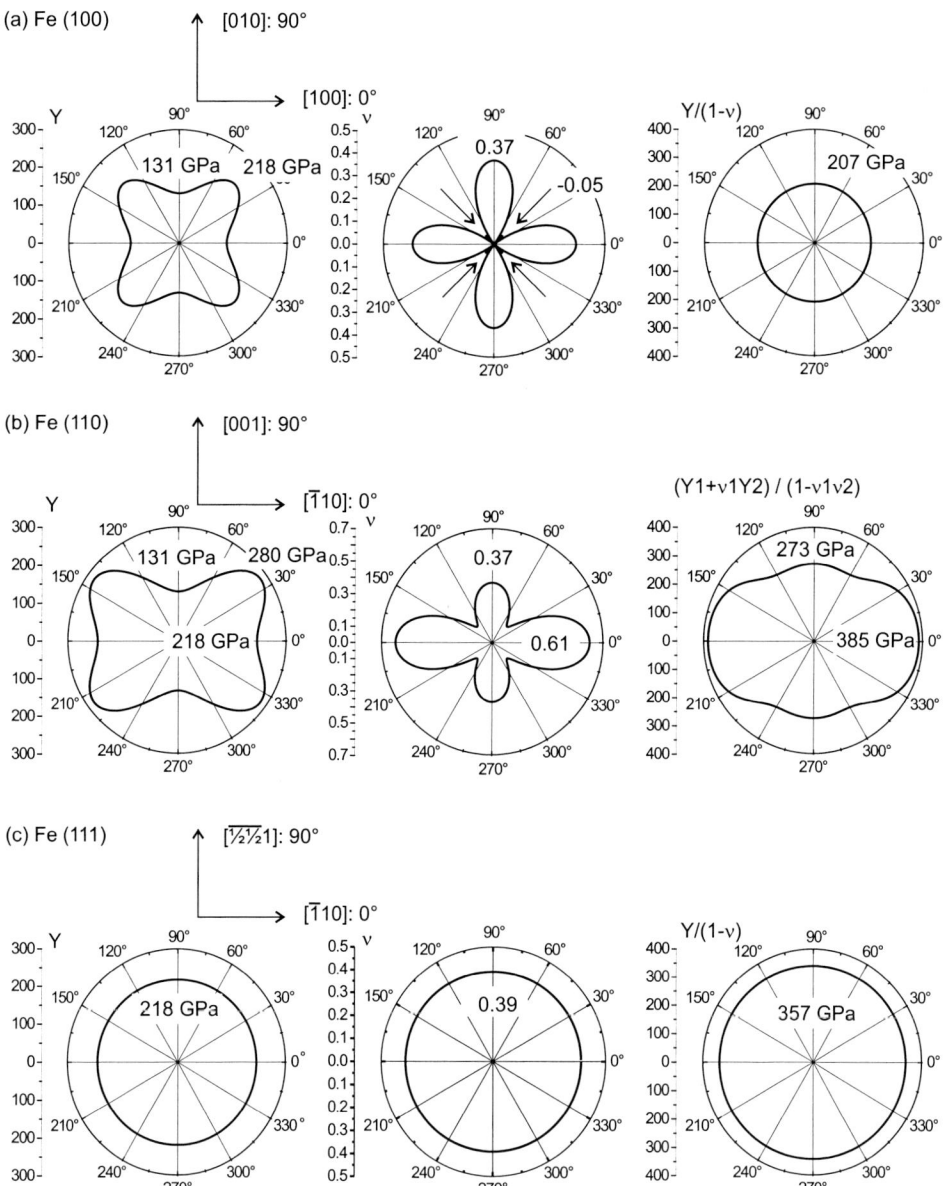

Figure 5. Angular dependence of the Young modulus Y and the Poisson ratio ν in the (a) (100), (b) (110) and (c) (111) plane of Fe. Note, that both Y and ν are anisotropic in the (100) plane, but the biaxial stiffness $Y/(1-\nu)$ is isotropic.

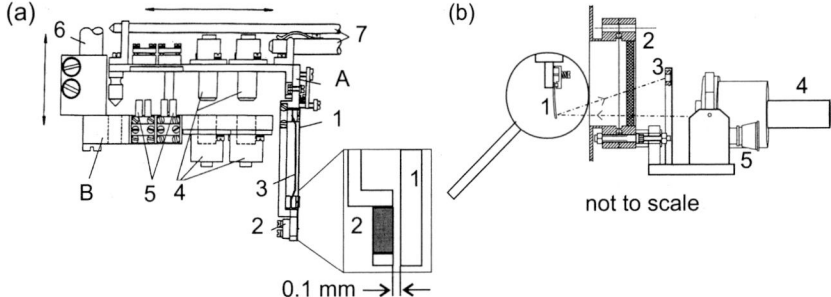

Figure 6. (a) Capacitive deflection measurement with transferable sample holder [18]. The sample 1 forms a capacitor with the plate 2 of the sample holder A, see inset. A can be connected and removed from the UHV baseplate B by relative movements indicated by the arrows. 3, sample heating; 4, floating electrical connections for capacitor; 5, filament and thermocouple contact; 6, manipulator rod; 7, transfer rods. (b) Optical deflection technique [34]. 1, sample inside of UHV chamber; 2, UHV window; 3, position sensitive detector; 4, laser; 5, gimbal mount.

2.2. Capacitive and optical deflection techniques

There are many approaches to measure stress-induced substrate bending. The foregoing discussion implies that imaging the substrate area to extract information about the curvature might be a worthwhile experiment. The image of a curved substrate could directly confirm the influence of clamping and of elastic anisotropy. Indeed, imaging techniques have been applied to measure magnetostrictive stress via a interference technique, which measured the biaxial curvature of the substrate [28]. Similar approaches have been described [29]. An alternative to the imaging is the use of an optical deflection technique with multiple beams, that are distributed over the sample length or scanned over the sample length [30–33]. The reader is referred to these references for further information.

The results of Fig. 3 suggest, that the effect of clamping is *not* a detrimental issue as long as one chooses the proper sample geometry and adopts the curvature detection scheme. The following examples for capacitive and optical deflection techniques represent an excellent compromise between simplicity and sensitivity.

When it comes to the detection of tiny displacements of cm sized samples, the application of capacitive techniques. A change of distance between a substrate and a reference electrode is detected via the resulting change of capacity, and outstanding sensitivity has been demonstrated . Examples include the measurement of magnetostriction of paramagnetic samples, where an extension as small as 0.01 Å has been detected [35]. The sensitivity for detecting changes of the gap distance can be increased by using the smallest possible gap distances leading to an high initial capacity.

Consequently, capacitive techniques have been employed by several groups to measure stress with high sensitivity [5,18,36–38].

A schematic of an UHV set-up with transferable sample holder is depicted in Fig. 6(a).

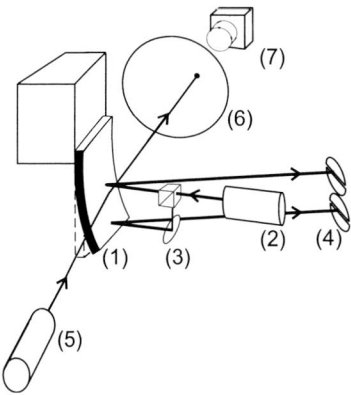

Figure 7. Two beam curvature measurement with simultaneous MEED measurement. (1): sample, (2): laser, (3): beamsplitter and mirror, (4): two position sensitive detectors, (5): electron gun, (6): LEED screen, (7): CCD camera.

This set-up has been used to measure changes of surface stress on Si and metal single crystal substrates due to gas adsorption and film growth [19]. The main drawback of this layout is the use of rather large samples.

The optical deflection technique shown in Fig. 6(b) can be realized with a free sample length of the order of 10 mm at a sample width of 2 mm. This reduces the cost and preparation efforts of the single crystal substrate and leads to a very compact design. This single beam technique measures the slope-related curvature at the sample end. It is superior to the capacitive technique, which measures the deflection-related curvature, if one is interested in approaching the $D = 2$ situation for reasons given in section 2.1 above.

The use of a two beam deflection technique was pioneered by [39]. It offers the additional benefit of measuring the local curvature near the sample end. The curvature can be expressed as the difference of slope between two points, divided by the distance between the points. The position signals of two reflected beams, that both hit the sample near the bottom end and are separated by a few mm on the sample, give the respective slope information. Calculating the difference of the slope, divided by the beam separation, gives the curvature. The original single beam optical deflection set-up of Fig. 6(b) was modified by the introduction of a beam splitter and a second position sensitive detector, as shown in Fig. 7. With this set-up, *direct* curvature measurements are performed, that represent the ideal situation of Fig. 3(c) with D=2 to an excellent degree.

Examples of stress measurements with the optical deflection technique are presented in the following two sections. Stress in Fe and Ni monolayers is discussed and the application of the technique to measure magnetostrictive stress is briefly explained.

3. Stress in Fe and Ni monolayers

The growth of epitaxial films on single crystal substrates has been reviewed before, e.g. see [40]. Here we concentrate on the application of stress measurements to monolayer growth. It is demonstrated that structural transitions in epitaxial layers can be detected by characteristic changes of film stress. It is shown that an adsorbate-induced change of surface stress can be the dominant factor that determines stress in the sub-monolayer regime.

The growth of Fe on W(110) is a well studied epitaxial system [41–45] and a rich variety of magnetic properties of the first two Fe layers has been described [46–50]. The system is characterized by a large lattice mismatch between Fe and W of 10.4 %, and all structural studies indicate that the first Fe layer grows pseudomorphically strained on the W surface. Continuum elasticity predicts a tremendous in-plane stress of 39 GPa along the [$\bar{1}$10] direction, and of 28 GPa along [001] [21]. Although the strain is isotropic, the resulting stress reflects the elastic anisotropy of the Fe film. The measurements presented in Fig. 8 reveal indeed an anisotropic stress, but the magnitude is roughly 50 % larger as expected from continuum elasticity, see Fig. 8(c). These stress values are almost an order of magnitude larger than the elasticity limits of high strength materials. The formation of misfit distortions already in the second layer of Fe is ascribed to this large film stress. The curvature measurements of Fig. 8 indicate the onset of the misfit formation by the kink in the stress curve at 1.2 ML. Additional Fe atoms are incorporated in non-pseudomorphic positions, and the horizontal section of the stress curve indicates that no additional stress is build up in the second and third layer.

For thicker Fe films, the reduced slope of the stress curve corresponds to a smaller stress of order 13 GPa. Diffraction experiments and STM studies reveal the formation of a regular misfit distortion network. Additional diffraction spots due to the distortion network are seen in low energy electron diffraction (LEED) experiments of Fig. 9(a). Regular faint distortion lines on top of Fe islands of the third and fourth layer are also imaged by STM in Fig. 9(b). The driving force for the formation of the misfit distortions is the reduction of elastic energy of the system. However, at room temperature kinetic limitations cause an only partial relaxation of the strain energy. At higher temperature, an almost stress-free growth of Fe is observed [21]. The strain relaxation is then ascribed to two mechanisms, (1) the formation of misfit distortion at the interface between first and second layer Fe, and (2) a relaxation of residual strain in Fe islands, which form at elevated temperature [21].

In addition to structural changes that are driven by strain relaxation, the influence of strain on the magnetic properties of Fe monolayers has been investigated [51,52]. It is found that even small residual strain in the sub-percent range changes the effective magnetoelastic coupling of Fe in sign and magnitude as compared to the respective bulk value. This aspect will be briefly discussed in section 4.

We conclude that for Fe coverage above 0.5 ML film strain is the dominant source of the measured stress. For a smaller coverage however, an anticlastic curvature of the W crystal is measured. The opposite slope of the curvature measurements in Fig. 8(a) indicate a saddle-like deformation of the W crystal, which is depicted in (b). The interesting result is, that W bends along [001] as if the Fe atoms are under a *compressive* stress, which is

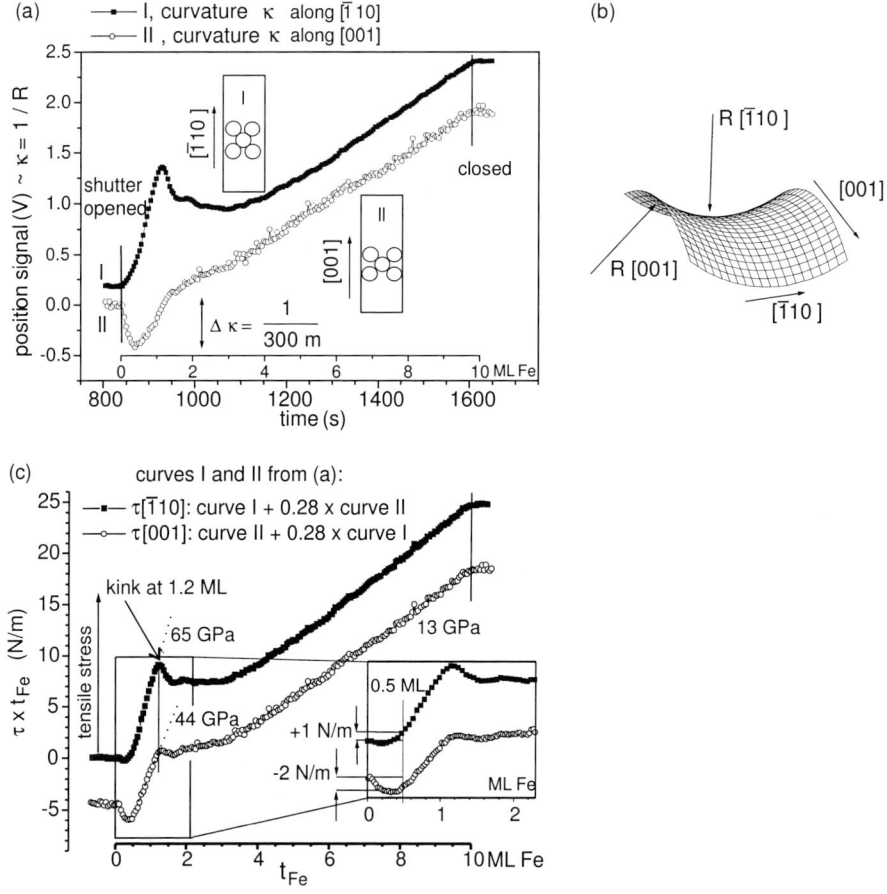

Figure 8. Biaxial curvature and stress measurement on Fe monolayers on W(110) [16]. (a) Curvature measurement along the [$\bar{1}$10] direction, curve I, and along the [001] direction, curve II. Note the opposite sign of curvature in the sub-monolayer range, leading to the anticlastic substrate curvature shown in (b). (c) The curvature data from (a) are combined with the Poisson ratio 0.287 of W to derive at the biaxial stress curves. The kink at 1.2 ML indicates the formation of misfit distortions in the second layer.

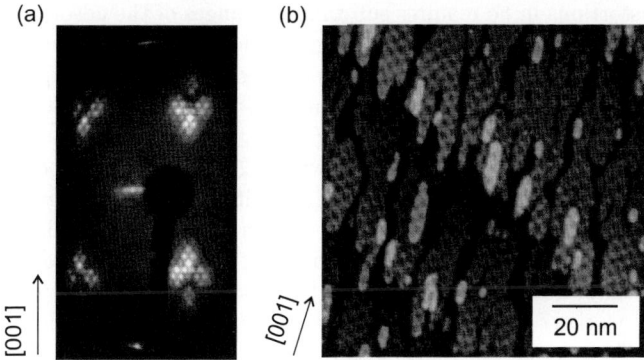

Figure 9. Misfit distortion network of 3 monolayer Fe on W(110) [21]. (a) LEED image of additional diffraction spots surrounding the (110) spots in diamond shaped areas. (b) STM image of islands of the third layer in grey with some islands of the fourth layer in lighter grey. A regular distortion line network is shown as weak lighter grey contrast.

opposite in sign compared to the *tensile* misfit. This unexpected result is ascribed to the dominant contribution of the change of surface stress of W due to the Fe adsorption [16]. Our results indicate, that this adsorbate-induced surface stress change is also anisotropic, and its value can be extracted from the inset of Fig. 8(c).

This result is an example for the often unexpected stress behavior for sub-monolayer coverage. Simple models, based on strain, surface free energy or electron transfer due to differences of electronegativity are not capable of explaining the anisotropy of adsorbate-induced stress [16,19]. Further experiments are necessary to elucidate the relevant processes.

The formation of misfit distortions in Fe requires substantial changes of the adsorption sites of the Fe atoms [54]. However, also very subtle changes of the atomic arrangements in the first layer can be detected with the stress measurement technique. An example is the the growth of Ni on W(110), which occurs in the Nishiyama-Wassermann mode [55] and results in fcc Ni layers already in the second layer [53].

The results of curvature measurements during the growth of Ni on W in connection with structural models are presented in Fig. 10. For the first half monolayer, a negative surface stress is found, before the positive slope indicates tensile in stress in the first layer. In situ LEED measurements are performed, and at each point of the stress curve the film structure has been identified as indicated by the labels in Fig. 10(a). We concentrate on the transition from the 8×1 to the 7×1 film structure, which occurs at the completion of the first layer. The hard sphere models of Fig.10 suggest a change a strain along [001] from +13 % to -1.3 %. The kink of the stress measurement reveal a compressive stress component at this transition. This indicates a repulsive interaction between Ni along [001] for the 7×1 structure, in qualitative agreement with the negative strain predicted in the hard sphere model. Thus, the stress change can be ascribed to a change of film strain in the first layer. The driving force for this transition is the higher packing density of the Ni atoms which result in a denser coverage of the W surface [56,57].

A similar anisotropy of the Ni-induced change of surface stress as compared to the adsorption of Fe on W(110) is observed in the coverage range up to 0.5. A two dimensional stress analysis reveals compressive stress along [001], in contrast to the large tensile strain of Ni, see Fig. 10(a). Along [$\bar{1}$10] a positive stress is measured [54].

This result supports the claim mentioned above, that adsorbate-induced changes of surface stress effects can be dominant for the stress behavior at surfaces.

The following section gives an example of the application of the highly sensitive curvature measurement technique to measure stress due to the magnetization of monolayers.

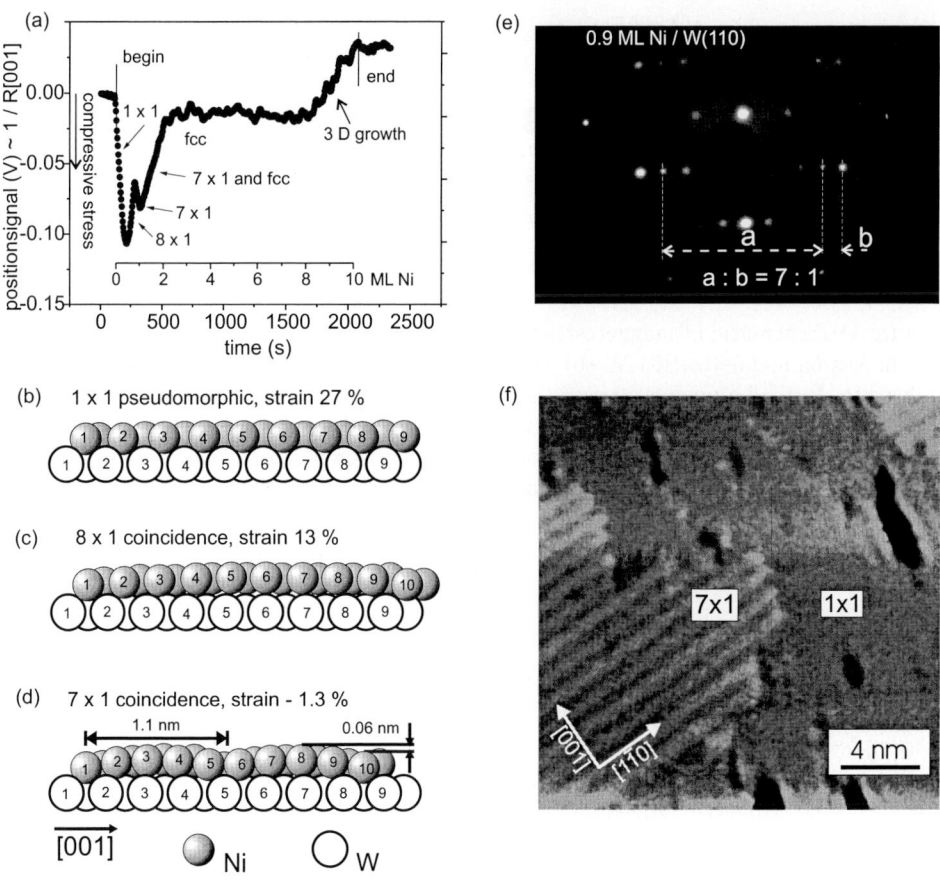

Figure 10. Stress and structure of the first Ni layer on W(110) [53]. (a) Curvature measurement along W[001]. The correlation between curvature and different monolayer structure as determined by LEED is indicated. Structural models using hard spheres for the pseudomorphic 1x1 state (b), the 8x1 coincidence state (c), and the 7x1 coincidence state (d). LEED image of the 7x1 state with characteristic satellite spots along [001]. (f) STM image of a partial 7x1 and 1x1 structure in the first Ni layer shown in grey. The 7x1 structure is identified by a regular spacing along [001] of dark-bright contrast lines, as expected from the vertical corrugation shown in (d).

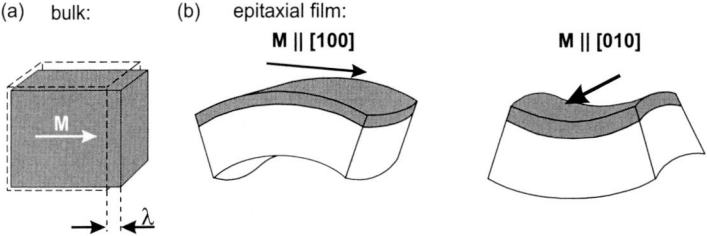

Figure 11. Schematic of magnetostrictive stress measurements. (a) Bulk samples show a strain λ upon magnetization M. (b) Films cannot expand freely as they are bonded to the substrate. Instead a magnetostrictive stress is created upon magnetization. The resulting curvature depends on the stiffness of the substrate and on the magnetization direction. Reorientation of the film magnetization from a direction along the sample length to the sample width induces a biaxial curvature change which is measured to extract the effective magnetoelastic coupling coefficients.

4. Magnetoelastic stress measurements

An exciting aspect of the study of the physical properties of monolayers is that strain states are accessible in epitaxial films, that cannot be achieved in bulk samples. Even with high strength materials, strain in excess of one percent is beyond the elasticity limits, and defects are formed. In epitaxial monolayers however, strains of several percent are measured and strongly modified physical properties are to be expected. The effect of strain on the modified magnetostrictive behavior of monolayers is discussed as one example.

Magnetostriction is the relative change of length of a sample upon magnetization [58]. Bulk Fe is found to expand upon magnetization, as sketched in Fig. 11(a), leading to a magnetostrictive strain λ of 2×10^{-5} for magnetization and strain measurement along the cubic axis. The underlying physical principle is the magnetoelastic coupling which leads to the formation of magnetoelastic or magnetostrictive stress upon magnetization. This interaction can be described by a contribution to the elastic energy of the system that is proportional to strain, in contrast to the elastic energy density, that is proportional to the second power of strain. Magnitude and sign of the magnetoelastic coupling constant B determine whether a bulk sample expands or contracts upon magnetization. A free expansion of a ferromagnetic film upon magnetization is not possible, as the film is rigidly bonded to the substrate. Instead, magnetostrictive stresses act in the film upon magnetization, and the resulting substrate curvature, as indicated in Fig. 11, is measured to determine the effective magnetoelastic coupling directly [21].

The easy direction of magnetization of a ferromagnetic films is determined by the magnetic anisotropy. A detailed knowledge of which is of key importance for the design of any magnetic or magneto-electronic device. An important contribution to the magnetic anisotropy is the magnetoelastic energy due to film strain. In simple terms, for a film with a positive magnetoelastic coupling constant B, the easy axis of magnetization is aligned

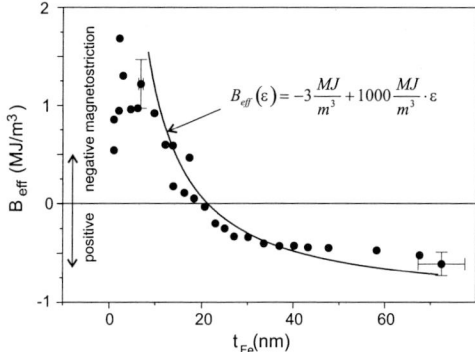

Figure 12. Magnetoelastic coupling as a function of Fe thickness on W(001) [21]. Note the zero crossing at 20 nm. The solid curve models the influence of film strain ϵ, which is derived from the stress measurements during growth, upon the effective magnetoelastic coupling B_{eff}. This model corresponds to a strain correction of the bulk constant B_1.

along the film direction with a negative strain. For a system with an in-plane tensile strain, the perpendicular strain is negative. This favors a film magnetization pointing out of the film plane, as in the case of Ni monolayers on Cu(001) [15]. For general cases, the epitaxial orientation and the direction of magnetization have to be considered and other anisotropy contributions have to be included [52].

Our measurement of the magnetoelastic coupling in Fe monolayers shown in Fig. 12 reveals that the effective magnetoelastic coupling constant changes with Fe film thickness. This is in contrast to the expectation of a bulk like behavior, which leads to a constant magnetoelastic coupling of -3.4 MJ/m^3. Even for thicker films of 70 nm, the magnetoelastic coupling deviates substantially from this bulk value.

The magnetoelastic coupling has a zero crossing at 20 nm and it changes its sign for thinner films. A detailed analysis which takes the thickness dependence of film stress into account, reveals that film strain and not film thickness is the dominant factor that leads to a non-bulk-like magnetoelastic behavior in strained epitaxial films [21,59,15,60,61].

The solid line in Fig. 12 represents a strain dependent correction of the magnetoelastic coupling and it gives a reasonable description of the experimental data for a thickness above 10 nm. A far too low magnetoelastic coupling is measured in thinner films, as compared to the strain model. A detailed understanding of this reduced magnetostrictive behavior in a few monolayer thin films is still lacking.

Recent first principles calculations provide a theoretical justification of strain dependent contributions to the magnetoelastic coupling [62].

To derive at this important result of a strain-dependent correction of the effective magnetoelastic coupling it was essential to have experimental data on both film stress and magnetoelastic stress from in situ curvature measurements.

5. Conclusion and Outlook

Curvature measurements are a well suited technique to measure stress in atomic layers and at surfaces. Radii of curvature of the order of 1000 m are easily detected, which translated to sub-monolayer sensitivity for adsorbate-induced stress even on relative thick (several 100 μm) and stiff substrates (e.g. W, Y:400 GPa). This numbers do *not* mark the limits of the technique. In cases where the stress induced effect can be repeated during the measurement time, averaging techniques and phase sensitive signal detection can be employed to detect radii of curvature as large as 100 km. This sensitivity has been proven when the tiny effects of magnetostrictive stress in ferromagnetic monolayers were studied.

What can be learnt from stress measurements in atomic layers and at surfaces? The measurements indicate that the classical stress-strain relations that are used in continuum elasticity to derive stress from strain and vice versa are not necessarily relevant in the thickness limit of single atomic layers. Here, the effect of surface stress changes due to the adsorbate can be dominant. Stress opposite in sign as expected from misfit arguments has been measured. Thus, stress measurements in this thickness range contribute to a better understanding of the still largely unknown processes that determine forces in the surface region of a solid.

Stress measurements are also very sensitive to subtle atomic rearrangements in monolayers. In contrast to diffraction techniques that require extended areas of coherent structure to obtain a characteristic change of the diffraction pattern, stress measurements can identify the *onset* of a structural change with sub-monolayer accuracy.

The independent experimental determination of both stress and strain promises new insight into the phenomena that govern forces and structural changes at surfaces and in atomic layers. To this end, the in-situ combination of curvature and surface X-ray diffraction experiments has been realized [45] to investigate stress-strain relations on an atomic scale, beyond the limitations of continuum elasticity.

Acknowledgment

This contribution was written during a stay as Professeur invité at the C.R.M.C.2 at Luminy. I benefited form many discussions with the colleagues there and I appreciate their hospitality and assistance. I thank two former PhD students, Axel Enders and Thomas Gutjahr-Löser for the fruitful cooperation on which many of the presented results are based. I acknowledge gratefully numerous discussions with Klaus Dahmen and Harald Ibach from the IGV, Jülich on the nature of curvature measurements. It is may pleasure to thank Jürgen Kirschner for his continuous vivid interest and generous support of this work.

REFERENCES

1. R. Hoffman, Phys. Thin Films 3 (1966) 211.
2. K. Kinosita, Thin Solid Films 12 (1972) 17.
3. M. F. Doerner, W. Nix, CRC Crit. Rev. Solid State and Materials Science 14 (1988) 225.
4. W. Nix, Metall. Trans. A 20A (1989) 2217.

5. R. Abermann, Vacuum 41 (1990) 1279.
6. R. Koch, J. Phys.: Condens. Matter 6 (1994) 9519.
7. J. Daughton, A. Pohm, R. Fayfield, C. Smith, J. Phys. D: Appl. Phys. 32 (1999) R169.
8. P. Feibelman, Phys. Rev. B 56 (1997) 2175.
9. I. Batirev, W. Hergert, P. Rennert, V. Stepanyuk, T. Oguchi, A. Katsnelson, J. Leiro, K. Lee, Surf. Sci. 417 (1998) 151.
10. X. Qian, F. Wagner, M. Petersen, W. Hübner, J. Magn. Magn. Mater. 213 (2000) 12.
11. V. Stepanyuk, D. Bazhanov, W. Hergert, Phys. Rev. B 62 (2000) 4257.
12. V. Stepanyuk, D. Bazhanov, A. Baranov, W. Hergert, P. Dederichs, J. Kirschner, Phys. Rev. B 62 (2000) 15398.
13. P. Marcus, X. Qian, W. Hübner, J. Phys.: Condens. Matter 12 (2000) 5541.
14. T. Gutjahr-Löser, Ph.D. thesis, Martin-Luther Universität Halle-Wittenberg, Mathematisch-Naturwissenschaftlich-Technische Fakultät (1999).
15. T. Gutjahr-Löser, D. Sander, J. Kirschner, J. Appl. Phys. 87 (2000) 5920.
16. D. Sander, A. Enders, J. Kirschner, Europhys. Lett. 45 (1999) 208.
17. S. Timoshenko, S. Woinowsky-Krieger, Theory of Plates and Shells, McGraw-Hill, Singapore, 1959.
18. D. Sander, H. Ibach, Phys. Rev. B 43 (1991) 4263.
19. H. Ibach, Surf. Sci. Rep. 29 (1997) 193.
20. P. Marcus, Surf. Sci. 366 (1996) 219.
21. D. Sander, Rep. Prog. Phys. 62 (1999) 809.
22. W. Young, Roark's Formulas for Stress and Strain, McGraw Hill, Singapore, 1989.
23. K. Dahmen, S. Lehwald, H. Ibach, Surf. Sci. 446 (2000) 161.
24. J. F. Nye, Physical Properties of Crystals, Oxford University Press, Oxford, 1985.
25. W. A. Brantley, J. Appl. Phys. 44 (1973) 534.
26. R. F. S. Hearmon, The elastic constants of non-piezoelectric crystals, Vol. 2 of Landolt-Börnstein Numerical Data and Functional Relationships in Science and Technology Group III, Springer-Verlag, Berlin, 1969.
27. R. F. S. Hearmon, The elastic constants of crystals and other anisotropic materials, Vol. 18 of Landolt-Börnstein Numerical Data and Functional Relationships in Science and Technology Group III, Springer-Verlag, Berlin, 1984.
28. R. Gontarz, H. Ratajczak, P. Šuda, phys. stat. sol. 6 (1964) 909.
29. G. Degand, P. Müller, R. Kern, Surf. Rev. Lett. 4 (1997) 1047.
30. D. Flinn, D. Gardner, W. Nix, IEEE Trans. Electron. Dev. ED-34 (1987) 689.
31. C. Volkert, J. Appl. Phys. 70 (1991) 3521.
32. A. Shull, H. Zolla, F. Spaepen, Mater. Res. Soc. Symp. Proc. 356 (1995) 345.
33. J. Floro, E. Chason, S. Lee, R. Twesten, R. Hwang, L. Freund, J. Electronic Mat. 26 (1997) 969.
34. D. Sander, A. Enders, J. Kirschner, Rev. Sci. Instrum. 66 (1995) 4734.
35. E. Fawcett, Phys. Rev. B 2 (1970) 1604.
36. E. Klokholm, IEEE Trans. Magn. MAG-12 (1976) 819.
37. M. Moske, Ph.D. thesis, Georg-August Universität Göttingen, Mathematisch-Naturwissenschaftliche Fachbereiche (1992).
38. M. Weber, R. Koch, K. Rieder, Phys. Rev. Lett. 73 (1994) 1166.
39. A. Schell-Sorokin, R. Tromp, Phys. Rev. Lett. 64 (1990) 1039.

40. D. King, D. Woodruff (Eds.), Growth and properties of ultrathin epitaxial layers, Vol. 8 of The chemical physics of solid surfaces, Elsevier, Amsterdam, 1997.
41. U. Gradmann, G. Waller, Surf. Sci. 116 (1982) 539.
42. H. Bethge, D. Heuer, C. Jensen, K. Reshöft, U. Köhler, Surf. Sci. 331-333 (1995) 878.
43. C. Jensen, K. Reshöft, U. Köhler, Appl. Phys. A 62 (1996) 217.
44. E. Tober, R. Ynzunza, F. Palomares, Z. Wang, Z. Hussain, M. van Hove, C. Fadley, Phys. Rev. Lett. 79 (1997) 2085.
45. H. Meyerheim, D. Sander, R. Popescu, J. Kirschner, P. Steadman, S. Ferrer, Phys. Rev. B subm.
46. H. Elmers, U. Gradmann, Appl. Phys. A 51 (1990) 255.
47. U. Gradmann, Magnetism in ultrathin transition metal films, Vol. 7 of Handbook of magnetic materials, Elsevier Science, 1993, Ch. 1, p. 1.
48. H. Elmers, J. Hauschild, Surf. Sci. 320 (1994) 134.
49. H. Elmers, J. Hauschild, H. Fritzsche, G. Liu, U. Gradmann, U. Köhler, Phys. Rev. Lett. 75 (1995) 2031.
50. D. Sander, R. Skomski, C. Schmidthals, A. Enders, J. Kirschner, Phys. Rev. Lett. 77 (1996) 2566.
51. A. Enders, D. Sander, J. Kirschner, J. Appl. Phys. 85 (1999) 5279.
52. D. Sander, A. Enders, J. Kirschner, J. Magn. Magn. Mater. 200 (1999) 439.
53. D. Sander, C. Schmidthals, A. Enders, J. Kirschner, Phys. Rev. B 57 (1998) 1406.
54. A. Enders, Ph.D. thesis, Martin-Luther Universität Halle-Wittenberg, Mathematisch-Naturwissenschaftlich-Technische Fakultät (1999).
55. E. Bauer, Appl. Surf. Sci. 11/12 (1982) 479.
56. C. Schmidthals, D. Sander, A. Enders, J. Kirschner, Surf. Sci. 417 (1998) 361.
57. C. Schmidthals, A. Enders, D. Sander, J. Kirschner, Surf. Sci. 402-404 (1998) 636.
58. C. Kittel, Rev. Mod. Phys. 21 (1949) 541.
59. T. Gutjahr-Löser, D. Sander, J. Kirschner, J. Magn. Magn. Mater. 220 (2000) L1.
60. R. Koch, M. Weber, K. Rieder, J. Magn. Magn. Mater. 159 (1996) L11.
61. G. Wedler, J. Walz, A. Greuer, R. Koch, Phys. Rev. B 60 (1999) R11313.
62. M. Komelj, M. Fähnle, J. Magn. Magn. Mater. 220 (2000) L8.

STM Spectroscopy on semiconductors

D. Stiévenard

IEMN, UMR8520 (CNRS), Département ISEN, 41 Bd Vauban, 59046 LILLE Cédex, France

After a short recall on the main principles of STM spectroscopy, we give illustrations on the microscopical studies on shallow and deep defects in semiconductors. For deep defects, we show that, in order to correctly understand the spectroscopic results, it is necessary to take into account the coupling between the energy level associated with the defect and the conduction or valence bands. This coupling is analysed in terms of emission and capture rates usually used for electrical characterization of the defects. Some new experimental suggestions are also proposed concerning the determination of the emission rate of a defect using the STM technique. A second example is devoted to the investigation of the electronic structure of a quantum box. We show that the observed standing wave patterns are associated with the probability density of the ground and first excited states. Finally, spectroscopic results of organic molecules adsorbed on silicon are presented. STM images are highly voltage dependent and we discuss the influence of alkyl chains on the adsorption process and on the appearance of the molecules in the STM images.

1. INTRODUCTION

Since the pioneering work of Binning et al. on gold surface [1] and mainly on the Si(111)-(7x7) reconstructed surface [2] with an atomic resolution, the Scanning Tunnelling Microscope (STM) is one of the most powerful experimental tool used to study surfaces, interfaces (via cleaved materials), point defects in semiconductors and more recently organic molecules on metallic or semiconductor surfaces. The tunnelling current varies approximately exponentially with the tip-surface separation, typically increasing by a factor of 10 for every 0.1 nm reduction in the separation. The current is mainly bound to the last atom on the tip, the closest to the sample, and this is a reason why atomic resolution can be obtained. Moreover, the tunnelling current is a convolution between the topography and the local density of states (LDOS) of the surface. Used carefully in the spectroscopic mode, the STM allows the determination of the LDOS.

The aim of this communication is to briefly recall the physical basis of the STM spectroscopy and to give illustrations on the following subjects: shallow and deep defects in semiconductors, semiconductor quantum boxes and organic molecules on silicon.

2. PHYSICAL BASIS OF STM SPECTROSCOPY

To perform STM measurements, a potential difference is applied between the tip and the surface. To get an STM image, the potential is fixed and the tip is swept over the surface. However, by varying this potential difference, eigenstates of the tip and sample become available for tunnelling and lend themselves for spectroscopic studies.
An approximate expression for the tunnelling current is [3]:

$$I \propto \int_0^{eV} \rho_s(E) D(E) dE \qquad (1)$$

where $\rho_s(E)$ is the local surface density of states (LDOS); $D(E)$ is the transmission probability, reduced for small voltages and energies at the Fermi level to $\exp(-2Ks)$ where s is the tip-surface separation (noted also z in the following part of the paper) and K the vacuum decay constant. R.M.Feenstra has shown that the normalized conductance is proportional to a normalized surface density of states [4]:

$$\frac{dI/dV}{I/V} \approx \frac{\rho_s(E)}{(1/eV)\int_0^{eV}\rho_s(E)dE} \qquad (2)$$

In such a case, eq. (2) is independent of s. Note that an easy way to get an I(V) curve independent of s is to apply a linear ramp voltage given by [5]:

$$\Delta s(V) = s_o (1 + \frac{|V|}{4\Phi}) \qquad (3)$$

where Φ is the average work function of the two electrodes (the semiconductor and the tip). The slope of the ramp is of the order of 0.5-1 Å/V. (Eq.(3) is a first order approximation with respect to $V/2\Phi$).

From (2), the normalized conductivity (dI/dV)/(I/V) is often said to be proportional to the LDOS. However, for large band gaps such as GaAs(110) surface, the method is not available since both the current and the conductivity approach zero at the band edges and their ratio tends to diverge. In such a case, some generalization of the I/V normalization approach is requested.

One method to suppress the divergence in the normalized conductivity is to broaden the function I/V by convolution with a suitable function, leading to the elimination of the zeros of I/V within the gap. Generally, an exponential (or gaussian) function is taken. The broadened I/V function is then given by [5]:

$$\bar{I}/\bar{V} \equiv \int_{-\infty}^{+\infty}(I_m/V') \exp(\frac{V'-V}{\Delta V}) dV \qquad (4)$$

Generally, ΔV is of the order of the band gap value in order to eliminate all the band edges divergences and noise within the gap. Away from the band edges, the other features in the spectrum are only slightly affected by the broadening procedure.

A second type of normalization method is to account for the different values of the tip-surface distance by using a multiplicative correction of the form $\exp(2Ks)$, where s are the known values of the relative tip-surface distances. The only parameter in this approach is K, which in general depends on both the tip-sample separation and bias voltage, $K = K(s,V)$. These dependences are usually small, so that, in many cases, it is sufficient to assume a constant K value.

However, in some cases, the exact form $K(s,V)$ has to be known. It is then necessary to measure the decay length simultaneously with the measurement of the tunneling spectra. R.M.Feenstra [6] has proposed an original approach, starting from spectroscopic data acquired with variable tip-sample separation and thus transformed, in a parameter free method, to constant separation.

As κ depends on both s and V, we denote it $\tilde{\kappa}$, which can be expressed as [6]:

$$\tilde{\kappa} = \frac{1}{2}\frac{d}{ds}\ln(\frac{\sigma(z_o,V)}{\sigma_m(z(V),V)}) \qquad (5)$$

where $\sigma(z_o,V)$ is given by:

$$\sigma(z_o,V) = I_m(z_o,V_o) g_m(z(V),V) \exp(\int_{V'}^{V} g_m(z(E),E) dE) \qquad (6)$$

where m is referred to a measured quantity, o to a given s distance and $g = (dI/dV)/I$.

An example of the variation of $\tilde{\kappa}$ is given in Figure 1 for a GaAs(110) surface. The dotted lines correspond to a tip with a weak work function, as detailed later.

Other examples of LDOS on n-GaAs, n-InP, p-GaSb, n-InAs, n-InSb are given in Figure 2, Reprinted from R.M.FEENSTRA, [4].

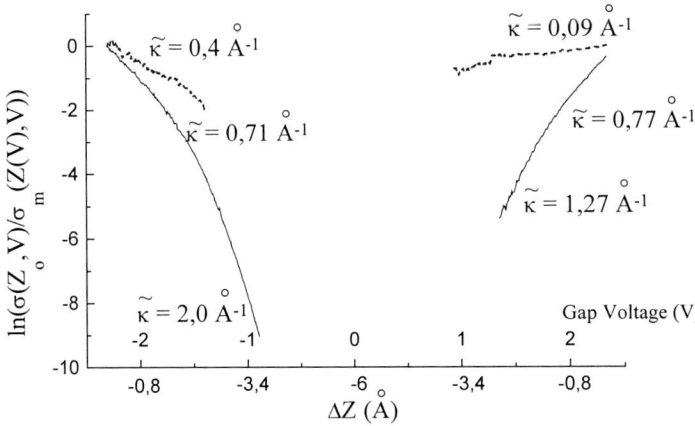

Figure 1: $\tilde{\kappa}$ for a GaAs(110) surface

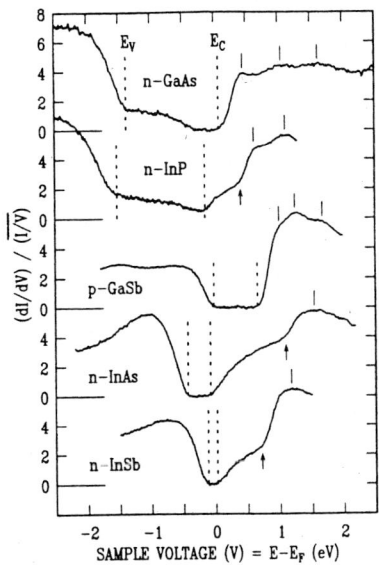

Figure 2: LDOS of various III-V semiconductors

Finally, one key parameter is the barrier height, which governs the tunnelling current and consequently the physical interpretation of the STM pictures. This barrier height (Φ) is associated with the work functions of the surface and of the tip. One way to control the barrier height is to work in ultrahigh vacuum (UHV) where the surface is well controlled and clean. In such a case, it is possible to focus on the nature and the role of the tip apex.

STM and STS (Scanning Tunnelling Spectroscopy) measurements were performed [7] with an UHV STM with a base pressure of 5×10^{-11} torr on GaAs (110) cleaved surfaces, residually p doped. The samples were cleaved *in vacuo* to yield atomically flat, electronically unpinned (110) surfaces cross-sectional to the (001) growth direction. The W tips were prepared as follows: after an electrochemical etching in NaOH, they were sharpened in UHV by mixed heating, field emission and ion milling processes. Very clean tips with a radius lower than 6 nm were routinely obtained. Finally, the presence of water on the tip or on the surface, known to decrease ϕ, is here excluded since the surface is heated up to 150 °C for 12 hours in UHV and the tip is also heated up to 1300°C during its preparation.

In a first step, ϕ were carefully determined. Starting with the Tersoff and Hamann description of the tunnelling barrier [8], ϕ can be deduced from the variation of the tunnelling current I versus the shift of the tip in the z direction perpendicular to the sample:

$$\phi = \frac{\hbar^2}{8m}\left(\frac{d\ln I}{dz}\right)^2 \quad (7)$$

in which ϕ and the inverse decay length κ of the electrons in the barrier are related by:

$$\kappa = \sqrt{\frac{2m\phi}{\hbar^2}} \quad (8)$$

Figure 3 [7] gives the normalized conductance $(dI/dV)/((\bar{I}/\bar{V}))$ of a GaAs(110) surface measured with two tips, labelled A and B.

To avoid that this conductance diverges at the band edges, an exponential function for broadening the conductance was used, with a broadening width equal to 1.4 eV. With the tip A (solid lines), a gap voltage of 1.45 eV is measured, in good agreement with the gap voltage of GaAs. As to the position of the Fermi level, it is given for a zero gap voltage. This zero gap voltage is closer to the valence band edge than to the conduction band edge, which corresponds to a lightly p-doped semiconductor, in agreement with the doping level of the sample (residually p doped). With the tip B (dashed lines), the gap is of the order of 2.1 eV and the Fermi level position seems to indicate a n-type semiconductor.

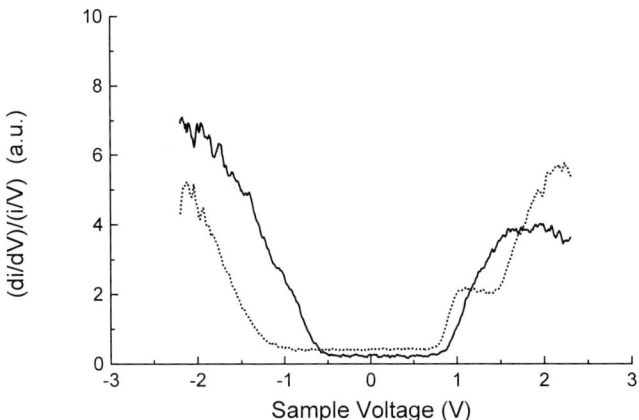

Figure 3: Normalized conductance $(dI/dV)/((\bar{I}/\bar{V}))$ of a GaAs(110) surface measured with two tips: tip A (solid line), tip B (dashed line)

Figure 1 shows the ratio σ_o/σ_m obtained from $I(z(V),V)$ measurements with a z ramp of 6 Å towards the surface when the voltage approaches zero. $\tilde{\kappa}$ was deduced using the relations (5) and (6). For the tip A, $\tilde{\kappa}$ varies from 0.77 to 1.27 Å^{-1} i.e. ϕ varying apparently from 2.5 to 5 eV. When the tip-surface distance decreases, $\tilde{\kappa}$ increases. This variation is due to the participation of electrons with an important $k_{//}$ to the STM current. $k_{//}$ is estimated to be 1.06 Å^{-1}, in good agreement with the value (0.94 Å^{-1}) of the wave vector at the edge of the Brillouin zone for GaAs. Finally, this precise determination of $\tilde{\kappa}$ shows an asymmetry between positive and negative voltage. This can be attributed to the contribution of the image potential. For the tip B, $\tilde{\kappa}$ varies from 0.09 to 0.4 Å^{-1} i.e. ϕ varies apparently from 0.03 to 0.7 eV. The tunnelling junction exhibits a very low barrier height, giving rise to artefact in the STM interpretation (gap, doping level).

For a low ϕ, it is easy to explain an apparent widening of the gap. Indeed, for a given gap voltage and a given tunnelling current, the lower ϕ, the greater z. So, for the same Δz, the variations of the conductance and of the current are smaller. As the band edges are associated with the lowest current (there are no surface states in the gap), they are automatically noisy. That leads to an apparent widening of the gap associated with a decrease of the current sensitivity. As to the apparent change of the doping level, it is directly related to the band bending induced by the tip. For a given doping level, this band bending decreases if the work function of the tip decreases, as in a MOS capacitor [9]. So, in STM spectroscopy, the corresponding positive gap voltage necessary to obtain a non zero current decreases: the apparent position of the Fermi level (zero gap voltage) seems to be shift toward the conduction band.

Now, the problem is to explain why the barrier height decreases. For that purpose, a non-equilibrium Green function formalism developed by Keldysh [10] was used, as applied by Caroli et al. [11], to calculate the tunnelling current for the tip-sample system. It provides an expression where non-coupled Green function of the tip and the substrate are linked by the tunnelling matrix. This formula includes the effect of the bias voltage and the tip-surface electronic coupling. The non-coupled Green functions of the tip and the substrate are calculated using the empirical tight-binding method together with the recursion and decimation techniques. The elements of the tunnelling matrix are those of Harrison [12], modified by a semi classical exponential factor at distances larger than the sum of the covalent radii [13]. The influences of the macroscopic electric potential on the surface density of states and of the space charge region in the semi-conducting sample are also taken into account following the technique detailed in ref. [14].

The simulated system is built of a semi conducting sample and a metallic tip. The sample is a (110) GaAs surface. The tip is a tungsten cluster of 14 atoms forming a pyramid (1-4-9) having a base of nine atoms, this cluster forming a protrusion below a sphere of radius 40 Å. The apex atom of the pyramidal protrusion is either a W atom, or Ga or As atoms. In this last case, the tight-binding calculation is made quasi self-consistent by adding a diagonal term to the Hamiltonian, adjusted to fulfil the local neutrality criterion [15].

The value of the current intensity I is then calculated and the barrier height Φ versus the tip-sample distance z is deduce from it, using (7).

For the calculation, the tip-sample distance is the space between the tip apex atom nucleus and the plane of the surface sample nuclei. Results are shown in Figure 4a for tip apex atoms W, Ga and a negative sample bias (-2 volts). The same type of behaviour is obtained for a positive sample bias. (Figure 4.b)

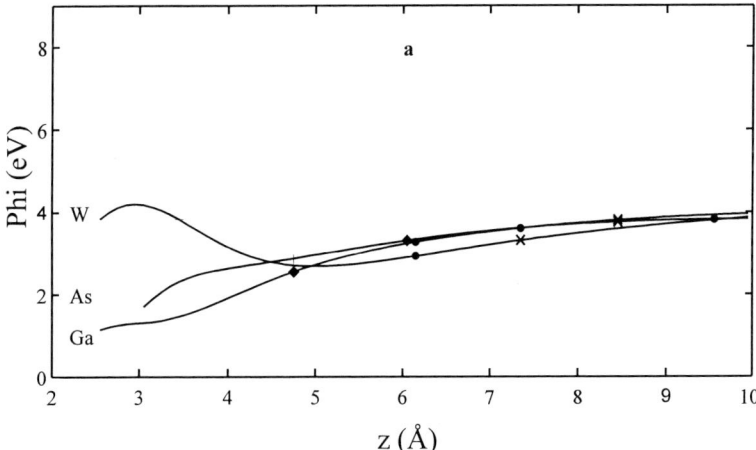

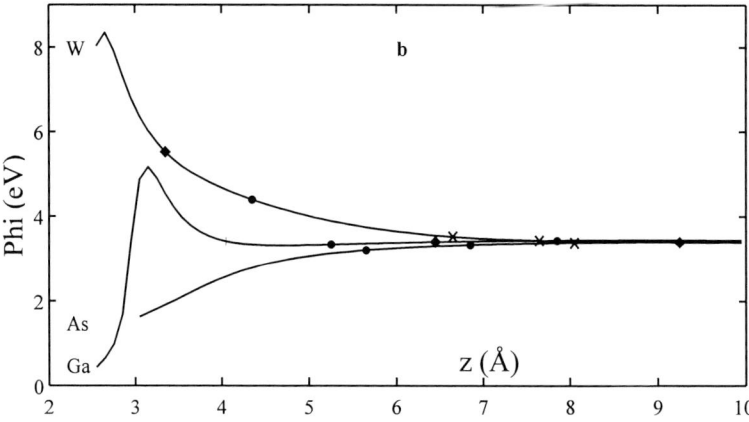

Figure 4: Calculated barrier height Φ versus tip to -sample distance z, for one W, Ga or As tip apex atoms and for a sample bias of -2 volts (Fig.4a) and + 2 volts (Fig.4b). Crosses show the value of z and Φ for a 1 nA tunnelling current. Each point corresponds to one decade of variation on the current amplitude. Distances are measured from the atom nuclei.

The form of the current is globally [10]:

$$I \approx \int_{E_{fs}}^{E_{fp}} n_s(E)\, e^{-kz\sqrt{\Phi+E}}\, n_p(E)\, dE \qquad (9)$$

where $n_s(E)$ and $n_p(E)$ are the local densities of states on the substrate and tip, k is a constant, Φ is the potential barrier height, and E_{fs}, E_{fp} are the surface and tip Fermi levels respectively. If the local densities of states present peaks at some energies, the integrand at these energies will be larger and the corresponding term $\sqrt{\Phi+E}$ will have more weight in the calculation of the barrier height Φ. So Φ will get close to $\Phi+(E_{peak}-E_f)$. Precisely the form of the local densities of states on the tip apex atom is very different with the nature of this atom: in the case of tungsten there is a peak at the Fermi level, whereas the peaks are far from the Fermi level in the case of Ga or As. This explains why the apparent barrier height varies with the nature of the tip apex atom. There are no more variations when the barrier gets wider, because the decrease of the exponential with energy becomes then to fast by comparison with the variation of $n_s(E)$ and $n_p(E)$, and it is the exponent $\sqrt{\Phi + E}$ of the larger integrand that predominates.

This effect is increased by the formation of an electric dipole on the tip due to charge transfer to the apex atom. Taking into account the calculated self-consistent potential, the variation of the barrier height is estimated to be of the order of 0.5 to 1 eV, but this value can increase if there are several As or Ga atoms at the tip apex. In Figure 3, a tip-sample separation is actually reached. With the experimental conditions for imaging (V=2 volts, I = 200 pA), the associated tip-sample separation is of the order of 1 nm. As a z-ramp of 6 Å was used, a minimum value of 3 Å used for the simulation is reasonable.

At present we have not taken into account the image potential in our calculation, and this is why, in figure 3, the barrier height takes in some regions pretty large values. In reality the image potential reduces in all cases the barrier height when the tip-sample distance is small, but does not modify the relative values of the different cases (W, Ga, As). Finally, these calculations show that for large z, ϕ becomes independent of the nature of the apex.

The influence of the shape of the tip has also been tested. The sphere radius on which the cluster of 14 atoms made the final protrusion has been changed from 40 Å to infinite (the plane limit). The variation induced on ϕ is less than 0.2 eV. The major influence on ϕ is due to the nature of the last atom of the pyramidal protrusion, as shown previously.

In the following part, we present some applications of STS on shallow dopants in GaAs.

3. SILICON DOPANTS IN GaAs

Dopants are shallow defects in semiconductors and they have been studied using STM. This new approach allows a visualization of the dopants at an atomic scale. Silicon dopants are the most characterized [16-19] but Be dopants [20,21] or Zn dopants [22,23] has also been observed. Subsurface Si_{Ga} donors appear as protrusion, in both the empty and occupied density of states images whereas subsurface Si_{As} exhibit a long range elevation in occupied state and a depression in empty state images. The observed contrasts depend on the location of the silicon atoms in the surface layer, second or third layer. Figures 4 and 5 show STM images and their associated cross-section for Si_{As} and Si_{Ga} respectively.

To explain the STM images (for example, observation of protrusion for donor), a simple approach has been proposed [17] based on electrostatic interpretation (see Figure 7). Due to tip-induced band bending, at a positive sample bias, the subsurface region under the tip is depleted of electrons. In this region, the shallow Si_{Ga} donors are ionised and are positively charged. Close to the core of a donor, its coulomb potential locally decreases the band bending and effectively increases the density of states available for tunnelling. To compensate the enhanced tunnelling current, and when a constant current STM mode is used, the STM tip moves away from the surface, producing a protrusion superimposed on the lattice corrugation. At negative sample bias, the tip-induced band bending causes electrons accumulation. The current is due to electrons tunnelling out of the valence band states and filled states near the conduction band edge in the accumulation layer. The current is mainly associated with states located near the semiconductor Fermi level, and thus, as in the precedent case, donors induce a local increase of the filled states and a subsequent increase

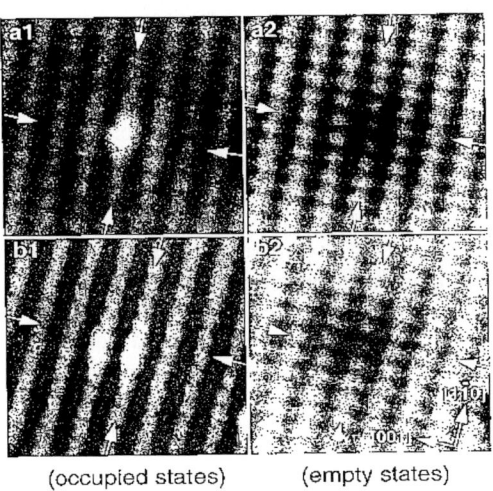

(occupied states)   (empty states)

Figure 5a: STM image for a Si_{As} dopant in GaAs [18]

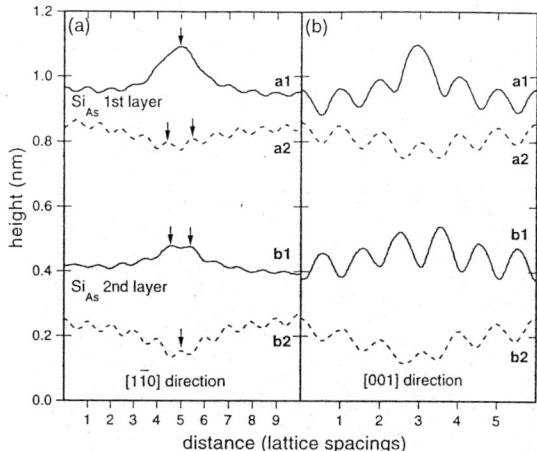

Figure 5b: Associated cross-section for a Si_{As} dopant in GaAs [18]

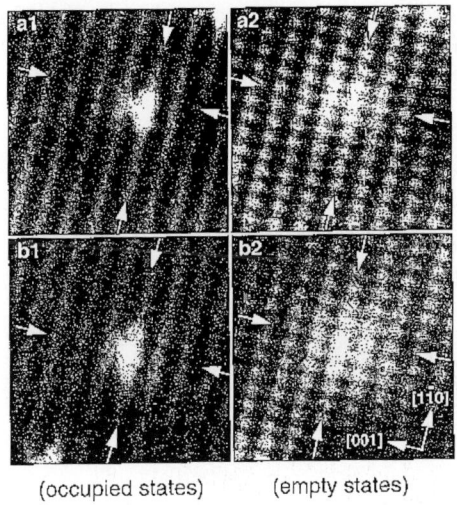

Figure 6a: STM image for a Si_{Ga} dopant in GaAs [18]

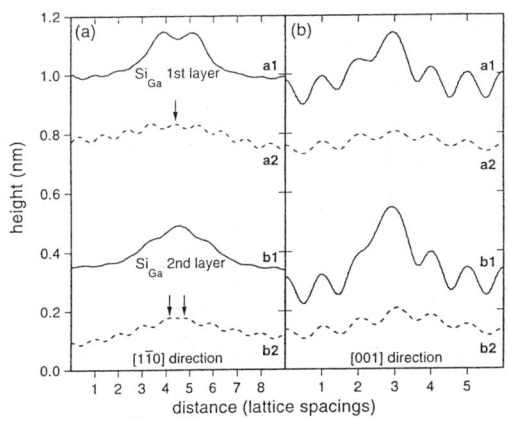

Figure 6b: Associated cross-section for a Si_{Ga} dopant in GaAs [18]

of the STM current. The screening length associated with the high concentration of electrons is always larger (4 or 5 nm) than the extent of the observed subsurface Si_{Ga} feature (2.5 nm) and therefore, screening does not affect the Coulomb potential perturbing the local bending.

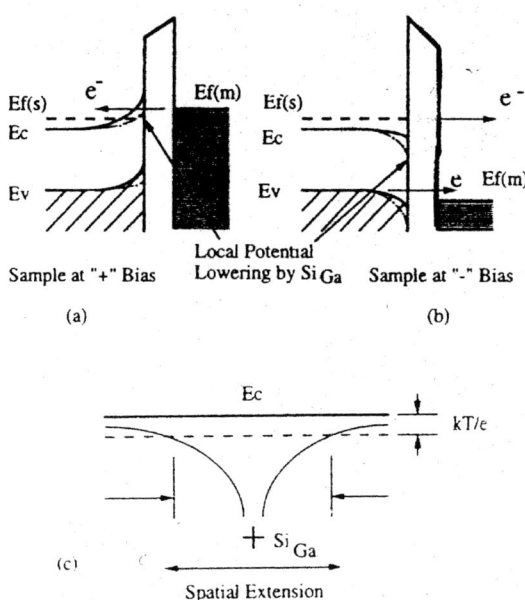

Figure 7: Energy band diagram illustrating the electron tunnelling process between metal tip and GaAs(100) surface in the presence of a tip-induced band bending: a) positive sample voltage, b) negative sample voltage, c) schematic illustration of the Coulomb potential around the Si atom [17]

These observations have been realized on bulk doped GaAs. Recently, the microscopic behaviour of silicon has been studied in silicon delta-doped layer [24]. The samples were grown using Molecular Beam Epitaxy with two growth temperatures: 480°C and 580°C. The δ-doped layer concentration is 0.03 M.L. The δ-doped layers are obtained by interrupting the growth of the GaAs host crystal closing the Ga shutter, opening the silicon one and leaving the As one open. To continue GaAs growth, the dopant shutter is closed and the Ga shutter opened again. Two AlGaAs marker layers (20 nm) are grown in order to help us to localize the delta-doped layers. Figure 8 gives XSTM images (44x44 nm^2) taken with a sample voltage of -2.0 V. The growth temperature is 480°C (Fig.8a) and 580°C (Fig.8b). In each image, bright hillocks (labelled D) with a diameter of 2.5 nm are observed and superimposed on the arsenic atomic lattice. For positive sample voltage, they also appear as bright hillocks, superimposed on the gallium atomic lattice (not shown here). At 480°C, some bright hillocks (labelled A in Fig.8a), with a diameter of 1 to 1.5 nm and surrounded by a dark region are observed. This dark region whose diameter is 4 nm is associated with a depleted region [17] around a negative charge center in a n-doped region. (For positive voltage, these defects A appear as simple depression superimposed on the gallium atomic lattice). All these observations are in agreement with the previous observations of donors (Si$_{Ga}$, D) and acceptors (SiAs, A) in bulk material made by J.F. Zheng et.al [17,22]. The silicon acceptors do not appear on the δ-doped layer grown at 580°C, but two other features are observed: some isolated vacancies (labelled V) and complexes (labelled C) associated with a donor and a vacancy.

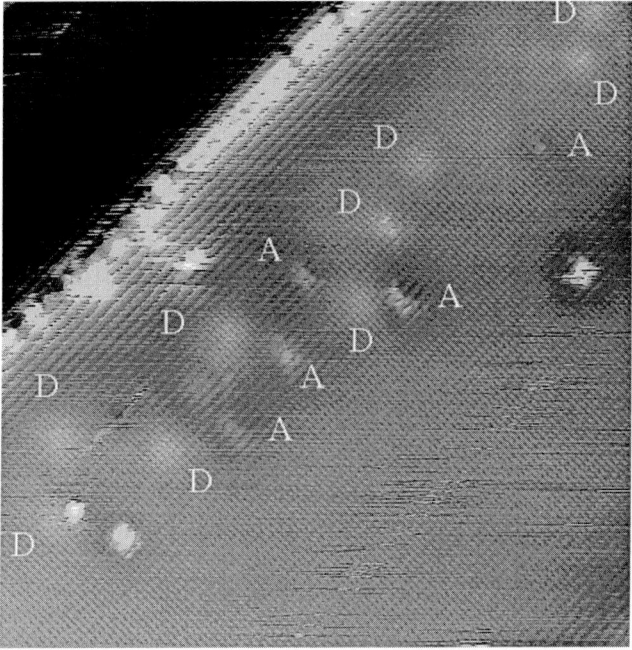

Figure 8a: STM image (45x45 nm^2) of the 0.03 M.L. delta-doped layer, grown at 480°C

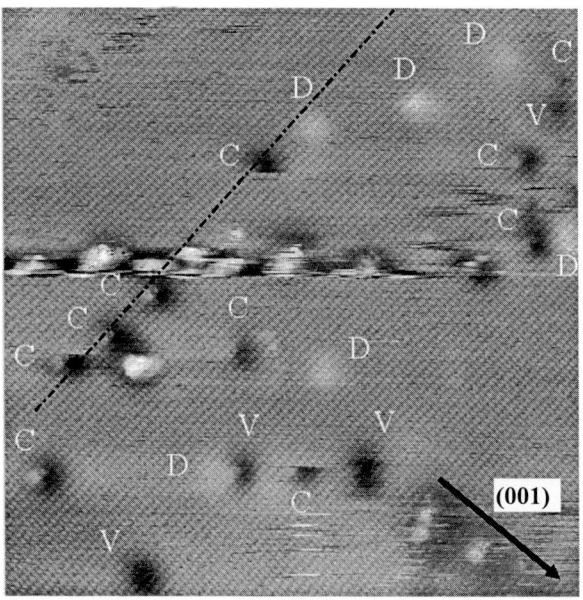

Figure 8b: STM image (45x45 nm^2) of the 0.03 M.L. delta-doped layer, grown at 580°C

The amphoteric character of silicon is shown (both donors and acceptors are detected in the same layer). Moreover, acceptors are created after donors in the [001] growth direction, as shown in Figure 9. This property found in delta-doped layer is in agreement with previous experimental results [18] observed in bulk Si doped material. Si is initially incorporated on Ga sites as donors [25]. Associated with that formation, the n-type doping of the layer increases, inducing a decrease of the energy formation of the acceptors, resulting in the incorporation of silicon on the arsenic sites [26]. Finally, the distribution of donors with a 580°C growth temperature is wide along the [001] growth direction. Including the position of the intended growth position of the delta-doped layer in the distribution, the authors propose that the main physical phenomenon associated with the extension of the layer is the segregation of silicon.

In conclusion of the STM results on shallow dopants, one important idea to keep in mind is that the STM images are associated with a local electrostatic perturbation of the bands around these defects and so, the STM images are due to an "indirect" effect, with a current which does not necessary go through the energy level of the defect. On the contrary, in the next paragraph, we present spectroscopic studies on deep defects and particularly on a famous deep level in GaAs, namely the As$_{Ga}$ antisite defect. A discussion is proposed about a basic problem: how tunnelling current occurs through deep defects?

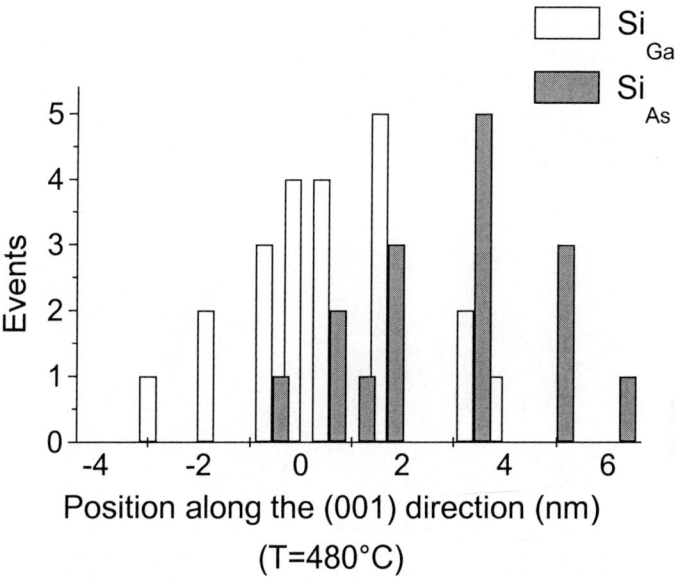

Figure 9: Spreading of the donor and the acceptors in the (001) direction at 480°C. The arrow gives the intended position of the delat-doped layer.

There is a necessary coupling between the energy level associated with the defect and the conduction or valence bands. This coupling is analysed in terms of emission and capture rates usually used for electrical characterization of the defects. Some new experimental suggestions are also proposed concerning the determination of the emission rate of a defect using the STM technique.

4. MICOROSCOPIC STUDIES OF THE As_{Ga} ANTISITE DEFECT

The As_{Ga} antisite defects has been extensively studied during the 15 years ago because it is a deep level associated with the famous EL2 defect [27], responsible of the compensation mechanism in semi-insulating GaAs layer. Antisite defects have been observed [28-30] in low-temperature-grown GaAs, which contain about 10^{20} cm^{-3} arsenic-related point defects. STM images (see Figure 10) reveal a central defect core with two satellites located about 1.5 nm from the core. The satellites are interpreted as antisite wave function tails extending along <112> surface directions. These images are in good agreement with theoretical STM simulation of antisite defect [31]. In fact, four types of structures (core + satellites), are observed but interpreted as belonging to the same type of defect, with the variation arising from the location of the defect relative to the surface plane. Local spectroscopy above the defect has also been performed with a variable position of the semiconductor Fermi level, from n^+ to p^{++} type material. The spectroscopy results are given in Figure 11 where one or two peaks appear, associated with the two charge states (0/+ and +/++) of the antisite defect.

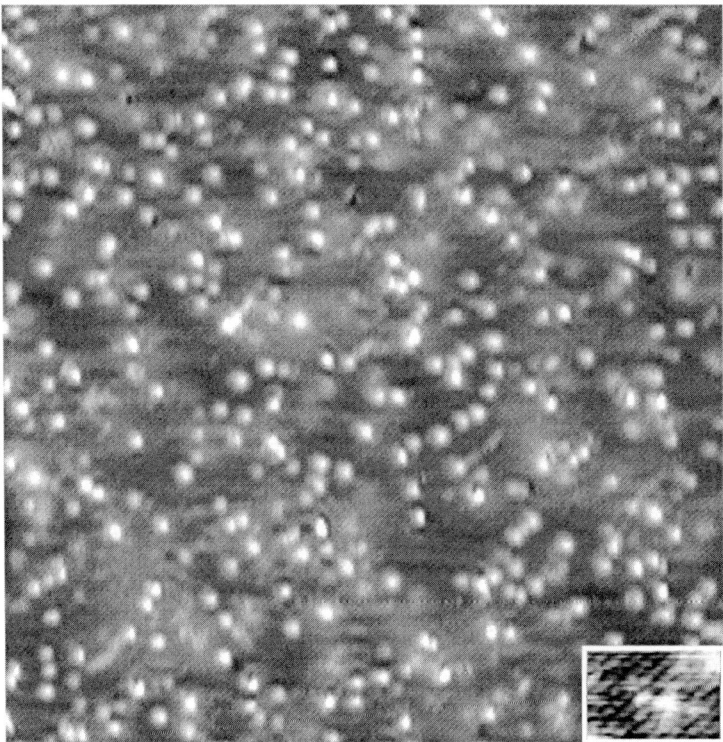

Figure 10: Constant current image (86 nm x 86 nm) of the (110) cleaved face of LTG-GaAs, acquired at a sample bias of –2.0 volts. Numerous point defects (arsenic antisites) can be seen [reprinted from 30]

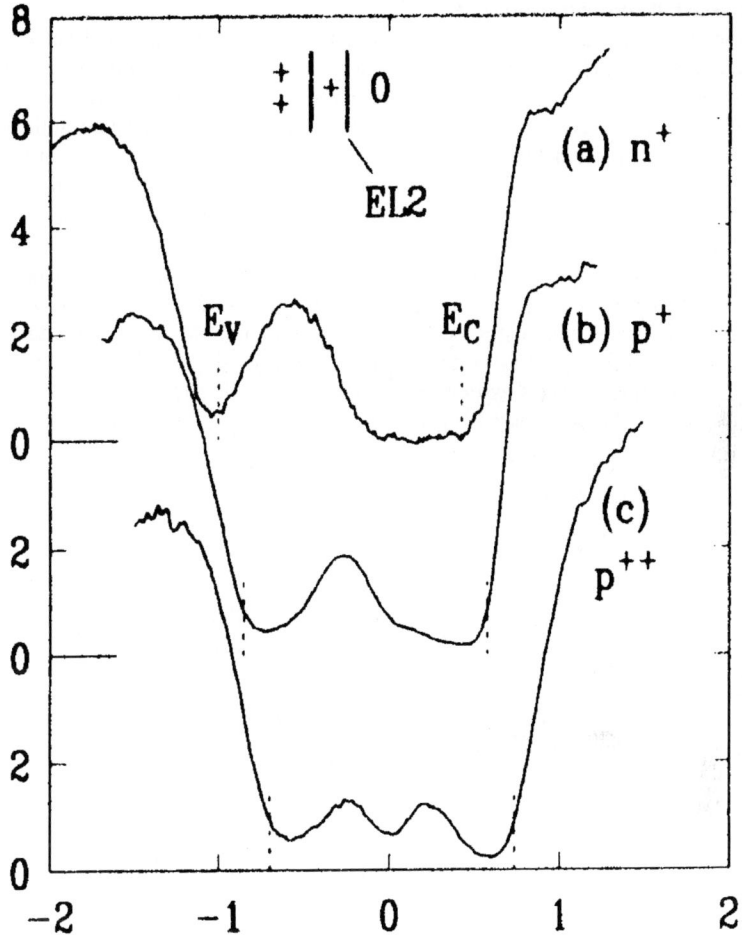

Fig. 11: Spectroscopic results on As$_{Ga}$ [28]

These spectroscopic results are discussed in the next part.

4.1. Defect assisted tunneling current

In the previous part of this paper, we have shown that the STM technique is a powerful technique, which allows studying defects at the atomic resolution. Moreover, in the spectroscopic mode, local density of states can also be measured with, *a priori*, the detection of the energy levels associated with defects, such as the arsenic antisite defect. However, one basic problem faces us: when a defect has an energy level localized in the gap of the semiconductor and when tunneling current is measured « through » the defect (resonant tunneling current), there is obviously a current from the tip to the energy level and therefore from the energy level to the bands of the semiconductor. In other words, the measured current is the result of a tunneling current combined with a current associated with the exchanges between the defect and the bands. In the following part, we analyze the influence of the thermal exchanges between the defect and the bands on the measured current. We show that, depending on the position of the Fermi level of the semiconductor with respect to the energy levels associated with the defect the tunneling current can highly decrease and thus, the interpretation of STM spectroscopy has to be carefully taken [30,32].

Calculation of the current

Figure 12 shows a schematic diagram of an energy level E_t laying in the semiconductor gap.

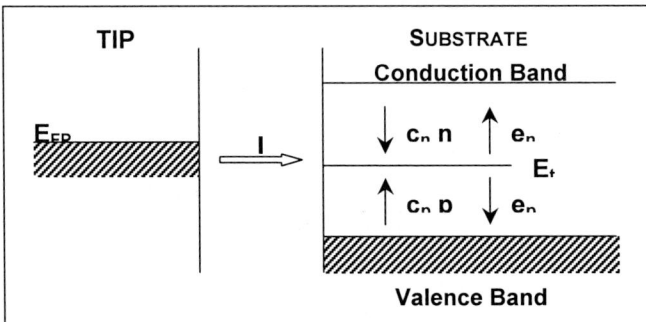

Fig. 12: Capture and emission rates associated with an energy level

The Fermi level of the tip is labelled E_{FP}. The emission and capture probabilities for electrons (holes) are e_n (e_p) and c_n (c_p) respectively; and if n and p are the free carriers concentrations (electrons and holes respectively) with f the occupancy function of the energy level, the current I_{th} associated with the exchanges between the defect and the bands is given by:

$$I_{th} = -e\,[f\,(e_n + c_p\,p) - (1-f)\,(c_n\,n + e_p)] \tag{10}$$

On the other hand, the tunnelling current I_{tu} from the tip to the energy level is:

$$I_{tu} = -e(1-f)W \qquad (11)$$

where W is transfer probability of an electron from the tip towards the surface and e is the absolute value of the electron charge. Under a permanent regime, these two currents are equal and, by eliminating f between (10) and (11), we obtain the final form for the measured current I, as:

$$I = -e \frac{W(e_n + c_p\, p)}{W + e_n + c_p\, p + e_p + c_n\, n} \qquad (12)$$

One observes that when e_n or $c_p p$ are large, I tends towards $-eW$: there is no attenuation of the current. On the contrary, if W is large, I tends towards $-e(e_n + c_p p)$: the current can decrease. It is the lowest process which imposes the current. The variation of the current can be estimated. We have fixed the tunneling probability W to 0.1 nA/e=6.25 10^8 /s. The semiconductor is a p-type one (n is always small compared with p) and the capture cross sections of the defect for the electrons and the holes, σ_n and σ_p respectively are equal to a typical value measured for defects equal to 10^{-15} cm^2 ($c_{n,p} = \sigma_{n,p} v_{n,p}$ where $v_{n,p}$ is the thermal velocity of electrons and holes). Figure 13 gives the variations of I versus the doping level N_a and the position of the energy level in the gap measured by the quantity E_t-E_i where E_i is the value of the intrinsic Fermi level.

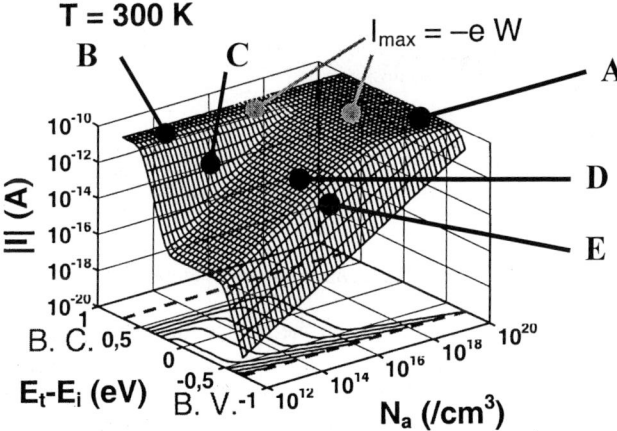

Figure 13: Variation of the current I versus the doping level and the position of the energy Et [32]

The general behavior is easily explained. When E_t is far from the band edges, i.e. ($-0,7\ eV \ll E_t-E_i \ll +0,7\ eV$), e_n and e_p are negligible with respect to $c_p\,p$ ($c_n\,n$ is always small as n is negligible). Then, the current I is equal to

$$-I \approx e\,\frac{W\,c_p\,p}{W + c_p\,p} \tag{13}$$

$I \approx -e\,c_p\,p \propto N_a$ if $W \gg c_p\,p$ (zone D), and $I \approx -e\,W$ if $W \ll c_p\,p$ (zone A). The electrons coming from the tip are emitted by the defect towards the valence band *via* the capture of a hole. When E_t is near the conduction band ($E_t-E_i \approx +0,7\ eV$), e_n increases until it becomes greater than $c_p\,p$. Thus, (zone C):

$$-I \approx e\,\frac{W\,e_n}{W + e_n} \tag{14}$$

(when e_n is larger than W, $I \approx -e\,W$ – zone B). The electrons coming from the tip are reemitted by the defect towards the conduction band. Finally, when E_t is near the valence band, ($E_t-E_i \approx -0,7\ eV$), e_p increases until it overcomes $c_p\,p$ and I given by:

$$-I \approx e\,\frac{c_p\,p\,W}{e_p} \tag{15}$$

rapidly decreases (zone E). Finally, let us remark that the maximum value of the current is equal to $-W$. The same kind of approach can be done in n type semiconductor.

Let us now applied this analysis to the case of the spectroscopy obtained on a defect with one energy level in the gap.

Spectroscopy analysis: case of one energy level

We have shown that, depending on the semiconductor doping value and the position of the energy level in the gap, the current through a defect can or can not exist. We now apply this analysis to spectroscopic studies in the case of one energy level in the gap, associated with a defect. Figure 14 explains the different possibilities. We consider a p-type material (the case of a n type material can be straightforward deduced), i.e. E_t is above the semiconductor Fermi level E_{FS}. In such case, the charge of the defect is governed by the position of the tip Fermi level E_{FP}. If E_{FP} is below E_t, (Figure 14a), the level is never charged and there is no current. Now, if E_{FP} is above E_t with e_n or $c_p\,p$ small (Figure 14 c and e), no current can occur, because the exchange from the defect to the bands are low. On the contrary, if e_n or

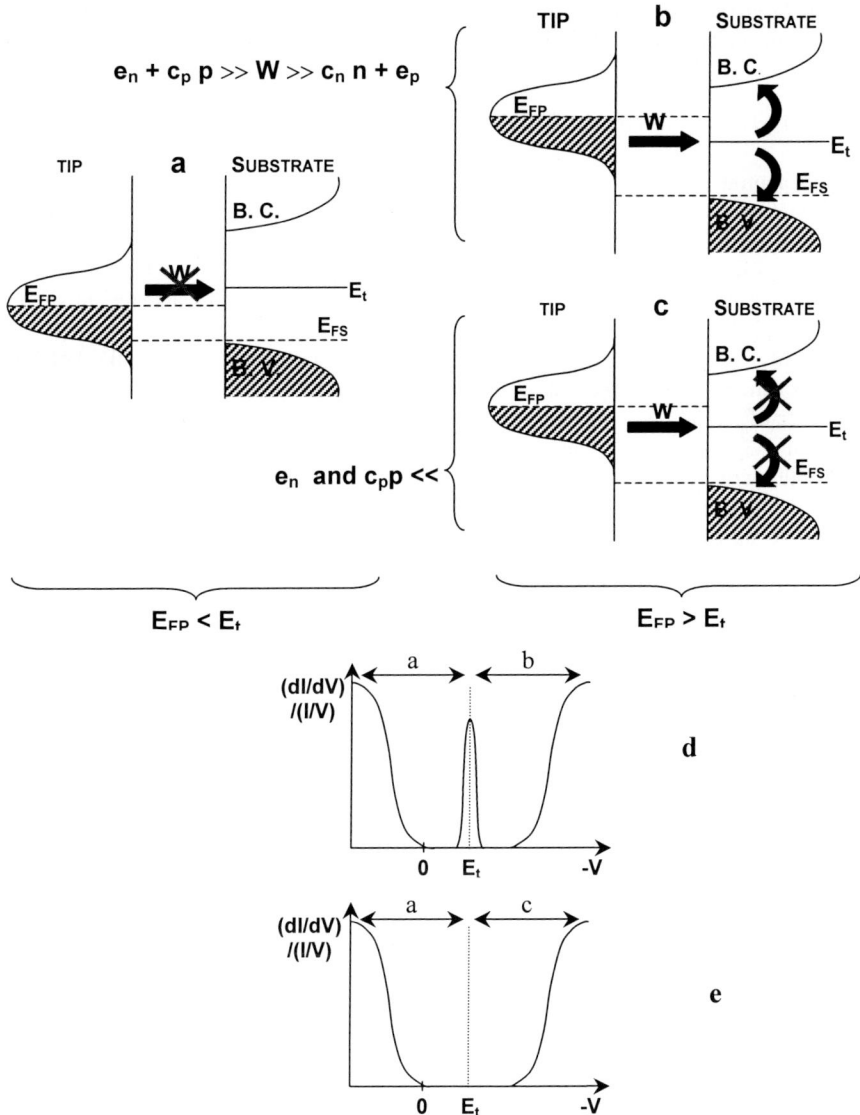

Figure 14: Spectroscopic studies associated with one energy level in the gap [32]

c_p p are large (with e_p and c_n n small), the exchanges of electrons with the bands are possible and a peak appears in the spectroscopic spectrum (Figure 14 b and d).

Case of two energy levels: the EL2 defect

This approach can be extended to the case of two energy levels. In such case, it is easy to show that two peaks are detected during spectroscopy measurements only when the Fermi level of the semiconductor lies between the two energy levels [32]. When the semiconductor Fermi level lies below or above the two levels, only one peak can be detected. This seems to be in agreement with the experimental data of R.M.Feenstra on the arsenic antisite in GaAs materials [28].

For the EL2 defect, being primarily composed of an arsenic antisite, the hole emission rate is of the order of 5×10^{-4} s^{-1} at room temperature [33]. The corresponding current is of the order of 7×10^{-23} A. Concerning the electron capture rate c_n n, it is of the order of 2 s^{-1} taking n~10^7 cm^{-3} and c_n equal to the product of the thermal velocity (~ 3×10^7 cm/s) times the capture cross section (~ 6×10^{-15} cm^2 at room temperature) [33]. The associated current is of the order of 3×10^{-19} A. Although the spectroscopic measurements were performed with a variable tip-sample separation and an initial current of 100 pA, the smallest measured current in the band gap is of the order of 100 fA. Therefore the measured current is much larger than the thermal current. The broad peaks observed in Fig. 11 cannot be explained by an exchange of carriers between the deep levels in the band gap and the bands. However, the LTG-GaAs layer contains a high concentration of arsenic antisites, which can give raise to a hopping conduction. From ref 32, the hopping rate W_{hopp} is related to the concentration N_T through the relation:

$$W_{hopp} = \frac{\sigma \Delta}{e^2 N_T^{1/3}} \quad (16)$$

where σ is the conductivity and Δ is the width of the band. In Fig. 11, the band of midgap states has a width of 0.5 eV. Knowing the defect concentration, the conductivity is estimated [34] to be $10^{-3} \Omega^{-1}$ cm^{-1}, leading to a hopping rate of the order of 7×10^8 s^{-1}. This value is close to the transfer probability W established for a tunnelling current of 100 pA (W=$6.25 \times 10^8 s^{-1}$). Thus we conclude that the tunnelling current results mainly from a hopping conduction rather than a thermal exchange between the midgap states and the bands. In Fig. 11, the appearance of a peak in the band gap is therefore possible due to the high concentration of arsenic antisites. This high concentration increases the probability to transfer free carriers from a subsurface defect to neighbour defects.

4.2. Determination of the emission or capture rates via the STM current

The emission and capture rates of a defect have been widely studied using, for example, the Deep Level Transient Spectroscopy technique. In such a case, the rates are measured over a great number of defects (typically 10^{14} to 10^{17}cm^{-3}) and the basic principle of the technique is an abrupt variation of the bias followed by the analysis of a transient (capacitance or current) associated with the emission or capture rates. Using the STM, we propose to determine these rates. There are two main differences: first, the rate will be measured on a single defect and second, the transient current will be induced by an abrupt change of the tip-surface distance. We consider the cases where the rates are large or small.

Case of large emission or capture rates

Let us assume that the emission rate is the dominant one and that the semiconductor is p doped. Thus, as shown in Figure 15, the method consists in selecting a bias voltage in order to the tip Fermi level E_{FP} is above E_t.

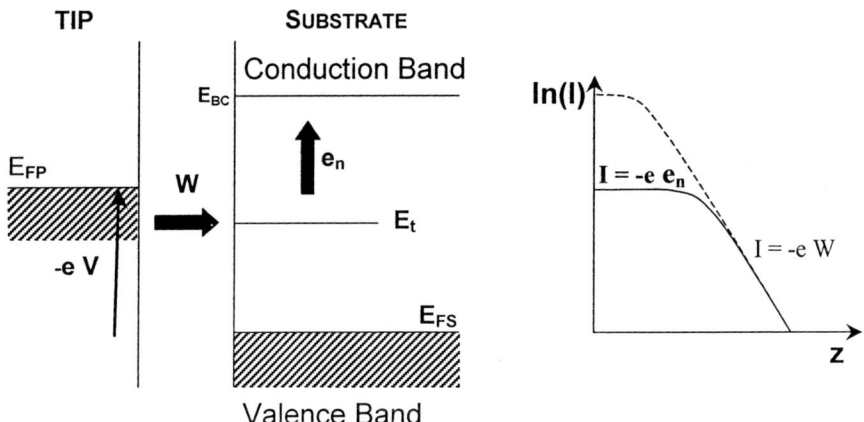

Figure 15: Case of large e_n measured *via* the tunnelling current.

Thus, as previously shown, the defect-assisted current is equal to:

$$-I \approx e \frac{W \, e_n}{W + e_n} \qquad (17)$$

This current is used to determine e_n. When the tip-surface distance z is "large", i.e. high enough to get W small, the current I is equal to $-eW$ and varies exponentially with z. But, when z decreases, W increases and I saturates to $-e\,e_n$. The measurement of the step in the I(z) curve should allow the determination of the emission rate. Experimentally, the measured current ranges typically from 1 pA to 1 nA, i.e. e_n and $c_p\,p$ ranging from $I/e = 10^7$/s to 10^{10}/s. At room temperature, with $\sigma = 10^{-15}$ cm^2, these conditions implies an energy level below 0.20 to 0.45 eV far from the conduction band, with a doping level ranging from 10^{14} to 10^{18} cm^{-3}.

Case of low emission and capture rates

In such case, the current is limited by the low exchanges between the defects and the bands, and it can be hardly measurable. So, we propose an other approach: now, the bias voltage is chosen in order to get the tip Fermi level E_{FP} at the same energy that the semiconductor conduction band. This implies the existence of a direct current I_1 from the tip to the conduction band, without current flowing *via* the defect energy level. As W is high and greater than e_n, the defect is necessary charged. Now, if we retract abruptly the tip far from the surface (the tip-surface distance increases from z_1 to z_2), W can become lower than e_n, and the defect keeps its charge during a time Δt ($I=I_2$) and then emits its electron towards the conduction band. The current turns into a new value I_3 equal to I_2 perturbed by the contribution (electrostatic effect) of the change in the defect charge. Δt is directly bound to e_n. If measurable, the current transient from I_2 to I_3 has an exponential behaviour with a constant time equal to e_n^{-1}. As in the precedent case, let us discuss on the electrical parameters of the defect and of the semiconductor. The first experimental limit is due to the needed time to retract the z-piezo, i.e. Δt must be greater than this retraction time. Typically, Δt must be greater than 1 ms, which implies that the rates must be lower than 10^3/s. At room temperature, with σ of the order of 10^{-17} cm^2, E_t must be at more than 0.5 eV below the conduction band (0.2 eV if T=110 K), with a doping level lower than 10^{13} cm^{-3} (10^{15} cm^{-3} at 110 K). These conditions are more restrictive than in the precedent case but are reasonably possible. In every case, the temperature is a free parameter that increases the possibility to work with comfortable experimental conditions.

In conclusion of this part, we pointed out that the defect assisted tunnelling current has to be taken into account for a good interpretation of STM images or spectroscopy results on deep levels. Finally, we propose new experimental approaches in order to determine emission (or capture) rate of one defect, using the defect assisted tunnelling current. In the next section, we present spectroscopic results obtained in InAs quantum boxes buried in GaAs.

5. STM SPECTROSCOPY OF InAs QUANTUM BOXES IN GaAs

Semiconductor zero-dimensional (0D) quantum structures, or quantum boxes (QB), exhibit a three dimensional confinement with a δ-function-like electronic density of states. In the past years, self-assembled QB have shown very rich spectroscopic signatures [35]. As their optical properties depend on the wave functions of the electron and hole confined in the box, the knowledge of the shape, the extent and the overlapping of the different wave functions is therefore of prime interest. So far, to glean details of the wave functions in such 0D structures, a growing number of theoretical works have been achieved. The electronic structure of semiconductor boxes with different sizes and facet orientations, embedded or not in an overlayer, have been calculated [36]. But the charge densities associated with the confined wave functions have not yet been resolved experimentally. Even though some experimental results, in electro luminescence and magneto-photoluminescence, have been obtained on the average spatial extent or symmetry of the hole or electron wave functions for an array of boxes [37,38], a detailed analysis of the wave functions in a single box is still missing.

In recent years, scanning tunneling microscopy (STM) and spectroscopy have provided unique means to characterize low dimensional structures. These techniques have allowed the determination of the energy level structure and the observation of the charge densities in artificial structures like the quantum corral [39] or on natural scattering centers like surface steps [40] and nanoscale islands [41]. Since the temperature imposes a limit on the spectroscopic resolution of the STM, most of the experiments were achieved at low temperatures. Indeed, at room temperature, the energy separation between electron levels must be in the range of a few kT, to resolve each of them individually. In the case of 0D semiconductor nanostructures, this requirement is fulfilled.

Such nanostructures can be formed by the controlled growth of InAs on a GaAs substrate in the Stransky-Krastanow growth mode. Recently, cross-sectional STM studies of InAs QB buried in GaAs have been achieved on the (110) face of cleaved samples to determine the shape and size of individual boxes at an atomic resolution [42]. Here, we investigate the conduction band (CB) states of InAs quantum boxes embedded in GaAs with the spectroscopic ability of the STM [43]. We observe at room temperature, for the first time, standing wave patterns in the InAs boxes associated with the lowest CB states. For comparison, within the single band effective mass approximation, we calculate the electronic structure of the box according to its shape given by the STM image. This calculation enables us to determine unambiguously the spatial distribution of the states in the (110) plane of a truncated box.

The InAs quantum boxes were grown by molecular beam epitaxy on a (001) oriented GaAs substrate, with a residual p-type concentration. The active part of the samples consists of 15 arrays of InAs boxes separated by 15 nm GaAs barriers. The whole structure is covered by a 140 nm GaAs overlayer. In order to build each box array, 2.3 monolayers of InAs were deposited on the GaAs layer within 20 s at a temperature of 520 °C. They were immediately buried with GaAs. Samples cut from the wafer were cleaved *in situ* at a pressure below 5 x 10^{-11} Torr. Polycrystalline tungsten tips were prepared by electrochemical etching. The tips were then cleaned by heating and sharpened by a self-sputtering process in UHV. Topographic STM images were acquired with a 120 pA current and positive sample biases.

In figure 16, we display a cross-sectional topographic STM image of a stack of self-aligned

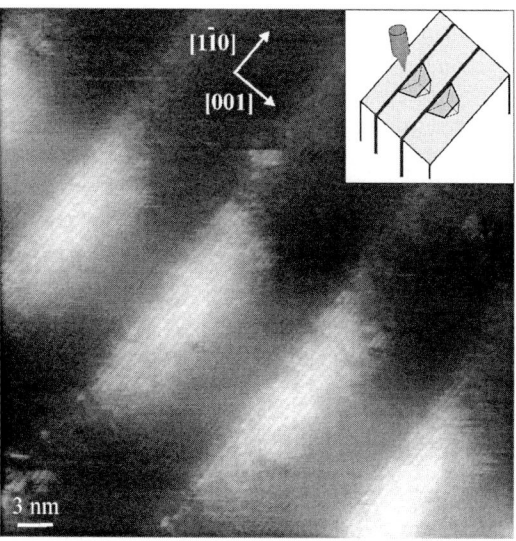

Figure 16: STM image of the (110) face of an InAs box-stack layer in GaAs. The image was acquired at a sample bias of 1.86 V. Inset: schematic diagram of the cleave.

QB along the [001] growth direction. The QB appear bright and the GaAs layers dark. The four boxes are lying on bright layers, which correspond to the wetting layers. We may expect the boxes, that show the largest sizes and the highest contrast, to be cleaved near the dot center, leaving one half of the boxes underneath the cleavage plane. We will now focus on such boxes, where the confined electron states can easily be resolved. In general, these boxes have a length around 20 nm and their height ranges from 4 to 5.6 nm.

As the sensitivity of the spectroscopy measurements depends strongly on the tip-sample separation, we first focus on the contrast mechanism of topographic STM images. To estimate the relative contributions of the electronic effects and the strain relaxation on the contrast, we have performed computations on cleaved InAs boxes embedded in GaAs. A truncated pyramid like InAs box, with a 20 nm [100] x [010] square base and {110} faces, was considered in the calculation. The truncated pyramid lies on a 0.4 nm thick wetting layer. The calculated box was cleaved perpendicular to the [110] direction, along the main diagonal of the square base. Strain relaxation at the cleaved edge was calculated with a finite difference method within continuum elasticity theory, following the work of Grundmann et al. [44].

Figure 17(a) shows the calculated structural height variation at the cleavage surface due to the strain relaxation. For comparison, a contour line plot obtained from the central dot on the STM image in Fig.16 is plotted in Fig. 17(b). Both the computation and the experimental observation give a height variation of 4 Å. At voltages higher than +1.86 volt, the strain relaxation is thus the major source for the contrast between the box and the surrounding layer. This result is similar to the one found for InGaAsP heterostructures imaged at sufficiently large positive sample bias [45].

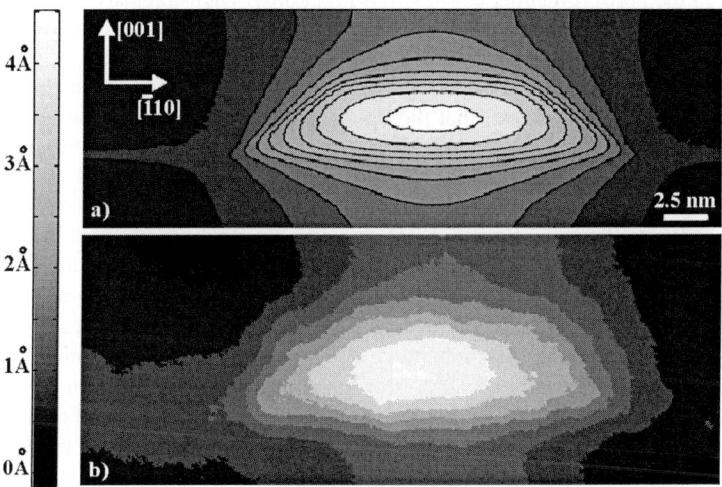

Figure 17: Comparison of the height variation between (a) a simulated topographical image and b) a STM image of the cleaved InAs box located in the center of the figure 16. A low pass filter was used to remove the atomic corrugation from figure 17b. For clarity, a contour line plot is displayed.

To investigate the electronic structure of a box, tunnelling current voltage curves were acquired simultaneously with the topographic image above individual quantum boxes. At every points of the image, obtained with a sample voltage of +2.15 volts, the feedback loop was switched off to measure I-V curves. During an I-V measurement, the vertical position of the tip is held stationary. A current image at a given voltage is then drawn by plotting in each point of the image the value of the current obtained at this point during the I-V measurement. The results are displayed in figure 18. Fig. 18(a) is the topographic image of a QB. This particular box has a base length of 20 nm and a height of 4 nm. Fig. 18(b) shows a typical series of tunnelling current voltage curves taken at different locations above the QB. Depending on the positions where the I(V) curves are taken, a significant voltage offset can be seen. The current images are displayed in figure 18(c) and 18(d). They were obtained for an applied voltage of +0.69 V and +0.82 V respectively. In these images, regions of high current are bright and regions of low current are dark. While the Fig. 18(a) outlines the box contour, the current image in Fig. 18(c) shows a standing wave pattern in the center of the box. The intensity of this feature varies with voltage: it becomes clearly visible at a voltage of +0.63 V and its intensity increases up to +0.74 V. At a voltage of +0.74 V, the standing wave pattern suddenly changes. Two new features are now apparent surrounding the central feature in the [1$\bar{1}$0] direction, as shown in Fig. 18(d). Their intensity increases with voltage up to +0.9 V. For sample voltages larger than +0.9 V, the box becomes brighter and brighter with no other distinct feature visible in the box. For such large voltages, we may expect to tunnel into the empty states of the wetting layer and the GaAs conduction band.

We have previously shown that the tip-sample separation is insensitive to the electronic effects for a voltage of +2.15 volts. Therefore, the variation of the tip-sample separation over the quantum box can be considered significantly small and cannot explain the measured 2 orders of magnitude in the current images.

The wave patterns in figures 18(c) and (d) are obtained without the interference of mechanical contributions and so reflect the spatial distribution of the lowest electron states confined in the box. As only the states lying between the Fermi levels of the sample and of the tip contribute to the tunneling process, the standing wave pattern at a voltage of +0.69 V corresponds therefore to the probability density of the electron ground state in the box, whereas at a voltage of +0.82 V, the standing wave pattern can be regarded as a combination of the probability densities of the electron ground and first excited states [46].

To confirm the origin of the observed standing wave patterns, we performed electronic structure calculations on a cleaved InAs box embedded in GaAs. The box has a similar shape to the one considered in the previous part. The conduction band states of the box were computed with a single band spatially varying effective mass approximation. A finite element method with piecewise-linear interpolation elements was used. The CB strain potential and effective mass tensors were calculated for each cell with an eight band k·p model [47,48]. The mean effective mass in the InAs box was found to be 0.039 m_0, in agreement with other calculations [48,49]. The piezoelectric potential was neglected, its effects being less important [44]. Energies are measured with respect to the conduction band minimum of bulk GaAs. All parameters are taken from ref. 49. A 4.07 eV high barrier was assumed between bulk GaAs and vacuum.

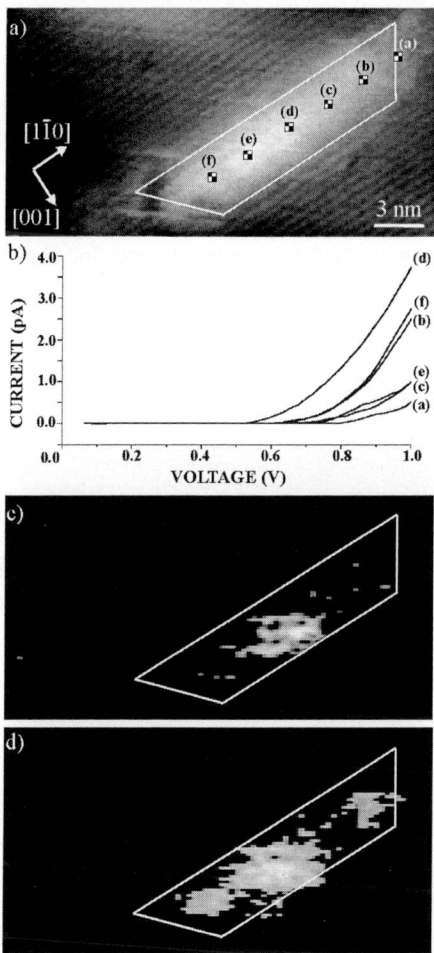

Figure 18: Simultaneously acquired topographic and current images of an InAs QB: a) STM topography with +2.15 V sample voltage, b) tunnelling current voltages curves for different locations in and near the box, c) and d) current images at samples bias of +0.69 and +0.82 V respectively. The grey scale ranges from 0.01 pA to 0.8 pA and 1.5 pA for figures 18c and 18d respectively.

The squared ground state |000> [50] and first excited state |010> are plotted in Fig. 19. Their energy is respectively E = -307 meV and E = -190 meV. The ground state is "s-like" (no nodes) and the first excited state is "p_y-like" with one nodal plane. Both states are pushed towards the surface by the strain potential which abruptly decreases at the cleaved edge [51]. Indeed, due to the strain relaxation, a « triangular » well appears along the [110] direction, whose depth with respect to the innermost parts of the box is about 500 meV. This triangular well increases the local density of states near the cleavage surface that is scanned by the STM tip.

Comparing the simulations with the experiments, we find that Fig. 19(a) matches Fig. 18(c). Similarly, a linear combination of Fig. 19(a) and Fig. 19(b) gives the probability amplitude distribution of Fig. 18(d).

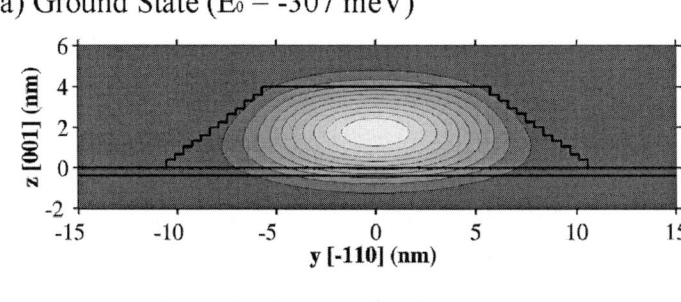

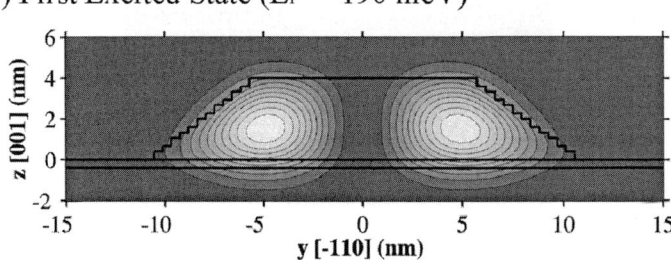

Figure 19: Isosurface plots of the charge densities of the electron ground and first excited states just below the cleavage face of a cleaved InAs box.

Therefore, the calculated density for the ground state and first excited state appears consistent with the STM current images. As the geometry of the box remains to a large extent unknown, others geometries (position of the cleaved surface, shape of the base,...) were investigated. In composition variations were also considered (In rich core with flanks down to $Ga_{0.7}In_{0.3}As$) [52]. Although the eigen energies may depend on the detailed geometry and alloy composition of the box, we always find a s-like ground state separated from the next p_y-like bound state by about 100 meV, which is consistent with our STM observations. In the case considered here, two other bound states exist with respect to the wetting layer (E = -31 meV): the $|100\rangle$ (E = -131 meV) and $|020\rangle$ (E = -60 meV) bound states. However, the $|100\rangle$ state may be hardly visible, the lobe pointing towards the surface being almost non-existent.

In conclusion, we have used the STM to study the probability amplitude distribution of conduction states in cleaved InAs/GaAs QB. Tunneling spectroscopy performed on a low dimensional semiconductor structure provides evidence for size quantization at room temperature, due to the large energy splitting. We have observed two bound states: the ground state shows an s-like envelope function whereas a p_y-like envelope function is found for the first excited state. The shape of these envelopes is consistent with our electronic structure calculation of cleaved InAs QB. Our experiment demonstrates that the spatial variation of the probability densities in the electron states of a semiconductor QB can be resolved. It should not be restricted to electron states but could open up the possibility of obtaining the probability densities of the hole states. Such a direct observation of the wave functions could bring strong evidence for the presence of a permanent dipole in InAs/GaAs QB [53].

The last part of this paper shows spectroscopic results obtained on organic molecules adsorbed on a silicon surface.

6. ORGANICS MOLECULES ADSORBED ON SILICON(100) SURFACE

π-Conjugated oligomers are subject to intense research activity due among others to their potential use as molecular wires in future molecular electronic devices. [54] For example, the characterization of the electron transfer through short chain oligothiophenes connected to metallic electrodes has been reported [55]. As microelectronics technology is based on the use of silicon substrates, there is an increasing need to link organic molecules to the existing silicon technology.

The optimization of the electronic transfer between the molecules and the silicon surface implies a detailed understanding of the nature of the interface between the molecules and the silicon surface.

In recent years, adsorption of unsaturated organic molecules on the Si(100) in ultra-high vacuum (UHV) has been studied and has revealed the possibility to attach molecules to the surface in a controlled manner [56]. Indeed, the silicon (100) surface is made up of silicon dimer rows. The bonding between the two adjacent Si atoms of a dimer can be described in terms of a strong σ and a weak π bond. Due to the ease of the π bond cleavage, unsaturated organic molecules can chemisorb to the silicon surface by reactions similar to the reactions of cycloaddition, involving purely organic molecules in organic chemistry. These reactions lead to the creation of strong covalent bonds between the molecules and the silicon surface, resulting in the formation of a well-defined interface.

As shown in recent work, thienylenevinylene oligomers (nTVs) form a new class of π-conjugated oligomers of particular interest as molecular wires since they exhibit the largest effective conjugation and hence smallest HOMO-LUMO gap among extended oligomers with chain length in the 10 nm regime [57].

On the other hand, since nTVs contain unsaturated double bonds, cycloadditions reactions with the Si(100) surface may be expected and would allow a good connection of these organic chains with the silicon surface.

Due to the low intrinsic solubility of rigid conjugated chains, the synthesis of nTVs up to the hexadecamer stage required the introduction of alkyl chains at the 3 and 4 positions of the thiophene ring in order to increase the solubility. Whereas such a substitution has limited effect on the electronic structure of the oligomers [57,58], it exerts a strong effect on the interactions of the molecules with their physical and chemical environment. Indeed, it has been shown recently that whereas the cation radical of the unsubstituted tetrathienylenevinylene (4TV) (Fig. 20a) reversibly dimerizes, substitution of the 3 and 4 positions of the thiophene ring by hexyl chains (4TVH) (Fig. 20b) inhibits the dimerization process [59].

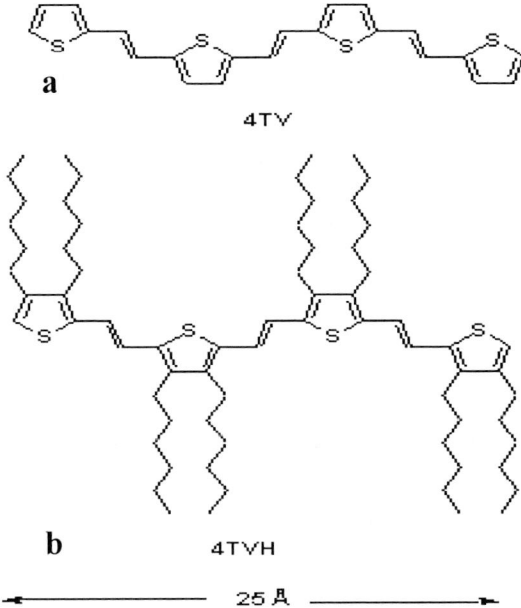

Figure 20: Chemical structure of a 4TV oligomers (a) and 4TV-H oligomers bearing hexyl chains at the β position of the thiophene rings (b).

On this basis, the structure of the nTV chain may be expected to exert a significant effect on the adsorption of the conjugated system on the Si(100) (2x1) surface. In order to test this hypothesis, we have used the scanning tunneling microscope to observe the different arrangements of the 4TV and 4TVH molecules after their adsorption on the Si(100) (2x1) surface. To understand the influence of the alkyl chains on the electronic structure of the adsorbed 4TVH, we performed voltage dependent images. We show that the alkyl chains lead to a weak interaction between the backbone of the molecules and the surface.

Experiments were performed with a scanning tunneling microscope (STM) in an UHV system. The n-type (5 Ωcm) Si(100) wafers were resistively heated to 600 °C for 12 h to degas the sample, cleaned by heating at 900 °C and flashing at 1280 °C for several seconds. Prior to the adsorption of the molecules on the Si surface, the surface was observed in scanning tunneling microscopy (STM) to check its cleanness and the defect densities, which has to be small in comparison with the dose of molecules adsorbed on the surface. The 4TVs oligomers, synthesized as already reported [57], were deposited into a Mo crucible, which was transferred into a homemade evaporator in UHV. Before the deposition process, the oligomer powder was outgassed by heating for several hours. The evaporation process was controlled by a quartz crystal microbalance. During deposition, the distance between the evaporation source and the substrate was 3 cm. The source temperature was measured by a (Tungsten-Rhenium) thermocouple in close contact with the crucible. The evaporation temperature of 4TVH and 4TV ranged from 150 °C to 180 °C. The substrate temperature was kept at room temperature during the deposition. The base pressure in the evaporation chamber was around 2×10^{-9} Torr during the deposition and the deposition process lasted only a few seconds to get submonolayer coverages.

At the end of the deposition, the sample was immediately transferred to the STM chamber, where the base pressure is below 5×10^{-11} Torr. After the transfer of the sample, a mass spectrometer was used in the evaporation chamber to check the absence of decomposition products such as thiophene moieties or alkanes.

As both molecules are made up of single C-C or double C=C bonds, the knowledge of the reactions of these bonds with the Si(100) 2x1 surface constitutes the basis for the understanding of the adsorption of the thiophene based oligomers on the silicon surface. As a starting point, we performed ab-initio calculations based on the local density approximation (LDA) [60] using the DMOL code [61] to predict the reaction products of ethane and ethylene on Si(100) 2x1. For the computation, we use a double numerical basis set and the spin-density functional of Vosko et al [62]. The surface is modelled with a silicon cluster containing 13 Si atoms. This cluster consists of 3 layers and includes two Si=Si dimers in the top layer. Bonds to missing subsurface atoms are terminated with hydrogen atoms. For the calculations, only the Si=Si dimers and the molecules were allowed to relax. The other atoms of the cluster were kept fixed using the Lagrange penalty constraint methods.

Figure 21 shows the reaction products of ethane ethylene molecules on the Si(100) surface. Regarding ethane, the molecule does not react with the surface, as depicted in Fig. 21(a). Ethane tries to reach a minimum potential away from the surface. Thus, due to their saturation, alkanes form an insulator layer, when they adsorb on the silicon surface.

Alternatively, ethylene undergoes a cycloaddition with a Si=Si dimer as shown in Figure 21(b).

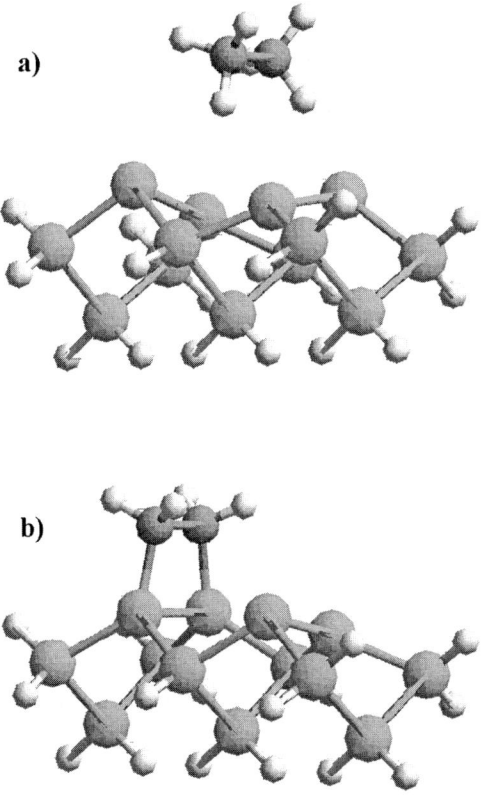

Figure 21: Optimised product geometries of adsorption of (a) ethane and (b) ethylene on the Si(100)

This result is in agreement with previous experimental [63] and theoretical [64] works. Upon adsorption, the C-C bond of the molecule stretches to 1.54 Å. The Si-C bond length between the molecule and the surface reaches 1.93 Å, which corresponds to the length of a covalent bond between a Si and a C atom.

As the 4TV oligomers contain C=C bonds in the thiophene rings and between the thiophene rings, they are likely to react with the Si=Si dimers in a way similar to the one described in Fig. 21(b). Figure 22 shows the Si(100) (2x1) surface after the adsorption of 4TV. The silicon dimers are visible, forming rows of gray bean shaped. On top of these rows, two different types of bright features can be seen. Feature A can be described as a rounded protrusion, whereas feature B has an oval shape with the direction of elongation parallel to the Si=Si dimer bonds. Counting statistics show that the A features comprises 75 % of the entire population of the adsorbates and the B features 25 %.

Figure 22: STM image of the Si(100) surface after deposition of 4TV oligomers. The image was acquired with a sample bias of –2.0 V and a tunnelling current of 60 pA. Two different types of protrusion, labelled A and B, are observed.

To characterize more accurately the adsorption sites of both features, we show, in Fig. 23, a high resolution STM image of the Si(100) surface obtained after the adsorption of the molecules. In this image, the Si atoms of the dimers are resolved separately. Their precise location allows the determination of the A and B feature adsorption sites. The type A feature appears to be centered over a single Si atom of a dimer. The B type feature is localized over a single dimer unit. Although adsorbates associated with the B features can be seen isolated, they tend to form arrays extending in a direction parallel to the dimer bonds, as shown more clearly in Fig. 22.

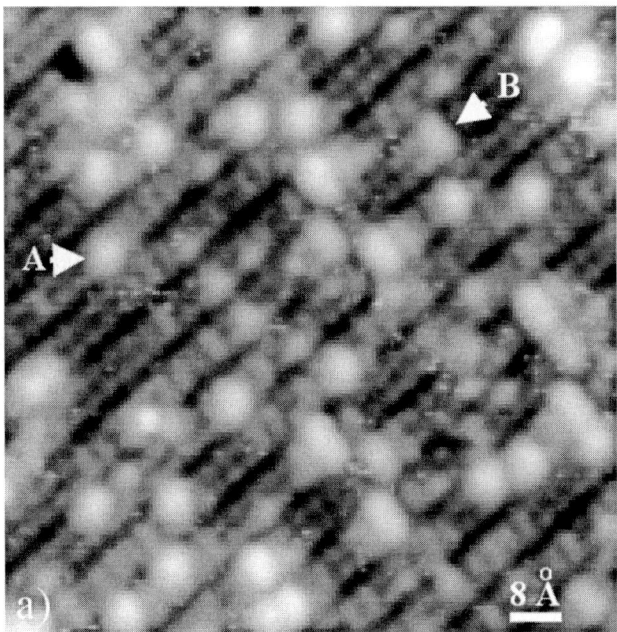

Figure 23: High resolution STM image of the Si(100) surface after deposition of 4TV oligomers. The image was acquired with a sample bias of –2.0 V and a tunnelling current of 40 pA.

From the dimension of the 4TV given in Fig. 20, it is clear that the features A and B do not have the expected size of the oligomer. Thus, the molecules are either broken or vertical on the surface. The gray scale in Fig. 22 is 3.2 Å and would favor a breaking of the molecules during the evaporation process. While the adsorption of 4TV oligomers gives features with a small size, the adsorption of 4TVH oligomers is quite different. A low coverage STM image of the Si(100) surface after the deposition of 4TVH is shown in Figure 24. Fine rows with a small corrugation can be seen extending along the main diagonal of the image. They correspond to the Si dimer rows. On top of the rows, the oligomers appear as elongated features. From this figure, they seem to be adsorbed randomly on the surface with

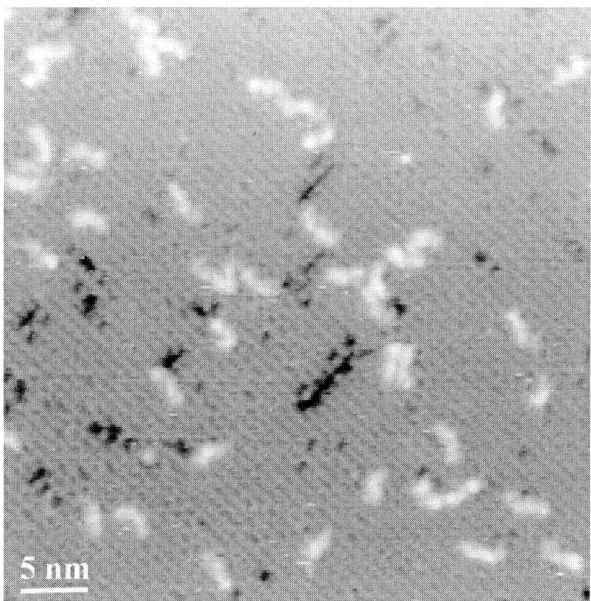

Figure 24: STM image of the Si(100) surface after deposition of 4TV-H oligomers. The image was acquired with a sample bias of –2.7 V and a tunnelling current of 60 pA. The grey scale ranges from 0 (black) to 2.6 Å (white).

no peculiar orientation in regard to the Si dimer rows. Even though many conformations are observed, the features have all almost the same lengths. Using the dimer rows as a template, we find a length of 25 ± 3 Å, close to the 22 Å calculated distance between the outermost side carbon atoms of the conjugated chain. As the hexyl chains are much shorter (~9 Å), the features observed in Fig. 21 correspond to the backbones of the 4TVH oligomers. This result is in agreement with previous STM investigations of alkylated oligothiophenes adsorbed on graphite, where only the oligothiophene backbone was imaged [65]. Our calculation has previously shown that the hexyl chain do not react with the Si(100) surface. Additionally, from photoconductivity experiments of self-assembled monolayers of long alkanes chains, deposited on the silicon surface, it was found that the highest occupied molecular orbital (HOMO) of alkanes containing between 12 and 18 carbon atoms was 4 eV below the top of the silicon valence band [66]. At a negative sample voltage V_0, electrons tunnel from the valence band states of the semiconductor and the states of the molecules, if these states are lying in the energy range eV_0. As the difference between the top of the valence band and the HOMO of the alkanes increases when the alkane length decreases [67], this state for a hexyl chain is lying much below eV_0, at a voltage of -2.0 volts. Therefore, it does not contribute to the tunneling current. The hexyl chains cannot be imaged in the voltage range commonly used while tunneling on semiconductors.

Although the alkyl chains do not appear in the STM images, it was shown that they played a role in the arrangement of decithiophenes adsorbed on graphite [65]. In our case, it is clear from the comparison between Fig. 22 and Fig. 24, which were acquired with the same sample voltage, that 4TV and 4TVH do adsorb in a different manner. While the 4TV appear as small adsorbates on the surface, the 4TVH are lying on the surface. Since the difference of structure between both molecules comes from the substitution of alkyl chains to the thiophene rings, we thus conclude that the difference of appearance in the STM images is caused by the alkyl chains.

To better understand the influence of the alkyl chains on the molecular arrangement of 4TVH, we have acquired STM images with different negative sample biases. Figures 25(a) and (b) were acquired simultaneously with two different voltages. A brief comparison between Fig. 25(a) and (b) reveals that the observation of the 4TVH oligomers in the STM images is highly voltage dependent. At sufficient high negative voltages, the molecules are visible, whereas at lower voltages, most of the molecules disappear and, only for a few molecules, their brightest part still remains visible. The calculated affinity and ionization energies of 4TV oligomers are respectively -1.75 eV and -5.79 eV [58]. The optical gap of 4TVH was found to be 2.4 eV [60]. Depending of the degree of coupling between the molecules and the surface, the HOMO of the 4TVH is thus positioned between 0.6 eV and 1.4 eV below the top of the silicon valence band, resonant with the occupied states of this band, as shown in figure 26. At high negative voltages, the electrons can then tunnel from the occupied states of the semiconductor and the HOMO of the oligomers to the empty states of the tip, thus allowing the appearance of the oligomers. Alternatively, at low negative voltages, only the states close from the Fermi level of the semiconductor can contribute to the tunneling current. These states, positioned in the silicon bulk gap, correspond only to the surface states associated with the Si dimers of the surface and

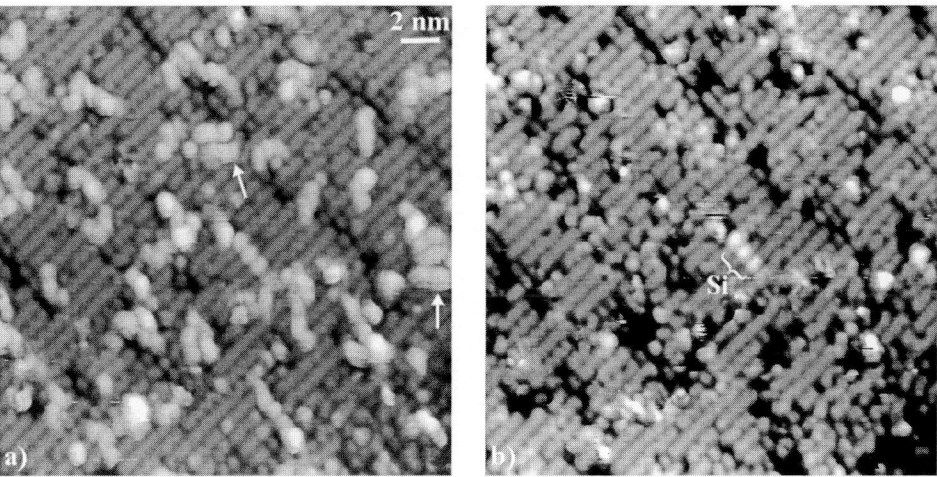

Figure 25: Voltage dependent images of the Si(100) surface after deposition of 4TV-H oligomers. The sample bias was in (a) –2.1 V and in (b) –1.3 V. The arrows indicate some fuzzy molecules. Three Si dimers positioned on top of the surface can be seen in the center of both images.

therefore an STM image acquired at low negative voltages shows mainly the Si dimers of the surface. To emphasize this phenomenon, we can focus on the three neighboring Si dimers seen in the center of Fig. 25(a) and Fig. 25(b). These dimers are lying on the surface and are thus positioned on the same plane as the molecules. They form a row of ad-dimers, [68] which is perpendicular to the Si dimers rows of the surface. These ad-dimers are bright in Fig. 25(a) and, though the reduced voltage in Fig. 25(b), keep their brightness, in clear contrast with what can be observed for the molecules.

The electronic interaction between a molecule and a conductive electrode leads to the extension of the electrode wave functions into the molecule. As a result, the molecular levels are broadened. This high degree of broadening allows the appearance of molecules, adsorbed on metals, at low voltages through a virtual resonance tunneling process [69]. As far as the Si(100) surface is concerned, surface states are lying in the gap of the material and

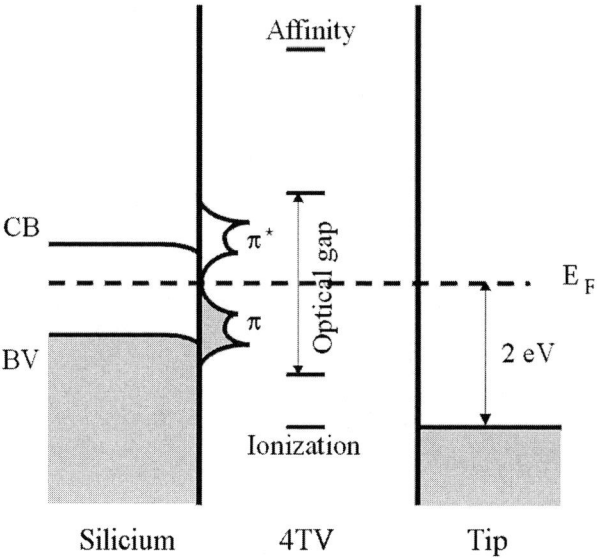

Figure 26: Schematic view of the energy diagram for semiconductor-molecule-vacuum-metal tunnelling. CB and BV correspond to the conduction and valence bands of the semiconductor. The semiconductor Fermi level is denoted E_F.

these states would allow a coupling with a broadened HOMO state. As most of the molecules do not appear at low biases, their HOMO state is therefore not coupled with the surface states in the band gap. We believe that the primary reason for the disappearance of the molecules, at low voltages, is caused by some alkyl chains. Indeed, as it was previously shown, ethane is weakly bound to the silicon surface. Thus, the dissociation of hexyl chains, which are saturated compounds like ethane, is not expected. As the hexyl do not bond to the Si(100) surface, steric hindrance prevents the oligomer backbone from reacting with the surface.

The weak interaction between 4TVH oligomers and the Si surface is supported by the observation of fuzzy oligomers, pointed by an arrow in Fig. 25(a). Such fuzziness can be attributed to the displacement of the oligomers with the STM tip, which indicates a small interaction between the oligomers and the silicon surface.

While most of the oligomers seem to be physisorbed, a few oligomer show a very bright part at both voltages. Their interaction with the surface may be stronger. In this case, the hexyl chains would not isolate the entire backbone from the surface, allowing the unsaturated part of the oligomer to react with the Si dangling bonds, in a similar manner as the one depicted in Fig. 21(b) for ethylene. As a result, the observation of different adsorption configurations suggests that the arrangement of the hexyl chains is random but sufficient to prevent the 4TVH oligomers from forming a well-ordered layer.

In conclusion, the molecular arrangement of 4TV and 4TVH oligomers adsorbed on the Si(100) surface has been studied by scanning tunneling microscopy and spectroscopy. While the result of the 4TV adsorption is ambiguous, 4TVH are observed lying on the Si(100) surface. This different behavior can be attributed to the steric effect of the hexyl chain, which prevents the direct interaction of the π-conjugated system with the Si atoms. Thus, substitution of the thiophene rings by alkyl chains provides a possible way to position the oligomer with its main axis parallel to the surface, while isolating the conjugated backbone. In the frame of using such oligomers as molecular wires, the adsorption of 4TVH is interesting since the hexyl chains allow the molecules to preserve extended π-electronic delocalisation. However the configuration adopted by the chains can vary from one oligomer to another and may make it difficult to selectively connect one end of the molecule to the surface. There is thus a need to substitute this end with functional groups capable of selectively react with the surface silicon atoms.

7. CONCLUSION

From some examples devoted to STM spectroscopy applied on semiconductors, we have shown that it is possible to get local spectroscopic information on defects, low dimensional nanostructures or organic molecules. However, as illustrated for defects or organic molecules, one has to be very careful for the interpretation of STM images or spectroscopic results. Finally, theoretical calculations are obviously needed to understand the STM data, but, as the systems are more and more complexes, the theory finds also its limitations in the size of the studied systems.

ACKNOWLEDGEMENTS

This work has been done in collaboration with B.Grandidier, B.Legrand, T.Mélin, and J.P.Nys for the experimental studies and X. de la Broise, C.Delerue, G.Allan, C.Krzeminski, Y.M.Niquet, C.Priester and M.Lannoo for theoretical calculations. InAs QBs were grown by J.M.Gérard and V. Thierry-Mieg from the Groupement Scientifique CNET-CNRS in Bagneux and nTVs have been synthetized by J.Roncali, P.Blanchard and P.Frère from IMMO (UMR 6501, CNRS) in Angers.

REFERENCES

1. G. Binning, H. Rohrer, and C. Gerber, Phys. Rev. Lett., vol. 49, pp. 57 (1982).
2. G. Binning, C. Rohrer, and C. Gerber, Phys. Rev. Lett., vol. 50, pp. 120 (1983).
3. C.B. Duke, "Tunnelling in Solids", Suppl. 10 of *Solid State Physics*, (F. Seitz and D. Turnbull, ed.), Academic Press, New York, (1969).
4. R.M. Feenstra, Phys. Rev.B, vol. 50, pp. 4561 4570 (1994).
5. P. Mätersson and R.M. Feenstra, Phys. Rev.B39, 7744 (1989).
6. C. Shih, R. Feenstra and G. Chandrashelkhar, Phys. Rev.B43, 7913 (1991).
7. D. Stiévenard, B. Grandidier, J.P. Nys, X. de la Broise, C. Delerue and M. Lannoo, Appl. Phys.72,569, (1998).
8. J. Tersoff and D.R. Hamann, Phys. Rev. Lett. 50, 1998 (1983).
9. S.M. SZE, Physics of semiconductors devices, Wiley, 2nd edition (1981).
10. V. Keldysh, J. exptl. theoret. phys. (U.S.S.R.) 47, 1515-1527 (1964).
11. C. Caroli, R. Combescot, P. Nozieres and D. Saint-James, J. phys. C 4, 916 (1970), J. phys. C 4, 2598 (1971) and J. phys. C 5, 21 (1972).
12. W.A. Harrison, Phys. Rev.B24, 5835, (1981).
13. L. Landau and E. Lifchitz, « Mécanique quantique », Mir, Moscou, p191 (1966).
14. X. de la Broise, C. Delerue and M. Lannoo, J. Appl. Phys.82, 5589, (1997).
15. M. Lannoo and F. Friedel, « Atomic and electronic structure of surfaces - Theoretical foundations », Springer-Verlag, p 64, 113 (1991).
16. J. Wang, T.A. Arias, J.D. Joannopoulos, G.W. Turner and O.L. Alerhand, Phys. Rev.B, vol. 47, pp. 10326-10334 (1993).
17. Z.F. Zheng, X. Liu, N. Newman, E.R. Weber, F. Ogletree and M. Salmeron, Phys. Rev. Lett., vol. 72, pp. 1490-1493 (1994).
18. C. Domke, P. Ebert, M. Heinrich, and K. Urban, Phys. Rev.B, vol. 54, pp. 10288-10291 (1996).
19. C. Domke, P. Ebert, and K. Urban, Phys. Rev.B, vol. 57, pp. 4482-4485 (1998).
20. M.B. Johnson, O. Albrektsen, R.M. Feenstra, and H.W.M. Salemink, Appl. Phys. Lett., vol. 63, pp. 2923-2925 (1993).
21. M.B. Johnson, P.M. Koenraad, W.C. van der Vleuten, H.W.M. Salemink and J.H. Wolter, Phys. Rev .Lett., vol. 75, pp. 1606-1610 (1995).
22. Z.F. Zheng, M.B. Salmeron and E.R. Weber, Appl. Phys. Lett., vol. 64, pp. 1836-1838 (1994).
23. P. Ebert, M. Heinrich, K. Urban, K.J. Chao, A.R. Smith and C.K. Shih, J. Vac. Sci. Technol. A, vol. 14, pp. 1807-1811 (1996).
24. B. Grandidier, D. Stiévenard, J.P. Nys and X. Wallart, Appl. Phys. Lett., vol. 72, pp. 2454-2456 (1998).
25. J.E. Northrup and S.B. Zhang, Phys. Rev.B, vol. 47, pp. 6791 (1993).
26. R.W. Jansen and O.F. Sankey, Phys. Rev.B, vol. 39, pp. 3192 (1989).
27. E.R. Weber, H. Ennen, U. Kaufman, J. Windscheif, J. Schneider, and T. Wosinski, *J. Appl. Phys.*, vol. 53, pp. 6140-6145 (1982).

28. R.M. Feenstra, M. Woodall, and G.D. Pettit, Phys. Rev.Lett., vol. 71, pp. 1176-1179 (1993).
29. R.M. Feenstra, A. Vaterlaus, J.M. Woodall, and G.D. Pettit, Appl. Phys. Lett., vol. 63, pp. 2528-2530 (1993).
30. B.Grandidier, X. de la Broise, D.Stiévenard, C.Delerue and M.Lannoo, Appl. Phys. Lett.76, 3142 (2000).
31. R.B. Capaz, K. Cho, and J.D. Joannopoulos, Phys. Rev. Lett., vol. 75, pp. 1811-1814 (1995).
32. X. de la Broise, C. Delerue, M. Lannoo, B.Grandidier and D.Stiévenard, Phys. Rev.B, 61, 2138 (2000)
33. G.M. Martin, A. Mitonneau, D. Pons, A. Mircea, D.W. Woodward, J. Phys. C **13**, 3855 (1980).
34. M. Kaminska, E.R. Weber, 21st International Conference on the Physics of Semiconductors, ed. Ping Jiang and Hou-Zhi Zheng (World Scientific, Singapore, 1992), 357 (1992)
35. K. Brunner et al, Phys. Rev. Lett. 73, 1138 (1994) ; J.Y. Marzin et al, Phys. Rev. Lett. 73, 716 (1994) ; D. Gammon et al, Science 273, 87 (1996) ; R.J. Warburton et al, Phys. Rev. Lett. 79, 5282 (1997), Landin et al, Science 280, 262 (1998).
36. J. Kim, L.-W. Wang, A. Zunger, Phys. Rev. B 57, R9408 (1998) ; L.-W. Wang, J. Kim, A. Zunger, Phys. Rev. B 59, 5678 (1999) ; O. Stier, M. Grundmann, D. Bimberg, Phys. Rev. B 59, 5688 (1999).
37. D. Bimberg, N.N. Ledentsov, M. Grundmann, N. Kirstaedter, O.G. Schmidt, M.H. Mao, V.M. Ustinov, A. Egorov, A.E. Zhukov, P.S. Kopev, Zh. Alferov, S.S. Rumivov, U. Gösele, J. Heyrenreich, Jap. J. Appl. Phys. 35, 1311 (1996).
38. S.T. Stoddart, A. Polimeni, M. Henini, L. Eaves, P.C. Main, R.K. Hayden, K. Uchida, N. Miura, Appl. Surf. Sci. 123/124, 366-370 (1998).
39. M. F. Crommie, C. P. Lutz, D. M. Eigler, Nature 363, 524 (1993).
40. P. Avouris, I.W. Lyo, Science 264, 942 (1994) ; Ph. Hofmann, B. G. Briner, M. Doering, H-P Rust, E. W. Plummer, A. M. Bradshaw, Phys. Rev. Lett. 79, 265 (1997).
41. J. Li, W-D. Schneider, R. Berndt, S. Crampin, Phys. Rev. Lett. 80, 3332 (1998)
42. W. Wu, J.R. Tucker, G.S. Solomon, J.S. Harris, Appl. Phys. Lett. 71, 1083 (1997); B. Legrand, J.P. Nys, B. Grandidier, D. Stiévenard, A. Lemaitre, J.M. Gérard, V. Thierry-Mieg, Appl. Phys. Lett. 74, 2608 (1999) ; B. Lita, R.S. Goldman, J.D. Philipps, P.K. Bhattacharya, Appl. Phys. Lett. 74, 2824 (1999); H. Eisele, O. Flebbe, T. Kalka, C. Preinesberger, F. Heinrichsdorff, A. Krost, D. Bimberg, M. Dähne-Prietsch, Appl. Phys. Lett. 75, 1060 (1999).
43. B.Grandidier, Y.M.Niquet, B.Legrand, J.P.Nys, C.Priester, D.Stiévenard, J.M.Gérard and V.Thierry-Mieg, Phys. Rev. Lett.85, 1068, 2000.
44. M. Grundmann, O. Stier, D. Bimberg, Phys. Rev. B 52, 11969 (1995).
45. R.M. Feenstra, Proceedings of International Conf. on Defects in Semiconductors, to appear in Physica (2001).
46. Coulomb effect can be neglected. Indeed, after the tunneling of the electrons into the empty eigenstates of the QB, their annihilation occurs mainly through a radiative

recombination process. For such boxes, the radiative lifetime of the tunneling electrons is of the order of 1.3 ns, which is short enough to avoid further charging of the box for a tunneling current of 1 pA.

47. H. Jiang, J. Singh, Phys. Rev. B 56, 4696 (1997).
48. C. Pryor, Phys. Rev. B 57, 7190 (1998).
49. M.A. Cusack , P.R. Briddon, M. Jaros, Phys. Rev. B 54, R2300 (1996).
50. The states are labeled |lmn>, l, m, and n being the number of nodes in the x ([110]), y ([1$\underline{1}$0]), and z ([001]) directions.
51. Y.M. Niquet, C. Priester, H. Mariette, C. Gourgon, Phys. Rev. B 57, 14850 (1998).
52. N. Liu, J. Tersoff, O. Baklenov, A.L.Holmes Jr., C.K. Shih, Phys. Rev. Lett. 84, 334 (2000).
53. P.W. Fry *et al.*, Phys. Rev. Lett. 84, 733 (2000).
54. R.E. Martin, F. Diederich, Angew. Chem. Int. Ed. 38 1350, (1999).
55. C. Kergueris, J.-P. Bourgoin, S. Palacin, D. Esteve, C. Urbina, M. Magoga, C. Joachim, Phys. Rev. B 59 12505, (1999).
56. J.S. Hovis, H. Liu, R.J. Hamers, Surf. Sci. 402-404 (1998) 1 ; J.S. Hovis, H. Liu, R. Hamers, J. Phys. Chem. B 102 6873, (1998).
57. J. Roncali, Acc. Chem. Res. 33 (2000) 147 ; I. Jestin, P; Frère, N. Mercier, E. Levillain, D. Stiévenard, J. Roncali, J. Am. Chem. Soc. 120, 8150 (1998).
58. C. Krzeminski, C. Delerue, G. Allan, V. Haguet, D. Stiévenard, P. Frère, E. Levillain, J. Roncali, J. Chem. Phys. 111 6643, (1999).
59. E. Levillain, J. Roncali, J. Am. Chem. Soc. 111, 8760 (1999).
60. P. Hohenberg and W. Kohn, Phys.Rev. 136 (1964) 864 ; W. Kohn and L.J. Sham, Phys. Rev. 140 A1133, (1965).
61. Cerius2 User Guide. San Diego: Molecular Simulations Inc. (1997).
62. S.J. Vosko, L. Wilk, and M. Nusair, Can. J. Phys. 58, 1200 (1980).
63. H. Liu, R. J. Hamers, J. Am. Chem. Soc. 119 (1997) 7593.
64. Konecny, D.J. Doren, Surf. Sci. 417 169, (1998).
65. T. Kischbaum, R. Azumi, E. Mena-Osteritz, P. Bäuerle, New J. Chem. 241, 23 (1999).
66. C. Boulas, J.V. Davidovits, F. Rondelez, D. Vuillaume, Phys. Rev. Lett. 76, 4797 (1996).
67. D. Vuillaume, C. Boulas, J. Collet, G. Allan, C. Delerue, Phys. Rev. B 58, 16491 (1998).
68. Y.W. Mo, J. Kleiner, M.B.Webb, M.G. Lagally, Phys. Rev. Lett. 66, 1998 (1991); Y.W. Mo, J. Kleiner, M.B.Webb, M.G. Lagally, Surf. Sci. 248, 313 (1991).
69. J.K. Gimzewski, T.A. Jung, M.T. Cuberes, R.R. Schlitter, Surf. Sci. 386, 101 (1997).

Spatially resolved surface spectroscopy

J. Cazaux[a] and J. Olivier[b]

a DTI, UMR 6107, Faculté des Sciences, BP 1039, 51687 Reims Cedex 2, France.
b Laboratoire Central de Recherches, Thomson-CSF, 91401 Orsay Cedex, France.

The present state of the art of spatially resolved AES (Auger Electron Spectroscopy) and XPS (X-ray Photoelectron Spectroscopy) is considered here with the various technical solutions being used to improve their lateral resolution in prototypes or in marketed instruments and, for XPS, by using conventional X-ray sources or synchrotron radiation. A special attention is devoted to the conquest of the third dimension by in-depth profiling and the recent role of spectral and image processing. The correlated limitations in terms of acquisition time for the characteristic images with also the decrease of the signal-to-noise ratio are also mentioned with their consequence on the quantification procedure for elemental mapping. The limiting factors related to the geometry of the specimen (topographic effects) and to its composition (radiation sensitive and insulating materials) are also outlined.

1. INTRODUCTION

XPS and AES are the two most widely used electron spectroscopies for surface compositional analysis from basic point of view of surface science as well as from the technological point of view of industrial applications [1,2]. Since their early applications in the sixties, the improvements of their capabilities in terms of lateral resolution has been the object of continuous efforts and these efforts are presently stimulated with the recent emergence of nanotechnologies combined to the continuous development of, for instance, the miniaturisation of the electronic devices for the semiconductor industry or of the generalised application of dispersed binary catalysts for the petroleum industry. In the present review paper, the basic principles of the two spectroscopies of interest are briefly indicated (section 2). Next, the present state of the art in terms of spatially resolved spectroscopy is established in section 3. Problems to be solved and some possible remedies are discussed in section 4. Photoelectron microscopes based on the use of visible or UV photons are considered beyond the scope of the text.

2. BASIC PRINCIPLES OF XPS AND AES

The aim of the present section is just to give a short overview of conventional XPS and AES. For more details, the interested readers are refereed to classical textbooks in this field [1,2] but also to historical papers of P. Auger for the Auger effect [3] and K. Siegbahn for XPS [4].

The left hand side of fig. 1 illustrates the excitation process involved in XPS and in e⁻ AES (e⁻ : electron induced). In XPS, monochromatic x-ray photons of energy hν are absorbed (following

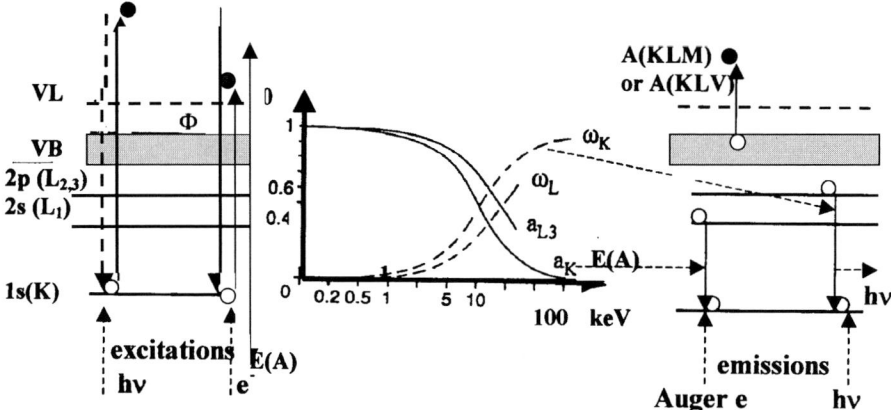

Fig.1 : Basic mechanisms involved in XPS and in e⁻AES. Left : Inner shell electron excitation with incident x-ray photons(and ejection of a photoelectron: XPS) and with incident electrons. Right hand side : schematic representation of the two de-excitation processes in competition, Auger electron emission and characteristic photon emission. Their respective probabilities ω_{ij} (ω_K or ω_L) and a_{ijk} (a_K or a_{L3}) as a function of the initial binding energy $E(A_K)$ is shown on the middle. VL : vacuum level, VB : valence band.

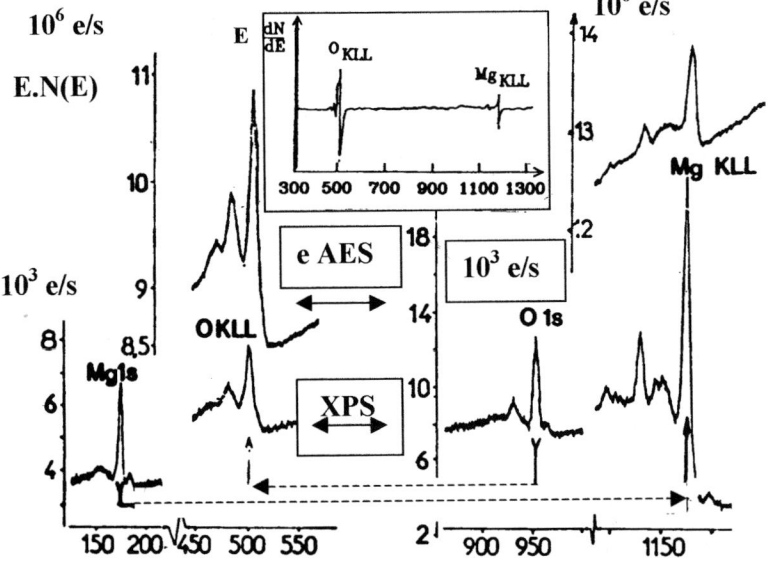

Fig.2 : XPS (Al Kα radiation) and e⁻AES spectra of MgO in the EN(E) mode. The dashed arrows correlate the photoelectron lines, O1s and Mg1s, to the corresponding Auger lines, O(KLL) and Mg(KLL).In the insert:e⁻AES spectrum of MgO obtained in the derivative mode.

the main x-ray absorption mechanism for x-ray photons in the keV range) causing the photo ionisation of the atom (species A or B or ..) and the ejection of an inner-shell (K) electron of atom A (binding energy : $E(A_K)$). The energy analysis of the photoelectron, E_K, permits the determination of the binding energy $E(A_K)$ from the relationship:

$$E_K = h\nu - E(A_K) - \Phi \qquad (1)$$

with Φ:work function (in eV). The identification of the excited atom A, is performed by comparing the value of $E(A_K)$ deduced from equation (1) to the tabulated values of the binding energies of all the possible elements [5].

The initial ionisation process is followed by two de-excitation processes (see right hand side of Fig.1): x-ray photon and Auger electron emissions but the Auger emission probability, a_{ijk}, is greater than the fluorescence yield, ω_{ij}, when the initial binding energy is $E(A) < 10$ keV (see middle part of Fig.1). (X-ray induced) Auger lines are also present in a XPS spectrum and the kinetic energy of the emitted Auger electrons, also nearly obeys to :

$$E(A_{KLL}) \approx E(A_K) - E(A_L) - E(A_L) \qquad (2)$$

In fact, the same Auger mechanism is used in e⁻AES where the initial atomic ionisation results from the bombardment of the specimen by a fine focused electron beam (operated in the 10 keV range). From the tabulated values [6] of all the Auger lines position, except H and He, the elemental identification is also possible in e⁻AES in a way similar to that of XPS or of EPMA (Electron Probe MicroAnalysis) where the characteristic photon energies are detected for such an identification. Then the elemental identification of XPS and of e⁻AES is based on the acquisition of an electron spectrum (below the hv value for XPS and in the 50 - 2500 eV range for e⁻AES) emitted from the specimen as the result of a x-ray irradiation or of an electron irradiation, respectlively. Examples of XPS and of e⁻AES spectrum acquired in the E N(E) mode are given on Fig. 2 for a MgO specimen while for e⁻AES the derivative mode of acquisition is illustrated in the insert.

The short values, in the 0.3-3 nm range, of the inelastic mean free path (IMPF), of these characteristic Auger and photoelectrons (carrying the elemental information) explains the surface sensitivity of XPS and of AES (a surface sensitivity which also requires operation under ultra high vacuum conditions for surface analysis). Figure 3 shows the IMPF values

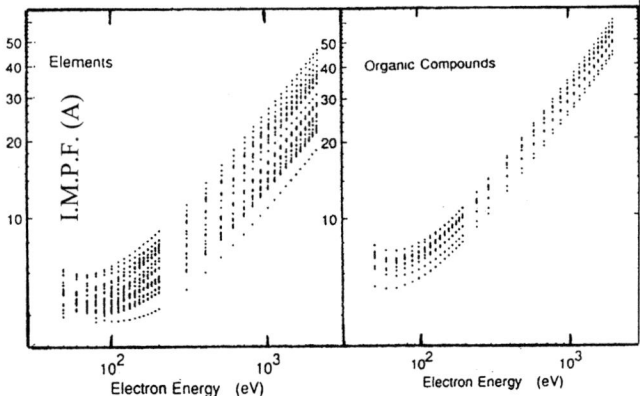

Fig.3. IMPF of electrons for 27 elements (left) and 14 organic compounds (right) as a function of electron kinetic energy (from Tanuma, Powell and Penn [7]).

calculated by Tanuma, Powell and Penn for a group of 27 elements (left) and for 14 organic compounds (right) [7]. This plot as a function of electron kinetic energy shows that the maximum surface sensitivity occurs in the 100 eV range but it also shows that the IMFP depends on the specimen composition (which is, in principle, unknown).

Besides this common point related to the surface sensitivity, there are obvious differences between XPS and e⁻AES. Involving only one atomic level, the spectra of the emitted photoelectrons are composed of rather narrow photoelectron lines reflecting the convolution of the natural width of the initial atomic level by the energy width of the incident radiation (which can be reduced by a monochromator set between the x-ray source and the specimen) while the e⁻AES spectra are much more complicated because, being the result of a three level process, different lines (close to each other) may be associated to a given initial level. The result is that the chemical shift effect (a shift of a fraction of eV up to a few eV which is a function of the electronic and chemical surroundings of the excited atom) is systematically exploited in XPS [1,2,4]. This is far from being the case in AES even if the corresponding chemical shift is sometimes far larger than in XPS (~ 8 eV instead of less than ~ 4 eV for Si in SiO_2 with respect to pure Si) and even if interesting chemical information may be extracted from the shape of the CCV (core-core-valence) and of the CVV Auger lines [8-10]. On the other hand, the advantage of AES over XPS results from the fact that incident electrons are easier to focus than incident x-rays and elemental mapping in Scanning Auger Microscopy (SAM) is more easy than in XPS.

Consequently, one goal of XPS is to improve its lateral resolution but with a good energy resolution to keep the advantage of the chemical shift effect over e⁻AES.

3. LATERAL RESOLUTION: PRESENT STATE OF THE ART.

3.1 Definitions

In optics, the lateral resolution, $l.r.$, characterises an instrument and is defined as the minimum distance between two points for which the two points are resolved. In principle it is then independent from the specimen composition. This resolution can be deduced from the point spread function (PSF) of the instrument (or from its Fourier transform: the transfer function) using next either the Rayleigh criterion (26,5 % deep between the 2 imaged points) or the Sparrow criterion(no deep between the two imaged points) [11]. From the experimental aspect, the PSF function, or more precisely, the line spread function (LSF) may be deduced from the first derivative of a profile taken across an abrupt interface (knife edge profile); such a profile corresponds to the edge spread function (ESF) and the criterion for the lateral resolution may be next applied to evaluate the minimum distance between two identical profiles. The advantage of this definition and this experimental procedure is that it can be easily transposed to depth resolution (of Auger depth profiles for instance) and to the energy resolution of spectrometers. Unfortunately, in most of the cases, the lateral resolution is deduced from the acquired knife edge profile directly because the derivative procedure amplifies the noise. Next arbitrary criteria are often applied to the experimental ESF : the distance between points giving 10 % and 90 %, or 20 % and 80 % or 25 % and 75 %. The result is that the exact performance of different instruments from the point of view of their lateral resolution is difficult to compare. In this context, the situation of SAM was subject to a long debate because, in AES, the PSF is composed of two contributions (see Fig. 3 inspired from ref [12]) : the sharp contribution of the incident beam and the contribution of the back scattering halo. The question was : is the lateral resolution governed by the incident beam

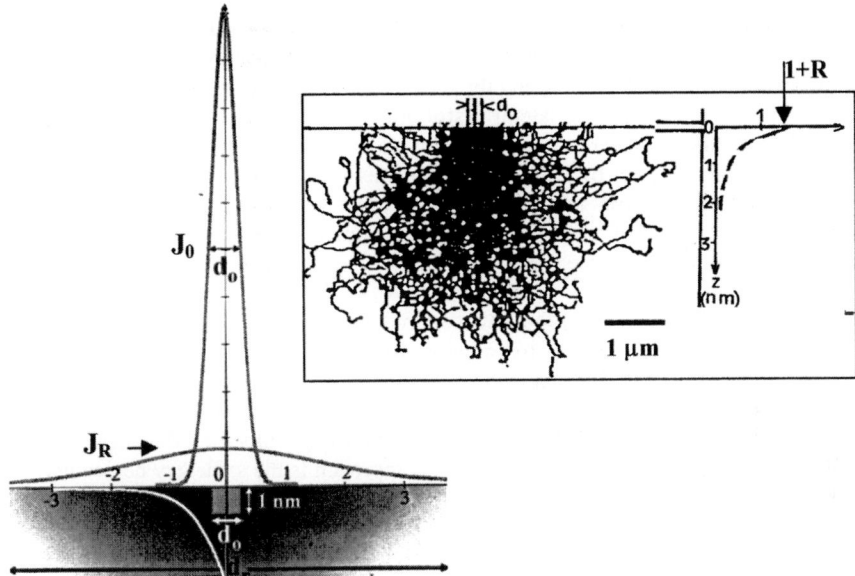

Fig.4. PSF of Scanning Auger Microscopy with the radial emission of Auger electrons induced directly by the incident beam (diameter d_0), J_0, and that induced by the backscattered electrons, J_R of relative weight R (here R ~ 0.4). The insert represents the random walk of a 20 keV electron beam into Al_2O_3. Auger electrons are generated everywhere into the irradiated volume but the escape probability of Auger electrons is more or less of the exponential form (effective escape depth : λ ~ 1 nm) shown on the right part of the insert or in white on the figure. The grey levels represent this volume when weighted by the lateral excitation probability and the escape probability (note the difference in the horizontal and vertical scales). The analysed volume is rather large but close to 50% of the emission originates from a volume of diameter d_0 and of height λ.

diameter or by the lateral dimension of the halo? When applied to SAM, the definition based on the minimum distance leads to the fact the lateral resolution is only related to the incident spot diameter alone (*l.r.* ≈ d_0) even when the Auger backscattering factor, R, is close to the unity .The present consensus on this point is demonstrated by the use of the narrowest possible probes in the scanning techniques (see below).

The performance of an instrument is not characterised only by its lateral resolution. Among many others parameters, there is also the minimum dimension, d_m, for a detail to be detectable (or the elements composing it to be identifiable). d_m is often confused with *l.r.* while the in-depth resolution is confused with the minimum buried thickness being detectable.

In fact d_m obviously depends of the signal-to-noise ratio and consequently of the incident beam dose and of the elements composing the specimen : then d_m can be either larger or smaller than *l.r.* (or d_0) [11]. The lateral resolution also differs from the minimum dimension, d_q, for a detail to be quantifiable. As illustrated on figure 3 for SAM, the lateral non uniformity of the response function of high resolution instruments requires change of the usual quantification procedures.

3.2 Microspectroscopy or Spectromicroscopy: Scanning or not Scanning?

For nearly all microanalytical techniques, elemental mapping may be obtained in two different ways. One way consists in focusing the incident beam on the specimen in order to acquire, first, a point spectrum. The lateral distribution of a given element is obtained by selecting the corresponding energetic window and by scanning the incident beam (or the specimen). An example is given by EELS (Electron Energy Loss Spectroscopy) in a STEM (Scanning Transmission Electron microscopy). The other approach consists in acquiring different characteristic images of a widely irradiated specimen by combining a magnification device to an energy filtering device. The corresponding example is Energy Filtering in a TEM.

Figure 5 gives a general view of these two approaches which illustrates the fact that the goal of the scanning approach is to obtain, first, a spectrum and next, images (spectromicroscopy) and the main device is the focussing element which has to give the finest incident probe as possible. The goal of the second approach is to obtain first, an image and next, spectra (microspectroscopy) and the lateral resolution is governed by the aberrations of the magnifying device. One advantage of this second approach is the same key device may be used for obtaining magnified characteristic images of the same specimen but being irradiated with different types of incident particles. In the context of surface microspectroscopy, an excellent example is given by the LEEM (Low Energy Electron Microscope) built by Bauer et al [13] which may be operated as a AEEM (Auger Electron EM) [14] or transformed into a XPEEM (X-ray excited Emission EM) by using a synchrotron radiation source [15] as well as into a SPLEEM (Spin-Polarised LEEM) or into a MIEEM (Metastable Impact Electron EM) [16]. Despite the fact that some of these instruments are recently available on the market [13], here, they will be considered as prototypes in the following subsections. SMART (SpectroMicroscope of All Relevant Techniques) is also a similar multipurpose microscope, presently in construction [17].

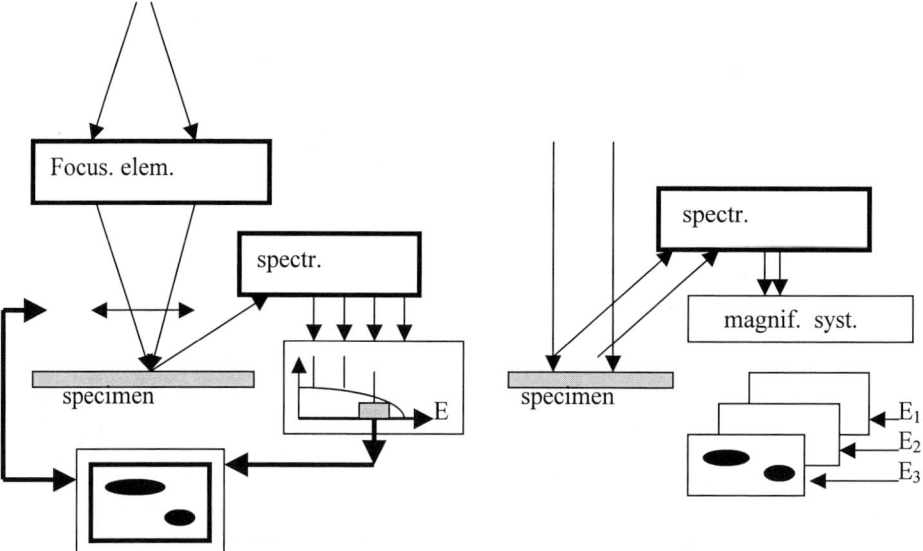

Fig.5. Schematic principle of spectromicroscopy (left) and of microspectroscopy (right).

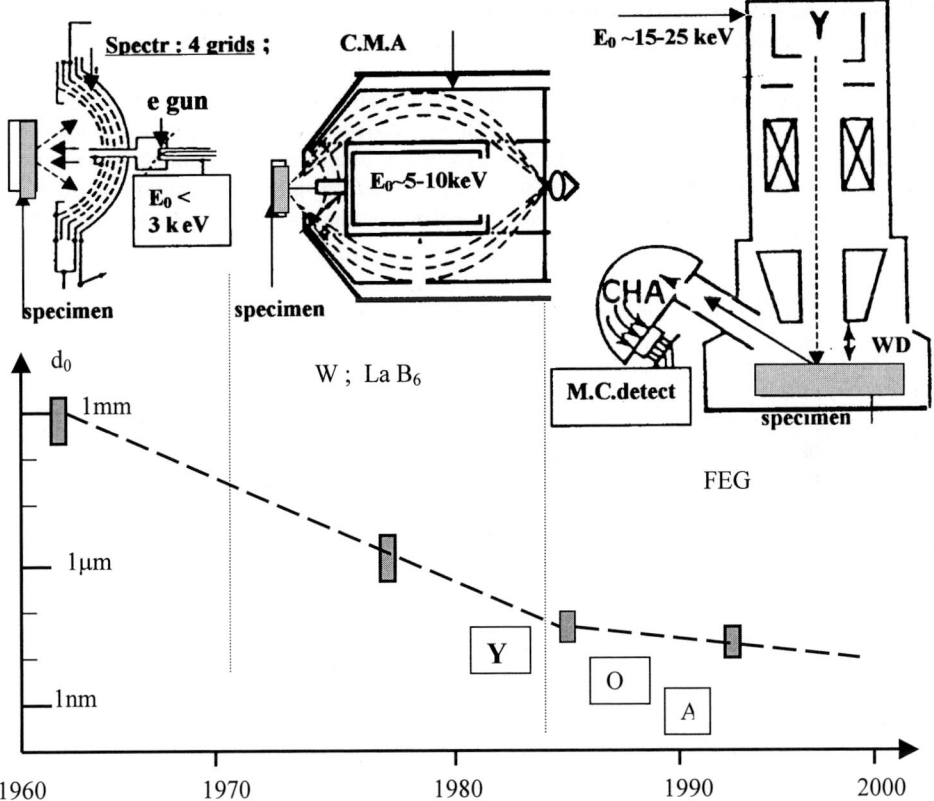

Fig.6. Top: approximate evolution of the architecture of the Auger instruments along the last four decades. Bottom: corresponding evolution of the probe diameter, d_0, for the commercial instruments (full symbols) and for some prototypes (open symbols with a letter; Y: York MULSAM at 20keV [22]; O: Orsay modified STEM [28-30] and A: ASU instrument, Venables and al. [23-26] at 100 keV.

3.3 Past and present state of the art of Spatially Resolved e⁻AES
3.3.1. Instrumental evolution of e⁻AES: commercial instruments.

From the early developments of the technique, the architecture of commercial Auger instruments has been extensively modified and fig. 5 illustrates such an evolution (in fact fig 5 only indicates the trends of the Auger equipment in terms of approximate periods, approximate incident spot diameters and examples of experimental arrangements). At the very beginning of AES, in the sixties, the spectrometer was a four-grid analyser derived from that used in Low Energy Electron Diffraction (LEED): fig. 6a. This instrument was operated at a primary beam energy below 3 keV and the incident spot diameter was in the mm range. Nowadays, for the marketed instruments there is a quite general consensus to build Scanning Auger Microscope (SAM) based on an architecture similar to that of a Scanning Electron Microscope (vertical column) under UHV conditions and with an electron analyser which is often of the CHA (Concentric Hemispherical Analyser) or of the CMA (Concentric Mirror

Analyser) type [1,2] but the two with a parallel detection system (see fig.6c). The SEM mode permits to obtain secondary electron images which permit to select the zone of interest for the surface analysis.

At the early developments of AES, the identification of elemental species was made using differentiated spectra acquired in an analogue mode with the help of a lock-in amplifier (see insert of fig.2 as an example), and incident beam intensities in the microampere range. The acquisition of the first derivative of the electron energy distribution, such as $\partial[EN(E)]/\partial E$ for example, remains very popular for semi-quantitave Auger analysis at moderate (> 1µm) lateral resolution because it presents some advantages, among them direct background suppression and the increase of peak visibility. In addition there is the easier identification of the chemical shift effect and of the line shape modification associated to the change of the chemical state of an atom [18,19]. Also in this derivative mode, spectral quantification is simple to obtain from peak-to-peak height measurements. However, to be effective, this procedure has only to be restricted to metallic mixtures where there are no spectral peak shape changes and for spectra recorded on one instrument [19]. Many of these problems disappear if the spectra are recorded in the direct spectral mode, N(E) or EN(E), for instance. These direct spectra are always acquired in a digital form by using a pulse counting technique when the incident beam intensity is in the nanoampere range. In the direct mode, a characteristic Auger intensity I(A) is directly related to the area of its corresponding peak obtained after removing a well defined background.

The counting mode has to be used for high lateral resolution SAM because it requires a lower primary beam intensity, therefore allowing the minimisation of the incident spot diameter. This minimisation is also obtained by increasing the primary beam energy so that, presently and in most of the instruments a Schottky field emission gun (FEG) is operated at a primary beam energy in the 15-25 keV range providing then a nA intensity in an incident spot of around 10-20 nm on the specimen.

3.3.2. Prototypes

The AEEM built by Bauer et al. is the unique example of an Auger instrument based on a non Scanning mode. Preliminary AEEM images of small Ag crystals on a Si substrate have been obtained at $E_0 \sim 2.5$ keV with a band width of 1 eV and with an edge resolution (15-85%) of only 200 nm but also with only a 15 s acquisition time [14]. The measured lateral resolution is worse than that predicted from theoretical estimates. Improvements will certainly be obtained in the near future making this approach attractive, and hence realising the possibility of obtaining various images of a given area of a given specimen.

All the other prototypes are based on the scanning mode and the principle of SAM was first demonstrated by Mac Donald and Waldrop [20]. An early practical instrument was built by Todd, Poppa and Veneklassen with a miniature field emission gun set inside a CMA and operated at 5 keV [21]. More recently two prototypes have been built in two universities. The different strategies being adopted illustrate very well the conflict between lateral resolution and quantification. At York University, the emphasis was placed upon the technique required to quantify images in order to map the variations of the chemical composition in the surface and spatial resolution was a second priority. Described by Prutton et al. [22] this instrument is referred to as MULTI spectral SAM (MULSAM). Two different Field Emission Guns (FEG) can be operated; their main characteristics are : $E_o = 20$ keV; $I_o \approx 6$nA; d_o(FWHM) ≈ 100 nm or are: $I_o = 10$ nA and $d_o \approx 20$ nm (at 5 keV). This instrument is also characterised by the variety of detectors being operated and among them, the fifteen detectors set at the exit of the CHA for a parallel detection of the Auger spectra.

At Arizona State University, the emphasis was placed upon the lateral resolution and the instrument is referred to as MIDAS (Microscope for Imaging Diffraction and Analysis of Surfaces) [23-26]. This instrument is based upon a heavily modified STEM operated at E_o = 100 keV, the incident spot diameter may reach the 1-2 nm range for a nA intensity [23]. The specimen is set in a magnetic immersion lens and a part of the structure of this lens acts as a beam paralleliser in such a way that almost all the emitted Auger electrons enter the CHA [27], optimising the collection efficiency of this analyser. A spatial resolution of ~ 3 nm in Auger peak images has been obtained on bulk samples and of 1 nm on thin specimens. Some Ag particles as small as 1-2 nm in diameter and containing as few as 10 Ag atoms have been also detected [24,26]. The advantage of using such an unusual high primary beam energy, results from the increase of brightness of the FE gun and the decrease of the background (because η does not significantly change between 20 and 100 keV and is spread over a larger energy range, then $\partial\eta/\partial E$ decreases).

Previously, in Orsay, the same primary beam energy was used in a marketed STEM but equipped with a CHA, in order to perform SAM experiments with the specimen set outside of the pole pieces. A sub-ten nm resolution was attained with a 8 nm probe size (and 8nA intensity) [28]. The elemental analysis of sub-ten nm Pd particles was performed and the number of Pd atoms being detected was estimated to be less than 4000 [29]. A minimum detectable concentration less than 2×10^{-3} was also attained by the detection of a silicon submonolayer buried in a GaAs matrix [30].

Fig. 5 includes the d_o values reached by some prototypes: that of York [22]; that of Arizona State University [23-27] and the STEM of Orsay equipped with an Auger analyser [28-30].

3.4. Past and present state of the art of spatially resolved XPS
3.4.1. Evolution of commercial instruments.

From the instrumental point of view, the main components of a XPS laboratory apparatus are: a x-ray tube (Al Kα radiation ~ 1486 eV; Mg Kα radiation ~1254 eV), a x-ray monochromator, the specimen with its environment and an electron analyser, all components being set in a ultra high vacuum (UHV) chamber. Like for other microanalytical (EPMA, EELS) or microscopical (TEM, SEM) techniques but in opposition to the significant changes in the Auger equipment over the last four decades, the basic architecture of laboratory XPS instruments looks similar with respect to the instruments initially built by Siegbahn et al [4] with, in particular, the quite systematic use of a spectrometer of the CHA type. In fact, many significant changes have been performed with a special attention to find solutions for improving the lateral resolution which was initially given by the projected area of the entrance slit of the analyser on the specimen surface and it was in mm range.

Presently all modern spectrometers operate with an input lens which, at a minimum, transfer a photoelectron image of the analysed area from the sample surface to the entrance slit of the analyser itself. This lens also acts as a retardation element for the electrons entering the analyser. More sophisticated versions of this input lens may project a magnified image on the entrance slit of the analyser, improving then the lateral resolution but with a corresponding loss in intensity. Then, for improving the lateral resolution it is possible to follow the two approaches considered in subsection 3-2, that is to say either the focusing approach or the imaging approach (by combining the electron optical properties of the spectrometer to that of additional electron lenses such as the input lens). In fact most of the commercial instruments combine a mixture of the two approaches.

The pure microprobe approach consists in using the focussing properties of an ellipsoidally shaped quartz crystal monochromator which provides a small x-ray spot on the specimen surface. By scanning a focused electron beam across the Al anode, the x-ray spot is then scanned on the specimen and elemental or chemical state maps may be obtained [31].

A pure imaging approach consists in playing with one of the possible modes for a sophisticated input lens. The imaging mode of the input lens(see fig 7A, right) provides a magnified photoelectron image of the sample in the form of a parallel photoelectron beam at the entrance slit of the CHA. The analyser introduces the energy dispersion but also the transfer of the image from the entrance slit to the exit slit. Another electrostatic lens may be set at the exit lens of the CHA to obtain, finally, a magnified filtered image on a two-dimensional detector. The spectrum mode may be operated by using the input lens to focus the emitted photoelectrons on the entrance slit of the analyser. The analyser is a stigmatic electron optical device which provides a point image of a monochromatic point object point so that the XPS Spectrum is normally displayed at the exit of the spectrometer [32] (see also, Tonner et al. for a similar approach but based on the use of the synchrotron radiation [33]).

A combination of the two approaches consists, for instance, in focusing the incident x-rays only in one direction in order to obtain, from the x-ray monochromator, an elongated x-ray spot on the sample. The combination of the input lens and of the spectrometer provides a line photoelectron spectrum or a E-x image at the exit of the spectrometer with the normal energy dispersion in one direction (in the plane of incidence which also contains the principal plane of the spectrometer) and the space direction, x, orthogonal to this plane of incidence [34,35].

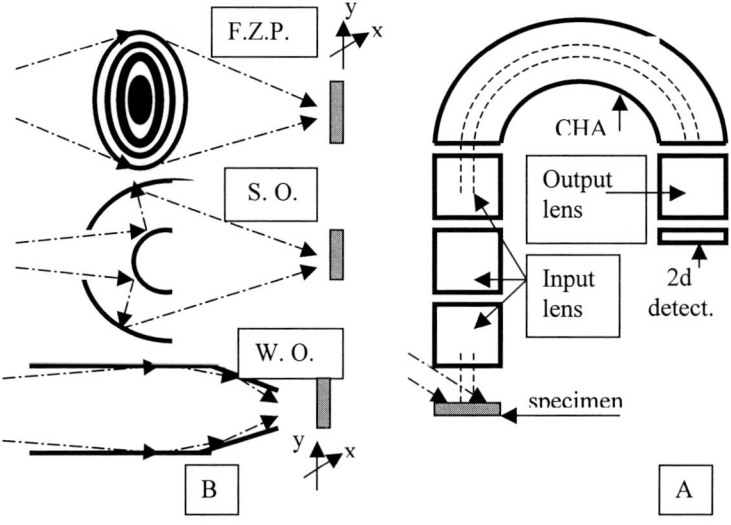

Fig.7.A (right): example of a combination of an input lens and of a spectrometer for image filtering in a marketed XPS instrument [32,33]. B(left): various methods used to focus incident x-rays on a mechanically rastered specimen in prototypes built near synchrotron radiation sources. FZP: Fresnel Zone Plates; S.O.: Scharwzchild objective; W.O.: Wolter objective.

Finally, the use of the imaging properties in two directions, x and y, of a spherical mirror analyser has been recently proposed at the place of the more conventional CHA[36].

Despite the wide variety of technical solutions for improving the lateral resolution of XPS, it seems that comparable values, situated in the 10 microns range, have been obtained or are claimed in the advertising brochures and in the technical sheets of the principal manufacturers of XPS instruments.

It is beyond the scope of the present paper to go deeper in this type of comparison because many other parameters have to be taken into account for the choice of an instrument (signal intensity, energy resolution, time of acquisition, ease of operation, additional attachments, etc). For instance, a lateral resolution in the ten microns range has been obtained from a long time but the simple method being used only applies to thin films attached on thin foil targets (with a back foil excitation of this target by a scanned incident electron beam [37]). The very close proximity of the x-ray source and of the specimen permits to obtain intense point spectra such as that shown on the bottom of fig. 2 for MgO [38].

3.4.2. Prototypes.

Many of the prototypes are operated in combination with synchrotron radiation sources and photon energies below 1 keV. One technical reason, among others, is that the harder the incident x-rays and more difficult they are to focus in the microprobe mode (or the higher the kinetic energy of the photoelectrons of interest and poorer is the resolution of the direct imaging mode). But if this rather low photon energy range presents some technical advantages for surface science, it hinders the use of it for elemental identification of compositionally unknown samples because of the risk of confusion between two elements having some outer shell electrons with close binding energies. Also, by tuning the synchrotron radiation, it is possible to obtain, in theory, characteristic photoelectron images without the use of an electron analyser. The method consists in subtracting two projected images acquired successively before and after tuning the incident photon energy across characteristic energy levels of the sample while the total photoelectron flux is collected [39]. This approach may be interesting for some specific applications but it cannot be of a general use because the low energy electrons are constituting the main contribution of the whole photoelectron flux so that the signal of the final characteristic image is just the difference between two very large signals while its noise results from the addition of the noise of these two initial images.

For characteristic x-ray photoelectron imaging in the filtered approach, two developments are pointed out, here (for other specific arrangements and applications the interested readers are referred to the special issue of the J. Electron Spectrosc. Relat. Phenom. 84 (1997) pp1-260). They are the photoelectron microscope built by Tonner et al [33] which was also mentioned previously and the LEEM operated in the XPEEM mode which provides energy filtered images by widely irradiating the specimen at a grazing incidence [15]. In the XPEEM mode of the LEEM and despite the chromatic and spherical aberrations of the cathode lens system, a resolution from several nm to several 10 nm is achievable when very small apertures are used (the usual compromise between intensity and resolution). For 65 eV photons, an energy resolution below 0.15 eV has been demonstrated in the local spectrum extracted from many filtered images [15].

Most of all the other prototypes are based on a microprobe approach where the sample is mechanically rastered to obtain characteristic photoelectron images in the scanning mode and now nearly all commonly use a CHA for acquiring the photoelectron spectra.

These focusing approaches differ from each other by the technical solution used for focussing incident x-rays on the sample solutions which also applies for transmission x-ray microscopy in the scanning mode. The experimental arrangements used to obtain x-ray micro beams have been recently reviewed [40,41] and some of them are illustrated on fig.7B.

Ade et al. (NSLS, Broockhaven) used a Fresnel zone plate to focus a 670 eV photon beam on the specimen [42]. The reported line scan of the knife edge test clearly show that a sub-micron lateral resolution was obtained (0.3 μm for the 25%-75% rise in intensity). More recently, using also a zone plate to focus ~ 500 eV photons, a 0.1 μm resolution has been recently claimed to be routinely reached by A. Barinov et al. (ELETTRA, Trieste) [43]. The corresponding characteristic images are acquired in parallel by using a battery of 16 detectors at the exit of the CHA and a great care is taken to avoid the collection of spurious photoelectrons being generated outside of the specimen - such as by the zone plate-.

Based on the total reflection of x-rays at grazing incidence, the advantage (over zone plates) of the Wolter type objectives are their larger working distances and the fact that they are achromatic (which permits to take the possible benefit of a tuneable x-ray source). These advantages are paid by a less resolution which is also strongly dependent of the accuracy of the machining of their (ellipsoidal) profiles [41]. A resolution situated in the micron range has been obtained at Lund with a radiation in the energy range of 15-150 eV [44]. Similar results have also been obtained in the total imaging mode at the Photon Factory (Tsukuba Japan) [45]. Based on the grazing incidence x-ray optics, the development of microcapillary (monocapillary concentrator, microchannel arrays, and polycapillary concentrator) seems very promising to obtain in a simple way sub-micron resolution in XPS [46,47].

In the soft x-ray region, a multilayer coated Schwarzchild objective has been used in the MAXIMUM photoemission microscope at the Wisconsin Synchrotron Radiation Centre. With less than 0.1 μm (90nm), the highest resolution in the world has been claimed to have been reached .The energy resolution is also very good, ~ 0.4 eV, but the photon energy was only of 95 eV [48,49]. Using also a Schwarzchild objective but operated in a table top equipment, the original approach of Kondo et al has to be pointed out despite more modest results: a lateral resolution of 0.5 μm and an energy resolution of 0.6 eV (for a photon energy of 250eV) have been obtained with the use of a Laser-Plasma x-ray source and a time of flight photoelectron spectrometer [50].

This last example demonstrates that original developments remain possible in this field. Besides the development of time of flight electron spectrometers for parallel recording of the spectra [51] there is the improvement of the surface sensitivity (via the background suppression). AES also uses, for instance, positrons as incident particles [52] and XPS uses x-rays at incidence angles below the critical angle for total reflection [53,54].

4.LATERAL RESOLUTION : CORRELATED PROBLEMS AND SOME REMEDIES.

When the dimension of a pixel (reported to the object plane), d_0, is decreased, its associated characteristic intensity is also decreased. The improvement of the resolving power (as defined in subsection 3.1) is not only related to the x-ray optics or to the electron optics of the instrument but has to be associated to its ability to obtain a detectable signal in a reasonable time of acquisition. The minimum dimension for a detail to be detectable results, then, from a compromise between the best resolution that is compatible with a sufficient signal intensity and this compromise leads to the notion of minimum dose (or fluence: number of incident particles per unit area) for a detail to be detectable. This minimum dose may be obtained with a low primary beam intensity and a very small d_0 dimension but the increased time of acquisition has to remain within acceptable limits. Also this minimum dose has to be compatible with the damage threshold of the specimen submitted to the particles. In addition, there is the problem of the accuracy of the quantification (the evaluation of the component concentrations) of a small detail when the signal intensities a subjected to very large statistical fluctuations. The cross-correlation between lateral resolution, signal intensity, radiation damage (including charging effects) and quantification procedure at high lateral resolution is the subject of the present section.

4.1. Minimum detectable concentration, x_m.

In conventional AES (point analysis) where only the change of the components' concentration as a function of depth is considered, the detected Auger signal intensity of element A, I(A) is related to its atomic concentration, $C_A(z)$, by [1,2]:

$$I(A) = I_o \, \text{cosec} \, \alpha \int_0^\infty [N°C_A(z)(1+R_{A/S})Q_A dz] a_{ijk} \, e^{-z/\lambda \sin\theta} T_A \quad (3)$$

where I_o is the incident intensity and α is the angle of incidence. The expression between brackets corresponds to the number of ionised atoms (electronic level i) per incident electron, where Q_A is the electron ionisation cross-section ; $R_{A/S}$ is the Auger backscattering coefficient, R, and the subscript A/S represents the reinforcement of the Auger signal of surface element A by the electrons backscattered by the substrate S ; N° is the atomic density (~ 5×10^{22} atoms/cm^3). Further, T_A is the collection efficiency of the analyser ; a_{ijk} is the Auger yield and the exponential term describes the attenuation of Auger electrons generated in the specimen at depth z when going towards the analyser (λ = attenuation length and θ = take off angle). The attenuation length, λ, is a functional parameter related to the inelastic mean free path (Fig. 2) and to the elastic mean free path of Auger electrons escaping into the vacuum [55-57]. When $C_A(z)$ is a constant over a depth of say, 3λ, the integration leads to

$$I(A) = I_o N°C_A (1+R_{A/S})Q_A \lambda \sin\theta \, a_{ijk} T_A \quad (4)$$

where now the normal incidence is only considered for the sake of the simplicity. The surface sensitivity of AES is increased when $\sin\theta$ is decreased but only 63 % of the signal come from the thickness $\lambda \sin\theta$ and the information depth is usually chosen to be $3\lambda \sin\theta$. This information depth is in the 1-5 nm range depending upon the kinetic energy of the Auger electrons of interest (and upon the take off angle θ).

For a pure element one may write eq. (4) in the form [58] :

$$I^\infty(A) = I_o . \beta(A) . T_A \quad (5)$$

where $\beta(A)$ is the quantum yield for the Auger electron production and is in the 10^{-4} range (with $Q_A \approx$ a few 10^{-20} cm^2 ; $\lambda \sin\theta \approx 1$nm, $a_{ijk} \approx 1$; N° $\approx 5.10^{22}$ cm^{-3}).

Similarly the background may be written in the form:

$$B = I_o \, \beta_B \, T_A \quad (6)$$

Using the Rose criterion [59] in the pulse counting mode, the minimum detectable concentration x_m (in atom/atom) is given by [58,60] :

$$x_m \approx \frac{3\sqrt{B.\tau}}{I^\infty(A)\tau} \quad (7)$$

where τ is the channel dwell time for the signal, $S \sim I\tau$, and for the background $B\tau$ (the two being expressed in counts). x_m is directly related to the S/N ratio [60] by $x_m = 3/(S/N)$ and for the major species it may vary from 10^{-1} to 10^{-4} as a function of the incident beam intensity I_o

and of time of acquisition of the spectrum [60]. The medium values (10^{-2}-10^{-3}) are obtained for $S^\infty(A) = I^\infty(A)\tau \approx 10^6$ counts (and $B\tau$ of about 10^7 counts as an order of magnitude). These medium values correspond to the usual sensitivity of e-AES : a fraction of monolayer [1,2].

Inserting eqs(5) and (6) into eq.(7) with also $I_o \sim J_o\, d_o^2$, one obtains the relationship correlating the incident dose (or fluence of the beam $F \sim J_o\, \tau_i$ with τ_i: irradiation time) to d_o and to x_m :

$$x_m \sim 3 / [k\, S^\infty(A)]^{1/2} \sim 3 / d_o[\, k\, c\, F\, \beta(A)\, T_A]^{1/2} \qquad (8)$$

where k is the signal-to-background ratio for the pure element of interest (k is between 0.1 and 1 in e-AES) and c is the ratio τ/τ_i between the dwell time per channel and the irradiation time (c is equal to the unity at maximum for parallel recording of the spectrum).

The minimum fluence F_m for a pure element to be detectable [when $S^\infty(A)$ is at least 3 times the statistical fluctuation of the background] is obtained by setting $x_m = 1$ in eq. (8) :

$$F_m \sim 10 / d_o^2\, [\, k\, c\, \beta(A)\, T_A] \qquad (9).$$

For incident probe intensities in the 1-10nA range, from eq.(7), the detected Auger signal intensity is expected to be in the 10^4 counts / s range for a pure element (with $\beta(A) \approx$ a few 10^{-4} and $T_A \approx$ a few 10^{-2}). In the sequential mode of acquisition of a (point) spectrum, at least 100 seconds are required for exploring only 100 channels with a dwell time of one second per channel. Then the minimum detectable concentration would only be of around a few percent. The required fluence would be, at minimum, 10^7-10^8 electrons per square Angstrom (or 10^4-10^5 C/cm^2) for a spot diameter of 10nm and the total time for the data acquisition of 64 pixels x 64 pixels images will be of around one hundred hours. This total time may be reduced down to one hour if the dwell time per channel is reduced to 10^{-2} s but the expected sensitivity (few tens %) would be very poor .This simple estimate explains the need of either using a detection of several channels in parallel at the exit of the analyser [22] or of optimising the collection efficiency, T_A, of the emitted signals [27]. In this field a major breakthrough of SAM will be obtained when the record of the full Auger spectrum will be possible over 1024 independent channels, like it is the case in EELS [61,62].

Eqs. (7) to (9) also apply for the Auger filtered image mode with the particular values of the specific operating conditions [14]. The reported acquisition time of 15 s, demonstrates the interest of this approach for the simultaneous acquisition of, at least 64x 64 pixels per image. The same equations also apply for XPS with the photoelectron intensity of a pure element A takes now the form [60] :

$$I^\infty_x(A) = \Phi\, d_o^2\, \beta_x(A)\, T_A \qquad (10)$$

where Φ is the incident flux (photons/unit area) in a pixel (in the object plane) of dimension d_o (less than the dimension of the projected slit of the analyser on the specimen surface); $\beta_x(A)$ is the quantum yield for the photoelectron production [$\beta_x(A) \sim C_A\, N^\circ\, Q^x_A\, \lambda\, \sin\theta$ with $C_A = 1$ for a pure element] and is also in the 10^{-4} range (with a photo ionisation cross section $Q^x_A \approx 10^{-20}$ cm^2, $\lambda\, \sin\theta \approx$ 1nm and $N^\circ \approx 5.10^{22}$ cm^{-3}). Eqs (8) and (9) also hold with a subscript for $k(k_x)$, $c(c_x)$, $F(F_x \sim \Phi\tau_i)$ and $\beta(\beta_x)$. For a given incident dose, the main difference between XPS and e-AES is the better signal to background ratio of XPS ($k_x \sim 10$).

Using the above numerical values for $\beta_x(A)$ and $\beta(A)$ and for T_A (a few %), fig 8 illustrates the correlation between sensitivity, x_m, incident dose, D, and pixel size, d_o, for XPS and for e-AES. The calculations are also based on $c \sim 1$ postulating a parallel recording for a point spectrum with a collection efficiency for the spectrometer of a few % and an energy window for the signal, ΔE (of a few eV), which covers its full width E_W for its full collection.

This type of calculation (or graph) allows the estimation of the acquisition time

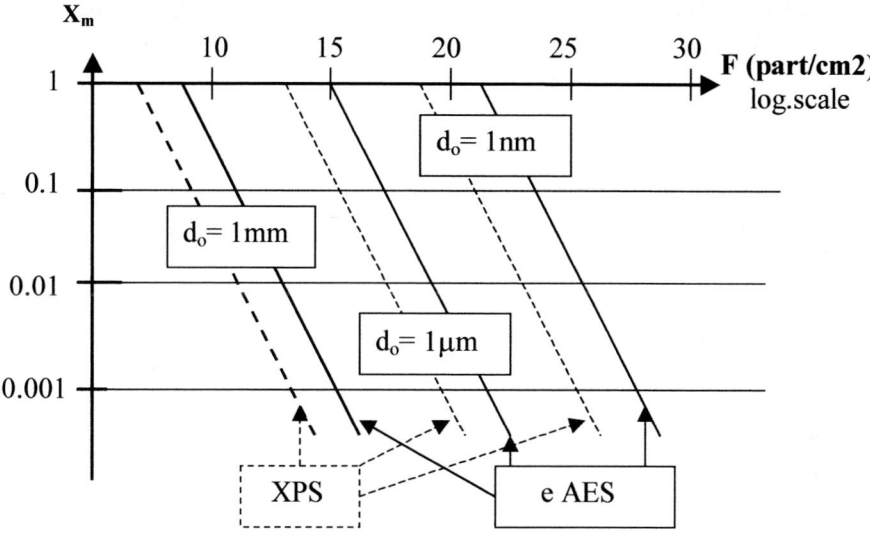

Fig.8. Correlation between sensitivity, x_m, incident dose, F (log scale in inc. part./ cm^2), and pixel size, d_o(3 values: 1mm, 1μm, 1nm), for XPS (dashed lines) and for e⁻AES (full lines): graph deduced from eq.(8) with $c \sim 1$, parallel recording of the spectra; $\beta(A)\ T_A \sim 10^{-5}$; k(or S/B) $\sim$ 10 for XPS and 0.1 for e⁻AES [60]. For $x_m = 1$, the dose values correspond to F_m the minimum dose for a pure element to be detectable.

required for a given incident probe diameter. For instance, the microscopy beamline at Hasylab provides about 10^8 photons/s in a focal spot of 1μm (photon energy in the 500 eV range) [41]. From the graph and for such a d_o value, F_m, the minimum dose for a pure element to be detectable in XPS is found to be of about 10^{13} phot.cm^{-2} and a 1 s acquisition time for one spectrum just allows us to reach only a few % sensitivity and 10 hours are needed for obtaining the corresponding characteristic map (64x64 pixels) in the scanning mode. Moreover, a parallel recording was assumed for the acquisition of the spectra. In the case of a sequential mode of acquisition, the required dose (as well as the acquisition time) has to be multiplied by the number of channels being explored (or by the factor c^{-1}). To benefit from the chemical shift effect of XPS, it is necessary to reduce the energy band width of incident photons and also to choose a better energy resolution (for the spectrometer); the two improvements lead to an increase of the dose (to be multiplied by $E_W / \Delta E$) but to a much more dramatic increase of the acquisition time except when the chosen specimen shows exceptionally large chemical shifts. This type of consideration also explains why the first XPS images have been obtained by back foil excitation: a photon intensity (Al Kα) of about10^8 photons/s was contained in a spot diameter of about 10 μm [38,60].

4.2 Problems related to quantitative elemental mapping.

In quantitative elemental analysis by XPS and AES, sophisticated procedures have been developed to improve the precision in the concentration evaluation from the acquisition of a good point spectrum (high S/N) [1,2]. The problem of the increased acquisition time when the typical pixel dimension is decreased make different the procedure to use for quantitative elemental mapping at high spatial resolution even if the intensity concentration relationship being used is the same: eqs (3) and (4) for e¯AES ; an eq. nearly similar to eq.(10) for XPS.

4.2.1 Background subtraction

For a good quantification, the first task is to remove the background from the detected signal and this task requires sophisticated procedures. In AES for instance, the main part of the background results from the tail of the secondary electrons, $\partial\delta(E)/\partial E$, and from the spectral distribution of the backscattered electrons, $\partial\eta(E)/\partial E$. If the whole spectrum is plotted on log-log axes the essentials of this background can be removed [63]. The remaining part results from inelastic interactions experienced by the emitted Auger electrons in AES and by the photoelectrons in XPS. It appears in the low kinetic energy tails and it is suppressed by linear or integral background subtraction methods [1] or better by the deconvolution method initiated by Tougaard [64]. In opposition to this approach, the characteristic signal intensity I(A) is just deduced from the subtraction of the contents of the two parent channels (one for collecting S+B, the other collecting B) in characteristic elemental mapping performed in a scanning mode (a noticeable exception to mention to this simple subtraction approach is that developed in York [63]). In the filtered image mode, the equivalent approach is the subtraction, pixel by pixel of two images acquired at kinetic energies (one on the peak and the other on the higher kinetic energy side of this peak). The correlated consequence of such a crude subtraction is that other information contained in the peak shape, which are used in point analysis, are often not considered in elemental imaging.

4.2.2. Backscattering factor effects in AES

Specific to AES is the problem of the reinforcement of the Auger signal by the backscattered electrons : the Auger backscattering factor, $R_{A/S}$, of eq. (3) which may vary from near 0 to the unity as a function of the substrate composition, of E_o and the characteristic line of interest. In Auger imaging, the substrate composition may change from place to place influencing then, significantly, the characteristic Auger signal intensity. Barkshire et al [65] resolved this problem by detecting simultaneously the electron backscattered current in a Scanning Auger Microscope. The backscattered (b.s.) electron detector is calibrated (for many elements) and the signal is related to the effective atomic number of the substrate. Next, this b.s. signal this is used to calculate the Auger backscattering factor from the expressions proposed by Shimizu et al. at selected angles of incidence ($\alpha = 0°$, $30°$ and $45°$) [66]. Based on the same philosophy (determination of η and next calculation of R), an alternative approach [67] consists in i) the determination of η based on the measurement of the specimen current and ii) the calculation of R by using an expression correlating the Auger backscattering factor R and the electron backscattering factor η [68]. Each of the two methods may also be applied to Auger in depth profiling where the change of the relative signal intensities is also strongly influenced by the change of substrate composition during the experiments [65].

4.2.3. Topographic effects

In SAM, when the incident beam is scanned along a surface, the change of the local slope changes the incident angle α and the take-off angle θ influencing, then, the detected intensities (via the influence of α and of θ in eq. 3). The discussion is often limited to the best weight to give at the measured background intensity (and sometimes to the best kinetic energy where the background has to be measured) for the best topographic correction [22,69-71]. This type of topographic effect certainly holds also in the Auger imaging mode as well as in the corresponding modes of XPS (because an equation similar to 3 may also be written) and the various approximate corrections proposed for SAM may be easily transposed to the

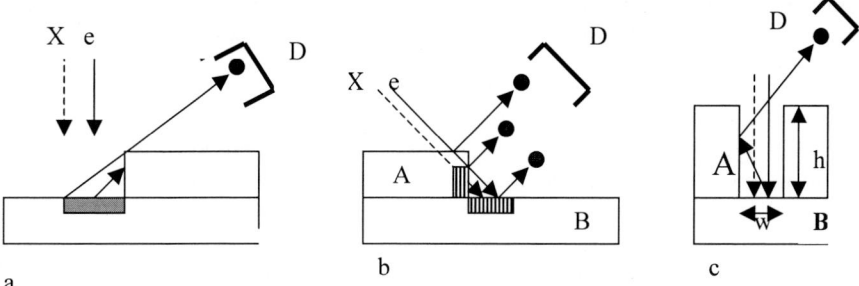

Fig 9: Topographic effects. a: shadowing effect of the detector by a step (in grey: shadowed zone). b: extra-contributions(hatched areas) from the riser and from the next step. c: combination of the shadowing effect for the signal issued from the bottom of the well and of the extra-contributions due to electrons backscattered by the curb of a well having a large aspect ratio h/w. When the signal issued from the bottom of the well cannot be detected, the analysis is impossible even when the dimension of interest, w, is far larger than the lateral resolution of the instrument. D: detector; A and B indicate a possible difference in composition between the irradiated zones.

imaging mode of Auger as well as to the XPS modes because there are based on the fact that, similarly to the signal, the background intensity is also influenced by topographic changes.
More specific to SAM is the use of several backscattered electron detectors for correcting rather large topographic effects [72]. The use of quadrants detectors at the exit of a CMA for instance has also to be pointed out [73,74].

The above considerations are related to rather smooth topographic change and more difficult to solve is the problem of large topographic effects (sharp edges or steps or wells) leading to additional contributions from the surroundings or to shadowing effects in the focused mode [75]. The shadowing of the detector by a step (fig. 9a) leads to a decrease of the characteristic signal which is a function of the position of the detector: this decrease may be very large at small take off angles. The undesirable extra contributions, (fig.9b), arise from places surrounding the point directly excited by the beam and these extra contributions obviously lead to false analysis when different compositions are involved in the direct and in the extra contributions. In SAM again, the use of Monte Carlo simulations permits a good understanding of the general mechanisms being involved in these processes but conflicting conclusions have been emitted on the best choice of the primary beam energy for minimising these effects [76,77]. It is clear that the weight of these spurious effects is a function of the angle of incidence and of the take off angle as well as of the specific topography of the specimen of interest. The worse situation nearly corresponds to that shown on fig 9c where the goal is to analyse the bottom of a well having a high aspect ratio, h/w. The shadowing effect prevent the direct contribution to be collected while the Auger electrons (excited by the backscattered electrons) issued from the curb of the well are easily detected. This example, very important for the semiconductor industry, illustrates the fact that the topographic effects may be sometimes the limiting factor for the applications of SAM for the analysis of highly integrated semiconductor devices (except cross sectioning them or thinning them by using "ion beam bevel section". The imaging approach (AEEM or XPEEM) certainly permits the extraction of the useful signal by the electric field of the cathode lens system but the lack of experimental results on specimen showing large topographic defects forbids us from going deeper into this subject.

4.2.4. Crystalline effects

The formalism used in the quantification procedures (eqs. 3, 4, etc…) only applies for amorphous specimen. In fact it is often incorrectly used for quantifying the composition of crystalline specimens despite the experimental evidence given by Photoelectron Diffraction (PhD) [78-80] and by Auger electron diffraction which also permits the direct imaging of surface and interface crystal structures [81-84].

It is clear that the probability for these effects to occur in polycrystalline materials increases with the improvement of the lateral resolution : a poor resolution averages the contribution of numerous microcrystals of different orientations while its improvement increases the risk to have only one microcrystal associated to one (or more) pixel. These channelling effects are a serious impediment to quantitative Auger analysis [85-87].

4.2.5. Radiation damage and charging effects (insulating materials).

From eq. (9) and fig. 8, it is clear that one limiting factor of the minimum detectable dimension, d_o, is not the performance of the instrument but is related to the specimen composition [60]. For radiation sensitive materials, the critical fluence of some organic specimens has been found to be as low as 10^{-3}-10^{-4} C/cm^2 (or 10^{16} part. / cm^2 or one electron per square angstrom) from experiments performed in electron microscopy [88]. Similar values have been reported in x-ray microscopy of biological specimens (1 x-ray photon / A^2) [89]. With the numerical value chosen for fig.8, these critical fluences imply that the corresponding specimens cannot be investigated at a resolution better than one micron in any mode of AES (taking into account the fact the surface is damage faster than the bulk and by the fact that the dose limit of eq.(9) corresponds to pure elements and not to compounds) or better than 0.1μm in any mode of XPS. Even when the critical fluence of less sensitive specimens is four order of magnitude higher, the possible decrease of the pixel size is only of two orders of magnitude.

The damaging effects are a function of the specific chemical composition of the specimen of interest and it is not the present purpose to discuss these effects in detail but it is interesting to observe that, for the energy range of incident particles, most of the radiation sensitive materials are insulating materials where the lack of conduction electrons prevent the rapid restoration of bond breaking between neighbour atoms. It is also interesting to point out that the initial mechanisms of radiation damage are closely related to mechanisms giving rise to the signal. Three initial causes of radiation damage may be considered and two of them are illustrated on Fig. 10 [90].

In AES, the first cause is the bond breaking between two neighbour atoms or the generation of electron-hole pairs (excitation of valence electrons towards the conduction band) either directly by the incident beam or indirectly via plasmon decay. For such a generation the energy lost by the incident electron is E(e.h.) (nearly equals to twice the energy band gap value). In XPS, there are the Auger electrons and photoelectrons generated by the absorption of the photon that play the role of the primaries of AES and that lose the energy E(e.h.) during their transport in the materials [91].

The number of electron-hole pairs, n(e.h.) generated by an incident electron of energy E_o or by a photon of energy hν is then given by :

$$n(e.h.) = E_o/E(e.h.) \quad \text{or} \quad n(e.h.) = h\nu/ E(e.h.) \quad (11)$$

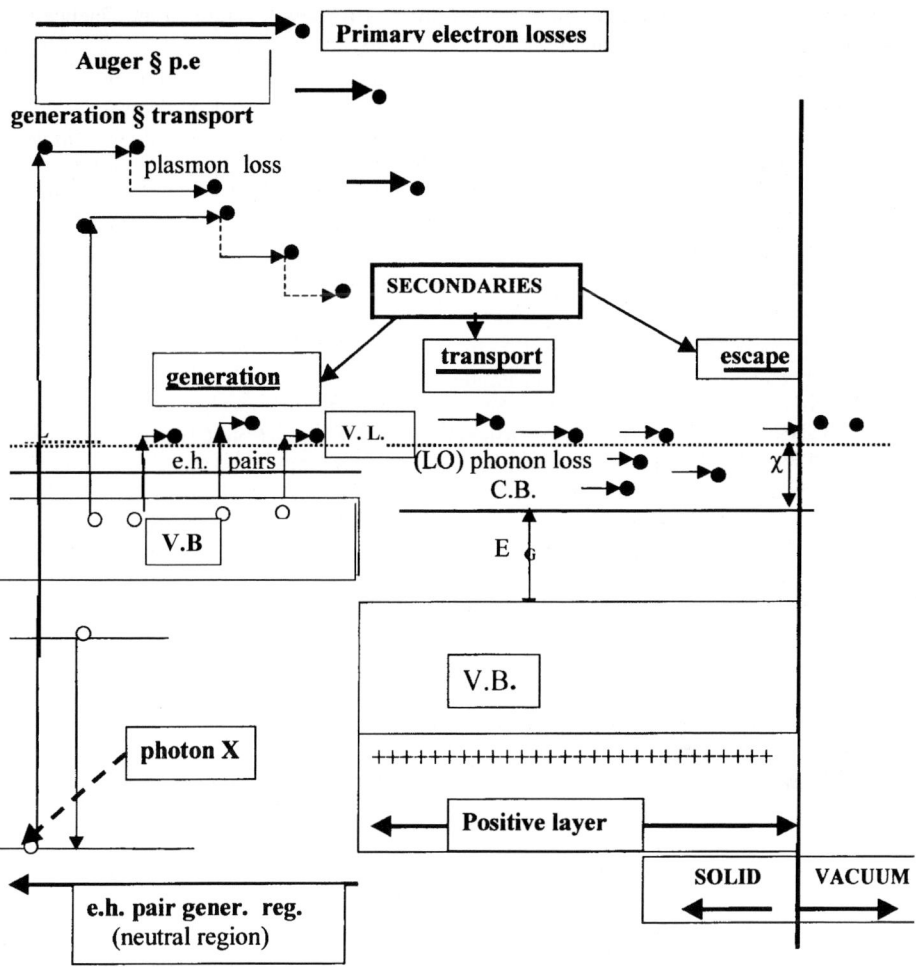

Fig.10. Electron–hole pair generation, transport and escape in insulators irradiated with electrons and photons (inspired from [90]). In e irradiation, the primary electrons (top left) mainly induces electron-hole (e.h.) pairs directly while, in x-ray irradiation, there are mainly the Auger electrons and photoelectrons generated in the absorption of a x-ray photon (bottom left and fig 1) which in turn, generate electron-hole pairs. A fraction of the electrons of the generated e.h. pairs are emitted into the vacuum leaving a positively charged layer and constituting the secondary electron emission from the specimen. The radiation damage units (rad or grey) includes implicitly the electron hole pair generation but that is to say the specific fact that the Auger process leaves the initial atom with two electrons missing while the secondary electron emission contribute to the (positive) charging of the specimen and to the internal electric field built-up which, in turn, increases the damage effects via the increase of the non recombined fraction of e.h. pairs.

It is next easy to correlate the absorbed dose, D (in Gy or J/kg), to the fluence of the incident beam F (inc. part./m^2) and to n(e.h.) :

$D(Gy) = F\, h\nu\, e\, \mu/\rho = F\, n(e.h.)\, E(e.h)\, e\, \mu/\rho$ (12) for incident photons or

$D(Gy) = F\, E_o\, e/\, R\rho = F\, n(e.h.)\, E(e.h.)\, e\, /R\rho$ (12') for incident electrons

where ρ is the mass density of the specimen, μ is its linear absorption coefficient for the photons of energy $h\nu$ ($h\nu$ in eV and then $h\nu e$ is in Joules with e =1.6 10^{-19} C) while R is the range of incident electron of energy E_o (E_o in eV and $E_o e$ in Joules).

Eqs 12 and 12' show that the absorbed dose is directly related to the number of e.h. pairs generated per incident particle. Then, for a given fluence of the beam, the advantage of XPS (over e⁻AES) is a less absorbed dose [$h\nu < E_o$; R (~1μm)< μ^{-1}(~10μm)] with a better S/N ratio. Another consequence of eqs 12 and 12'is that the units of radiation damage rad or Gy) are implicitly correlated to the electron-hole pair generation but these units do not take into account the second cause of radiation damage that is specific to the (electron- or photon-induced) Auger processes which leaves the initial atom with two electrons missing (for C, O, N, F) or four electrons missing (via the Auger cascade for Mg, Al, Si, Cl) while, in insulators, there are not enough conduction electrons to quickly restore the initial electronic environment (see the references included in ref. [61]. The correlated change of sign of halides (F⁻ F⁺; Cl⁻ Cl⁺) contributes to the desorption of the corresponding species (as well as the change of O^{2-} into O^o). In this situation the mechanism responsible for the signal is also responsible for the damage and for a given sensitive material the advantage of XPS over e-AES is only the better S/N ratio [60].

Finally, the third cause of radiation damage is closely related to the electric field up in insulators submitted to (e or x) irradiation. This electric field increases the non recombined fraction, Y, of the e.h. pairs, increasing then the damaging consequences. It also drives the migration of the mobile species such as Na ions in irradiated glasses [91].

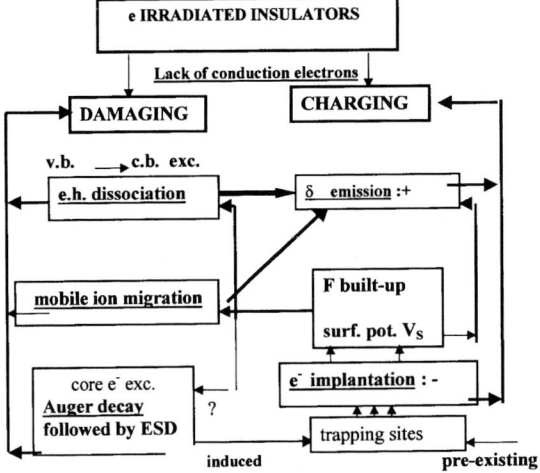

Fig. 11 (Inspired from [91]). Correlation between charging and damaging mechanisms in electron irradiated insulators (For x-ray irradiated insulators, the trapping of electrons, bottom right, has to be suppressed).

In fact, charging and damaging are closely correlated into insulators because the charging mechanisms in electron irradiated insulators involve a competition between the secondary electron emission from the specimen and the incident electron implantation (while, in XPS, the secondary electron emission is the main contribution to the charging of the specimen) and these secondary electron emissions correspond to the fraction of the electrons of the generated e.h. pairs that are emitted into the vacuum. Inspired from ref. [91], fig 11 summarises the strong correlation between charging and damaging in e-AES (and in XPS by suppressing the electron implantation).

In XPS the sign of charging is generally positive while it may be positive or negative for e-AES. Despite of the large values of the secondary electron yields from insulators (δ^X for incident x-rays, δ for incident electrons) the magnitude of the shift is often less than a few eV for positive charging because of the rapid freezing of the secondary electron emission while very large negative surface potentials may be obtained in e-AES. Despite this low magnitude, charging effects are spurious effects in XPS because the charging shift masks the benefit of the chemical shift effect. Based on the use of flood guns, various methods have proved to be efficient for compensating these effects in the case of wide illuminations (conventional XPS). When this emission is restricted to the very narrow area of a focused x-ray spot, the surface potential is a bell-shape function that induces not only a shift but also a distortion of the XPS spectra even if the surface is laterally homogeneous in composition [91]. Transposed to high resolution XPS it may be seen that the charging effects will play an increasing role with the increase of the incident flux on smaller and smaller incident spots. In theory the use of the XPS imaging mode (of XPEEM) may lead to less distortion of the spectra related to laterally homogeneous specimen but the use of additional flood guns in the cathode lens system may lead to technical complications.

For e⁻AES, charging effects often lead to very large energy negative shifts of the spectra (that sometimes forbid their acquisition) and to the shift in position of the incident beam (with the slowing down of incident electrons). The use of additional irradiation is often inefficient because of the difficulty of not only compensating the positive charges left by a high electron emission but also the negative charges implanted below the surface down to depths in the micron range. In conclusion for this point, one may say that most of the materials of the human environment are insulators and presently they are difficult to investigate in electron spectroscopy. In the future, these difficulties are expected to increase with the decrease of the pixel size(reported to the object plane).

4.3. Strategies

In the microprobe approach where analysis is more important than imaging, compromises may be found between conventional electron spectroscopy with large and intense incident probes on the one hand and two dimensional elemental (or chemical shift) imaging on the other hand (subsects. 3.3 and 3.4). These various intermediate strategies between a good resolution and a reasonable time of acquisition depend upon the goal to be reached but the first step, in AES, is obviously to obtain a good and high resolution SEM image which permits localisation of the region of interest. The implementation of a secondary electron detector into a XPS scanning probe instrument is subject to the same philosophy.

4.3.1. Point analysis.

If the region of interest is restricted to a point detail sitting on the surface (such as a dust grain which represents an important application in the failure analysis of integrated circuits), or a precipitate, a point spectrum with a probe size of the order of the detail to be investigated permits a rapid identification of the elemental components. In AES operated with submicron probes, the dimension of the emitting area remains in the micron range because it includes the backscattering halo and the standard quantification methods of AES or of XPS cannot be used when the dimension of the detail is less than the dimension of the area containing more than 90% of the signal. This difficulty, which does not hold in the imaging mode of AES, may be bypassed in the microprobe approach by taking the difference between two spectra, one acquired on the detail and the other outside it [14]. Such an approach applied to polycrystalline high temperature superconductors, has indicated that most of the grain boundary surfaces were deficient in oxygen and rich in copper compared to the bulk [92].

One specific aspect of AES in the microprobe mode remains that the analysed volume presents very blurred contours (see fig 4) with, laterally, a maximum ionisation probability inside the incident probe size and a very large faint halo combined to an exponential decay of the in-depth escape probability for the characteristic electrons. Then, inside the analysed volume, there is a cluster volume ($\pi d_o^2 \lambda \sin\theta/4$) from which 35-60 % of this signal is emitted because a significant fraction of the signal, $1/(1+R)$, comes from the area $\pi d^2_o/4$ and 66% of the signal comes from the depth $\lambda \sin\theta$ while the remaining contribution is diluted over a volume three times higher and covering a micron square area. For $d_o \approx 10$ nm, the cluster volume is of about 100 nm^3 and it contains only 5000 atoms (taking into account the usual atomic density of solids, 5×10^{22} at/cm^3). This order of magnitude explains the detection of a few silver atoms when the probe diameter reaches the nm value [24,26]. This result also illustrates the fantastic power of high resolution SAM in the detection of small amounts of matter (10^{-20}-10^{-21}g) but it also indicates the emergence of quantum limitations in

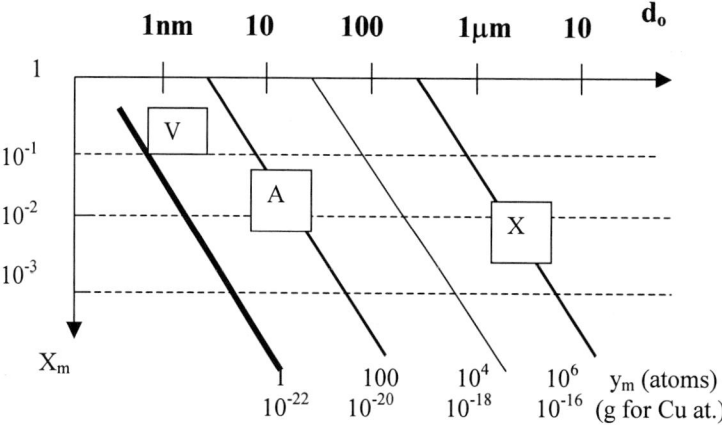

Fig.12. Correlation between x_m, d_o, and y_m in surface analysis (with an information depth of ~ 1nm). Typical results presently obtained with marketed instrument in AES (A) and in XPS (X) are shown with also the results obtained at ASU (V) [24,26]. The corresponding fluences for the beams may be obtained from the use of fig.8.

quantification, highlighting the need to apply specific procedures for quantifying the composition of so small a number of discrete emitters (see the appendix of ref [12]). It also illustrates the fact that the minimum number of atoms being detectable, y_m, is also a simple but useful parameter for characterising the performance of an instrument (besides lateral resolution and sensitivity). Despite the fact that, at the atomic scale, the concept of concentration itself becomes questionable (it strongly depends of the selected borders of the analysed volume [12]), fig. 12 shows a graph correlating x_m, d_o, and y_m for surface analysis ($\lambda \sin\theta \sim 1nm$) where some typical values presently attained are shown.

From the instrumental point of view the required stability for the specimen holder and for the incident probe has to be stressed when such fine probes are used.

4.3.2. Line scan profile

When the information of interest is only restricted to a change in composition along one direction on the surface, an Auger line scan profile parallel to this direction can be performed (for example, across a grain/grain boundary interface). In such a situation there are also some advantages to use a fine probe and a rather long dwell time for obtaining a first estimate of the concentration gradients, $C_A(x)$, $C_B(x)$, over short distances. These estimates may be next improved from a deconvolution from the observed profile by the PSF for instance [12]. The convolution effect of various PSF (given by eqs. similar to eq.(3)) on various types of sharp interfaces may be easily obtained [11] but the solution of the inverse problem is not so easy.

Nevertheless the scarce number of attempts in this direction is surprising with regard to the numerous attempts to solve a similar inverse problem with angle resolved XPS for non destructive in-depth profiling [93-96]. If the starting concentration values (far from the interface of interest) are accurately determined, the use of simple trying functions for $C_A(x)$, $C_B(x)$ etc (with one adjustable parameter) may certainly improve the first estimate of the concentration gradient changes with a lateral resolution of the order of the incident spot diameter. Standard deconvolution methods using Fourier transforms may also be used. For such approaches the key point is to know the radial distribution of incident particles in the beam (instrumental effect) and, for AES but not for high resolution XPS, to have an estimate of the radial distribution of the backscattered contribution (matrix effect). Similar to the investigation of small details at fixed probe, quantum effects have to be taken into account when sub-ten nm probes are use for obtaining an Auger line profile.

4.3.3 Signal and image processing

Like for many other scientific techniques, there is, now, the increasing role of signal processing in surface analysis. In AES, for instance, it is the increasing role use of Monte-Carlo simulations for estimating the signal intensities, the influence of the Auger backscattering factor and of the topographic effects [75] but also for simulating the background [97-99]. These simulations will allow the verification of the consistency between the experimental spectra and the estimated compositions of the detail and of its surroundings.

For the spectral imaging point of view, the techniques of interest here, consists in filling a tri-dimensional space: x, y, E_K where x, y are the usual co-ordinates of the specimen surface and E_K is the kinetic energy of the detected (Auger or photo) electrons. Then, the whole data set may be displayed in a space of axis Ox, Oy and E_K in which the contains of a given voxel (dimensions : $\Delta x . \Delta y . \Delta E_K$) is the intensity acquired at a point x_i, y_i and at the energy E_{Ki} (for the energy window ΔE_K: see fig.13). This data set may be acquired either in the scanning mode from the successive acquisition of spectra related to each point of the investigated surface or from the image filtering mode by acquiring successive images of the surface at different kinetic energies. The above principles are similar to that of EELS imaging or SEM-

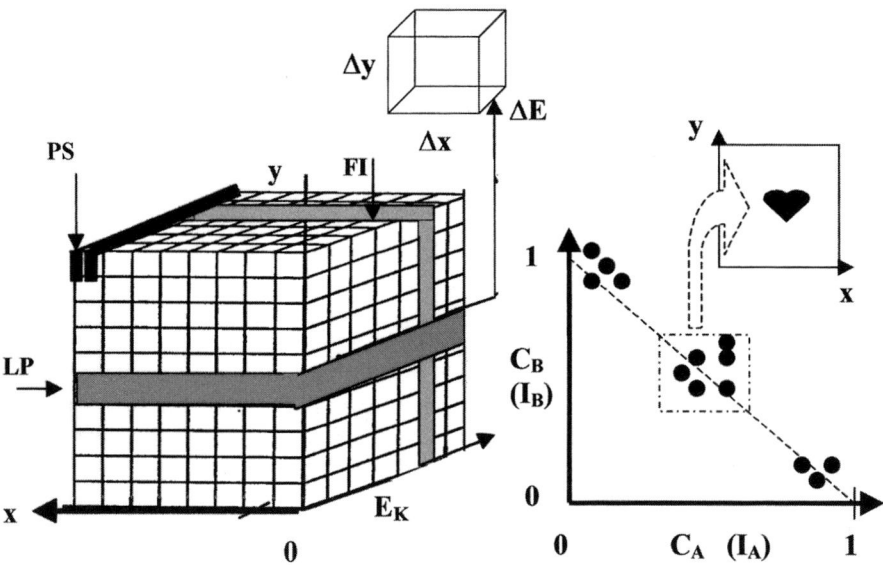

Fig.13. Left: principle of an image spectrum. Right : principle of the correlation (or scatter diagram) technique.

EDS imaging for the point of view of image processing [62,100]. In the two modes, the image-spectrum volume may completed by other signals such as the backscattered (and the secondary) electron signal η (x_i, y_i) [and $\delta(x_i,y_i)$] for instance. All the voxels' information being in a digital form it is next easy to manipulate this volume in order to extract an Auger point spectrum, or an Auger line profile (PS and LP respect. on fig. 13). Selecting a given energy E(A) it is also easy to display the image which corresponds to the change of S + B of element A as a function of the x, y co-ordinates. A simple background removal may be obtained from the subtraction, pixel per pixel, of the iso-energy images obtained on the peak to that obtained on the background. More sophisticated manipulations may also involve, in addition, other planes such as the η plane, η (x,y), for removing the Auger backscattering effects, for instance. An important development is, next, the use of correlation (or scatter) diagram technique that may be applied to any kind of spectromicroscopy or microspectrometry. Proposed by Browning [101] and widely popularised by Prutton et al. for SAM [100,102-104], this technique consists in displaying the concentration maps, $C_A(x, y)$ and $C_B(x, y)$ that have been obtained on a binary alloy, $A_X B_Y$, (for the sake of the simplicity) or by starting from the initial $I_A(x, y)$, $I_B(x, y)$ images. Next the characteristic intensities (or concentrations) of each pixel (of an image or a profile) are displayed in a I_A vs I_B (or C_A vs C_B) diagram (see fig. 13b). A simple inspection of such a diagram gives estimates of : the number of different phases, $A_x B_y$; their relative weights ; and to detect the existence of artefacts or of deviations from the statistical fluctuations of the number of counts. By windowing one cluster of pixels (related to a specific phase) it is also possible to find where they are located in the direct (x, y) space. This technique may be easily extended to multielement mapping [101] and the scatter diagrams may be displayed in three dimensions [103].

A clear conclusion of this subsection is that modern processing facilities permit to combine data from different sources in order to obtain more information than would be obtained from the separate processing of each image. Consequently there is the need, first, to collect the maximum of signals issued from the specimen when the incident beam is scanned.

5. EXPLORING THE THIRD DIMENSION

5.1 Non destructive sub-surface analysis

In XPS and AES, non destructive z-profiling is, in principle, possible (down to a thickness $t \approx 3\lambda$) by variation of the electron emission angle θ. From an equation similar to eq.3, it has been observed that the measured intensity $I(\theta)$ is the Laplace transform of the concentration $C(z)$. This concentration profile can be, then deduced from the use of the inverse Laplace transform of the measured angular intensities. This procedure is widely used in XPS [92-96] despite the amplification of the measurement errors in the solution of the inverse problem [19,105,106]. If a precise $C(z)$ profile is difficult to obtain, the variation of the take-off angle θ in XPS permits, at least, the distinction between a thin overlayer of an element A on a substrate B and an homogeneous surface of composition $A_X B_Y$. It also gives an estimate of the overlayer thickness t from the simple tilt of the specimen [19,92].

With standard spectrometers similar experiments are far less easy in AES because the tilt of the specimen also changes the backscattering factor R and the cosec α values of eq.(3). At fixed incidence angles, polar angle resolving experiments with parallel detection remain possible but they require the electron analyser to be adapted [73,82].

5.2 Destructive in depth profiling.

To go deeper than 3 λ in the in-depth distribution analysis, It is necessary to use a destructive technique either outside (ball cratering, chemical bevelling) or better inside the instrument. There are very few investigations carried out in surface spectroscopy that do not involve sputter etching either to remove surface contamination or in the production of the called depth profile. Depth profile is a combination of surface analysis (XPS or AES) and sputtering and Fig 14 illustrates such (Auger) profile which shows that sharpness of the different interfaces can reach the nm range for Fe/Cr multilayer structures grown by MBE on a (001) GaAs surfaces [107].

The recent progress in Auger (or XPS) depth profile (developed in conjunction with standard incident probes) results from technical developments related to "in situ" ion etching at oblique incidence with or without Zalar rotation system [108,109]. It also results from a careful analysis of the physical parameters involved in the sputter mechanism and the emission of Auger (or photo) electrons [82-84]. The third development of depth profile is related to the full exploitation of the collected data using factor analysis (FA) [18,110-112].

In a situation where successive spectra are acquired, this very powerful mathematical method, FA, is able to indicate the number of components (or factors) which are varying within the data set and to classify them as a function of their relative weight (down to the extra factors entirely related to noise). The shift in position and the change in the line shapes associated to changes in the chemical environment can be fully exploited by using FA in order to obtain, finally, depth profiles of the elements with respect to their chemical bonding states. Factor analysis is also useful in interpreting Auger or XPS spectra where peak overlap problems occur ; it also applies to any time series of Auger spectra such as that obtained during oxygen exposure (oxidation studies) or during nucleation and growth processes or that

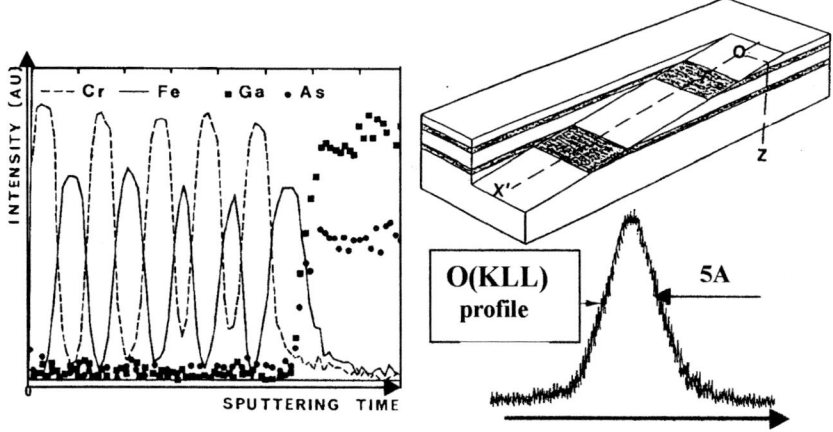

Fig. 14. Left: AES depth profile of five Fe(3 nm)/Cr(3nm) bilayer structure grown on a GaAs substrate (inspired from [107]). Top right : Principle of the ion beam bevel section where a Auger line profile along the bevel (x' direction) is transformed into depth analysis. Bottom right: Auger (O KLL) profile along an ion bevel section of a 0.5 nm thick oxide layer at a polysilicon /Si interface (from D.K. Skinner [120 b])

obtained during a line scan profile on a surface. The association of Auger depth profiling and FA is widely used for the investigation of magnetic/non magnetic superlattice structures [109] (see also [113-115] for some other examples of applications).

However excellent the use of FA is for the analysis of AES lineshapes, there is the requirement of standard spectra and the problem of deciding whether or not a factor of minor statistical importance should or should not be considered. For overcoming these difficulties, the use of neural pattern recognition has been recently suggested as an alternative route for post-processing AES spectra [116]. Independently from the data processing, three physical parameters influence the depth resolution from Auger sputter depth profiles. These parameters are : atomic mixing, surface roughness and information depth and the relative influence of each of them has been reported for Ni/Cr multilayers [117,118].

Finally, in the conquest of the third dimension one of the most (if not the most?) significant developments of AES for the past few years is the use of the ion beam bevel section developed in conjunction with finely focused incident electron or x-ray probes. This recent development transforms the good lateral resolution of modern SAM [119,120 a & b] and XPS [31] commercial instruments in an excellent depth resolution. The exception concerns most of the prototypes where the implementation of focused ion beams in the cathode lens seems sometimes difficult to realise. With this method, a finely focused ion beam is rastered in such a way as to sputter a very shallow bevel into the surface of the sample. Vanishingly small bevel angles are possible leading to magnification of one thousand or more between the vertical (depth) scale, z, and the nearly horizontal x' scale along which an Auger line scan is performed. Fig. 14 illustrates the principle of this technique and it gives also two examples of applications (inspired from the excellent review paper of Skinner dedicated to the role of AES in the semiconductor industry [120 b]). Other systems of interest for the study of magneto-optic recording media and giant magneto resistance effects have been explored by using this technique in conjunction with the use of Principal Component Analysis and Factor Analysis [121-123]. Computer controlled ion beam bevel sectioning may also be applied to the investigation of more heterogeneous systems where the ion microscalpel is used to remove a part of the materials above the detail or the interface of interest, in order to study next the revealed interface or detail with SAM [124-125].

5.3 Exploring the x, y, z space

Finally, it has been seen in the above subsection that the operation of ion beam bevel sectioning (or conventional ion etching) allows the removal of matter in order for an initially buried detail to become a surface detail. Such a detail (or region) of interest may be next investigated using either a fixed probe, or acquiring an Auger (or a XPS) line scan profile or even performing elemental mapping. In principle any kind of detail located somewhere in a solid may be then analysed with a depth resolution in the nm range and a lateral resolution in the 10 nm range in AES (or 0.1 to 10 µm in XPS).

One may also dream of obtaining successive characteristic x, y maps acquired at different depths in order to obtain some kind of spatial mapping or characteristic Auger (or XPS) microtomography of the region of interest in the specimen. For such purposes, the present limitations are related to the acquisition time required for obtaining good Auger (XPS) line scan profiles and good characteristic maps. This limitation will be partly overcome when a real parallel acquisition of the whole Auger or XPS spectrum will be possible. The other limitations are related to the physics of ion / matter interaction inducing atomic mixing and surface roughness.

Presently it remains that the exploration of the x, y, z directions is possible but only by selecting the direction(s) of interest and by finding the best compromise between sensitivity and spatial spreading of the required information. In other words, one may consider a hypervolume x, y, z, E_K composed of hypervoxels Δx, Δy, Δz, ΔE_K (extending the volume shown of fig. 13 to a fourth dimension) and by optimising the minimum time needed for acquiring the information related to a given hypervoxel with respect to the total acquisition time : t. The strategy is defined by the goal to be reached with the optimisation of the number of hypervoxels to be selected and the amount of information contained by each of them.

6. CONCLUSION

Two different approaches, the microprobe approach and the filtered imaging approach may be used for laterally resolved surface spectroscopy. Each of the two approaches presents its own advantages (first, the spectrum or first, the image) and limitations. For XPS (based on the use of a classical x-ray source or of the synchrotron radiation) and for AES, the nominal value of the lateral resolution that has been achieved by commercial instruments or by prototypes has been indicated in this review article. A few atoms have been detected with a prototype of an Auger instrument and, for XPS, the challenge is to obtain images at good lateral resolution but also at the best energy resolution as possible for providing chemical shift images not restricted to the large chemical shift of the Si./SiO$_2$ which may also be obtained easily in AES.

Besides the nominal values of the lateral resolution, some important limitations or improvements have also been outlined. The intrinsic limitations of the practical resolution are related to the specimen. They concern, for instance, its topography and also its possible modifications during its investigation (radiation sensitive materials) : the corresponding damaging effects are increasing in proportion to the inverse square of dimension of a pixel (reported to the objet plane).

The recent improvements involve modern processing facilities and modern methods for signal and images analysis (Multivariate Statistical Analysis and Correlation Diagrams) because they allow extraction of the maximum of information from the collected data.

Finally the conquest of the third dimension with the use of the ion beam bevel section in conjunction with finely focused incident probes is may be one of the most (if not the most?) significant developments of XPS and AES of the past few years but the corresponding attachment seems to be difficult to implement in most of the prototypes under operation.

REFERENCES

[1] D. Briggs and M.P. Seah in Practical Surface Analysis, Vol. 1, 2d Ed.; John Wiley and Sons, Chichester, 1990.
[2] J.C. Riviere in Surface Analytical Techniques, Clarendon Press, Oxford, 1990.
[3] P. Auger, Compt. Rend. 177 (1923) 169; 180 (1925) 65; Compt. Rend. 182 (1926) 773; Compt. Rend. 182 (1926) 1215.
[4] K.Siegbahn et al. in ESCA, Atomic, Molecular, and Solid State Structure Studied by Means of Electron Spectroscopy, Almqvist and Wiksells, Upsalla (1967) ; K.Siegbahn et al., ESCA Applied to Free Molecules, North Holland, Amsterdam (1969).
[5] J.A. Beardeen, Rev. Modern Phys. 39 (1967) 78.
[6] a) L.E. Davis, N.C. Mac Donald, P.W. Palmberg, G.E. Riach, R.E. Weber;Handbook of Auger Electron Spectroscopy, 2nd Ed,Physical Electronics Industries Inc,Min. (1976).
b) G.E. McGuire, Auger Electron Spectroscopy Reference Manual, Plenum NY (1979).
c) Y. Shiokawa, T. Isida and Y. Hayashi, Auger Electron Spectra Catalogue; A data Collection of Elements, Anelva Corp., Tokyo (1979).
d) T. Sekine, Y. Nagasawa, M. Kudoh, Y. Sakai, A.S. Parkes, J.D.Geller, A. Mogami and K. Hirata, Handbook of Auger Electron Spectroscopy, JEOL, Tokyo (1982).
[7] S. Tanuma, C.J. Powell, D. R. Penn, Surf. Interface Anal. 11 (1988) 577; 17 (1991) 911; 17 (1991) 927; 20 (1993) 77; 21 (1993) 165.
[8] P. Weightman, Rep. Prog. Phys. 45 (1982) 753; Microsc. Microanal. Microstruc. 6 (1995) 263.
[9] D.E. Ramaker, Physica Scipta T41 (1992) 77; J. Electr. Spectr. Relat. Phenom. 66 (1994) 269.
[10] Special issue of J. Electr. Spectrosc. Relat. Phenom. 72 (1995) 1-332.
[11] J. Cazaux, Surf. Interf. Anal. 14 (1989)354.
[12] J. Cazaux, J. of Surf. Anal., 3 (1997) 286.
[13] E. Bauer, Surf Rev Lett. 5 (1998) 1275.
[14] E. Bauer, C.Koziol, G.Lilienkamp, T.Schmidt; J. Electr. Spectrosc. Relat. Phenom. 84 (1997) 201.
[15] E. Bauer in Proceedings 12th Eur. Cong. Electr. Microsc., Brno, Ed. P.Tomanek and R. Kolarik, vol III (2000) I 27.
[16] G. Lilienkamp in Proceedings 12th Eur. Cong. Electr. Microsc., Brno, ibid vol III (2000) I 177.

[17] R. Fink et al (19 coauthors), J. Electr. Spectrosc. Relat. Phenom. 84 (1997) 231, W.Engel et al. (19 co-authors); in X-ray Microscopy and Spectromicroscopy, J. Thieme, G. Schmahl, D. Rudolph, E. Umbach eds Springer, Berlin III (1998) 55.
[18] J.T. Grant, Surf. Interf. Anal. 14 (1989).271.
[19] M.P. Seah in Practical Surface Analysis, ref [1], (1990). p. 201.
[20] N.C. Mac Donald and J.R. Waldrop, Appl. Phys. Lett. 19, (1971) 315.
[21] G.Todd, H. Poppa, L. Veneklassen ; Thin Solid Films 57 (1979) 213.
[22] M. Prutton, C.G.H. Walker, J.C. Greenwood, P.G. Kenny, D. Barkshire, I.R. Roberts and M. M. El Gomati, Surf. Interf. Anal., 17 (1991) 71.
[23] G.G. Hembree and J.A. Venables, Ultramicroscopy 47 (1992) 109.
[24] J. Liu, G.G. Hembree, G.E. Spinnler, J.A. Venables, Ultramicroscopy 52 (1993) 369.
[25] J.A. Venables, G.G. Hembree, J. Liu, C.J. Harland, M. Huang; Proceedings ICEM 13, Paris, July 1994, Eds : Les Editions de Physique, 1 (1994) 759.
[26] a) J. Liu, Microbeam Analysis 1995, Proceedings 29 th MAS Conf. Breckenridge 1995. E.S. Etz, VCH(1995)235.
b) J. Venables, G.G. Hembree, J. Liu J.S. Drucker, Inst. Phys. Conf. Series, 130 (1992) 415.
[27] P. Kruit, and J.A. Venables, Ultramicroscopy 25 (1988) 183.
[28] J. Cazaux, J. Chazelas, M.N. Charasse and J.P. Hirtz, Ultramicroscopy, 25 (1988) 31.
[29] J. Chazelas, A. Friederich, J. Cazaux, Surf. Interf. Anal. 11 (1988) 36, J. Chazelas, J. Cazaux, G. Gillmann, J. Lynch, R. Szymanski, Surf. Interf. Anal. 12 (1988) 45.
[30] J. Chazelas, J. Olivier, M.N. Charasse, R. Cabanel, J.P. Hirtz and M. Magis, J. Microsc. Spectrosc .Electron. 13 (1988) 371, J. Olivier, J. Chazelas and M. Razeghi, J. Microsc. Spectrosc .Electron. 13 (1988) 357.
[31] J.F. Moulder, S.R. Bryan, U. Roll, Fresenius J. Anal. Chem. 365 (1999) 83.
[32] P. Coxon, J. Krizek , M. Humpherson , I.R.M. Wardell, J. Electr. Spectrosc. Relat. Phenom. 52 (1990) 821.
[33] B.P.Tonner, D. Durham, T. Droubay, M. Pauli, J. Electr. Spectrosc. Relat. Phenom. 84 (1997) 211.
[34] U. Gelius, B. Wannberg, P. Batzer , H. Feller-Feldegg, G. Carlsson, C.G. Johansson, J. Larsson, P. Munger, G. Vegenfors; J. Electr. Spectr. Relat. Phenom. 52 (1990) 747.
[35] C. Coluzza , R. Moberg ; J. Electr. Spectrosc. Relat. Phenom. 84 (1995) 109.
[36] B.J. Tielsch, S.P. Page, D.J. Surman, E. A. Thomas, J.E. Fulghum , Abstract of 45th AVS meeting, Baltimore, nov. 1998 ;S.P.Page, European Patent, EP 0 458 498 B1 (1996).
[37] J. Cazaux, Ultramicroscopy, 12 (1984) 321.
[38] J. Cazaux, D. Gramari, D. Mouze, J. Perrin, X. Thomas, Inst. Phys. Conf. Ser. 61 (1982) 425.
[39] W. Swiech et al., J. Electron Spectrosc. Relat. .Phenom. 84 (1997) 171.
[40] P. Dhez, P. Chevallier, T.B. Lucatorto, C. Tarrio, Rev. Scient. Instrum.70 (1999) 1907.
[41] J. Voss, J. Electron Spectrosc. Relat. Phenom. 84 (1995) 29.
[42] H. Ade, J. Kirz, S.L. Hulbert, E.D. Johnson, E. Anderson, D.Kern, Appl. Phys. Lett. 56 (1990) 1841.
[43] A.Barinov, L. Casalis, L. Gregoratti, S. Gunther, M. Marsi , M. Koskinova in Proc.12th Eur. Cong. Electr. Microsc., Brno, Ed. P.Tomanek and R. Kolarik, vol III (2000) I 503. For technical details, see also L. Casalis et al.-20 coauthors- Rev. Sci. Instrum. 66 (1995) 4870 and M.P. Kiskinova, Surf. Interf. Anal. 30(2000) 464
[44] U. Johansson, R. Nyholm, C. Tornevik, A Flodstrom, Rev. Sci. Instr. 66 (1995) 1398.

[45] M. Hasegawa , K. Ninomiya , Rev. Sci. Instr. 66 (1995) 1361.
[46] D.H. Bildernack, D.E. Thiel, Rev. Sci. Instr. 66 (1995) 2059.
[47] U.W. Arndt, J. Appl. Cryst. 23 (1990) 161
[48] F. Cerrina ,A.K. Ray-Chaudhuri ,W. Ng, S. Liang, S. Singh, J.T. Welnek , J.P. Wallace C.Capasso, Appl. Phys. Lett. 63 (1993) 63.
[49] A.K Ray-Chaudhuri et al, J. Vac. Sci. Technol. A11 (1993) 2324.
[50] H. Kondo, T. Tomie, H. Shimizu in X-ray microscopy and Spectromicroscopy, J. Thieme, G. Schmahl, D. Rudolph, E. Umbach eds Springer , Berlin III (1998) 163.
[51] G. Schonhense in Proceedings 12th Eur. Cong. Electr. Microsc., Brno, Ed. P.Tomanek and R. Kolarik, vol III (2000) I 351.
[52] A.H. Weiss, S. Yang, H.Q. Zhou, E.Jung, S. Wheeler J. Electron Spectrosc. Relat .Phenom. 72 (1995) 305.
[53] J. Kawai, S. Hayakawa, Y. Kitajima, K. Maeda, Y. Gohshi J. Electron Spectrosc. Relat. Phenom. 76 (1995) 313.
[54] Y. Iijima, K. Miyoshi, S. Saito, Surf. Interf. Anal. 27 (1999) 35.
[55] A. Jablonski and C.J. Powell, Surf. Interf. Anal. 20 (1993) 771.
A. Jablonski in Springer Series in Surface Science, 18 (1989) 186.
[56] C.J. Powell, Surf. Sci. 299/300 (1994) 34.
[57] C.J. Powell, A. Jablonski, S. Tanuma, D.R. Penn, J. Electr. Spectros. Relat. Phenom. 68 (1994) 605.
[58] J. Cazaux, Surf. Sci. 140 (1984) 85.
[59] A. Rose, Adv. Electronics 1 (1948) 131.
[60] J. Cazaux, Scanning Electron Microscopy III (1984)1193; Ultramicroscopy 17 (1984)43.
[61] O.L. Krivanek, C.C. Ahn, R.B. Keeny, Ultramicroscopy 22 (1987) 103.
[62] C. Colliex, Mikrochemica Acta, 114/115 (1994) 71.
[63] J.A.D. Matthew, M. Prutton, M.M. El Gomati and D.C. Peacock, Surf. Inter. Anal. 11 (1988)173.
[64] S. Tougaard, Solid State Comm. 61 (1987) 547.
[65] I.R. Barkshire, M. Prutton and D.K. Skinner, Surf. Interf. Anal. 17 (1991) 213.
[66] R. Shimizu, Jpn. J. Appl. Phys. 22 (1993) 1631; S. Ichimura and R.Shimizu, Surf. Sci. 112 (1981)386.
[67] H. Benhayoune, O. Jbara, X. Thomas, D. Mouze, J. Cazaux, Surf. Interf. Anal. 20 (1993) 600.
[68] J. Cazaux, Microsc. Microanal. Microstruct. 3 (1992) 271.
[69] G. Todd, H. Poppa, J. Vac. Sci. Tech. 15 (1978) 672.
[70] M.M. El Gomati, J.A.D. Matthew and M. Prutton, Appl. Surf. Sci. 24 (1985) 147.
[71] C.J. Harland and J.A. Venables, Ultramicroscopy 17 (1985) 9.
[72] I.R. Barkshire, M. Prutton, J.C. Greenwood and P.G. Kenny, Appl. Surf. Sci. 55 (1992) 245.
[73] J. Cazaux, T. Bardoux, D. Mouze, J.M. Patat, G. Salace, X. Thomas, J. Toth, Surf. Interf. Anal. 19 (1992) 197.
[74] M.M.El Gomati, A. Gelthorpe, J. Dell in Proceedings 12th Eur. Cong. Electr. Microsc., Brno, Ed. P. Tomanek and R.Kolarik, vol III (2000) I 355.
[75] M.M. El Gomati, M. Prutton, B. Lamb, C.G. Tupprn, Surf. Interf. Anal. 11 (1988) 251.
[76] A. Umbach, A. Hoyer, R. Brunger, Surf. Interf. Anal. 14 (1989) 401.
[77] H. Ito, M. Ito, Y. Magatani, F. Soeda, Appl. Surf. Sci. 100/101 (1996) 152.
[78] J. Olivier, P. Alnot, F. Wyczisk, Physica Scripta. 41 (1990) 522 and references herein.

[78] J. Olivier, B. Bartenlian, M.N. Charasse, R. Bisaro, F. Wyczisk, J. Chazelas, J.P. Hirtz, Appl. Phys. Lett. 61 (1992) 1.
[79] J. Olivier, B. Bartenlian, R. Bisaro, F. Wyczisk, J. Vac. Sci. Technol. B10 (1992) 1835.
[81] T. Greber, J. Osterwalfer, D. Naumovic, A. Stuck, S. Hufner and L.L. Schlapbach, Phys. Rev. Letters 69 (1992) 1947.
[82] G.G. Frank, N. Batina, T. Golden, F. Lu, A.T. Hubbard, Science 247 (1990) 182.
[83] B. Akamatsu, P. Henoc, F. Maurice, C. Le Gressus, K. Raoudi, T. Sekine, T. Sakai, Surf. Interf. Anal. 15 (1990) 7.
[84] K. Marre, H. Neddermayer, A. Chasse, P. Rennert, Surf. Sci. 357-358 (1996) 233.
[85] F.E. Doern, L. Kover, N.S. McIntyre, Surf. Interf. Anal. 6 (1984) 282.
[86] H.E. Bishop, Surf. Interf. Anal. 15 (1990) 27.
[87] P. Morin, Surf. Sci. 164 (1985) 127.
[88] L. Reimer in Scanning Electron Microscopy, Springer Series in Optical Sciences 45 (1985) 124.
[89] G. Schneider in X-Ray Microscopy IV, V.V. Aristov and A.E. Erko editors, Chernogolovska, Russia (1994) 181, J. Cazaux, J. of Microscopy 188 (1997) 106.
[90] J.Cazaux, J. Electron. Spectrosc. Relat. Phenom. (2000) to appear.
[91] J.Cazaux, J. Electron. Spectrosc. Rel. Phen. 105 (1999) 155.
[92] D.M. Kroeger, A. Choudhury, J. Brynestad, R. K. Williams, R.A. Padgett and W.A. Coghlan, J. Appl. Phys. 64 (1988) 331.
[93] N. Iwasaki, R.Nishitani, S. Nakamura, Jap. J. Appl. Phys. 17 (1978) 1519.
[94] M. Pijolat and G. Hollinger, Surf. Sci. 105 (1981) 114.
[95] P.H. Holloway and T.D. Bussing, Surf. Interf. Anal. 18 (1992) 251.
[96] G.Y. Cherkashinin, Surf. Sci. 74 (1995) 67.
[97] T. Takeichi, K. Goto, V. Gaidorova, Appl. Surf. Sci. 100/101 (1996) 25.
[98] Z.J. Ding R. Shimizu, K. Goto, Appl. Surf. Sci. 100/101 (1996) 15.
[99] Z.J. Ding, T. Nagatomi,, R. Shimizu, K. Goto, Surf. Sci. 336 (1995) 397.
[100] M. Prutton,, Microscopy, Microstructures, Microanalysis 6 (1995) 289.
[101] R. Browning, J.L. Smialek, N.S. Jacobson, Adv. Ceramic. Materials 2 (1987) 773.
[102] M. Prutton, M.M. El Gomati, P.G. Kenny, J. Electr. Spectr. Relat. Phenom. 52 (1990) 197.
[103] P.G. Kenny, I.R. Barkshire, M. Prutton, Ultramicroscopy 56 (1994) 289.
[104] I.R. Barkshire, M. Prutton, G.C. Smith, Appl. Surf. Sci. 84 (1995) 331.
[105] C.J. Powell and M.P. Seah, J. Vac. Sci. Technol. A8 (1990) 735.
[106] S. Hofmann in Practical Surface Analysis (see ref 1) (1990) 143.
[107] P. Etienne, J. Chazelas, G. Creuzet, A. Friederich, J. Massies, F. Nguyen Van-Dau, A. Fert, J. of Cryst. Growth, 95 (1989) 410.
[108] A. Zalar, Thin Solid Films 124 (1985) 223, A. Zalar, E.W. Seibt, P. Panjan, Appl. Surf. Sci. 100/101 (1996) 92.
[109] S. Hofmann, Progress in Surf. Sci. 36 (1991) 35; Surf. Interf. Anal. 21 (1994) 673.
[110] E.R. Malinowski, D.G. Howery in Factor Analysis in Chemistry, Wiley NY 1980.
[111] S.W. Gaarenstroom, Appl. Surf. Sci. 7 (1981) 7 ; Appl. Surf. Sci. 26 (1986) 561.
[112] J.S. Solomon, Surf. Interf. Anal. 10 (1987) 75.
[113] T. Morohashi, T. Hoshi, H. Nikaido, M. Kudo, Appl. Surf. Sci. 100/101 (1996) 84.
[114] W. Bohne, F. Fenske, S. Kelling, A. Schopke, B. Selle, Phys. Stat. Solidi b194 (1996) 69.
[115] H.G. Steffen, S. Hofmann, Surf. Interf. Anal. 19 (1992) 157.

[116] C. Gatts, A. Zalar, S. Hofmann, M. Ruhle, Surf. Interf. Anal. 23 (1995) 809.
[117] S. Hofmann, A. Zalar, E.H. Ciplin, J.J. Vajo, H.J. Mathieu, P. Panjan, Surf. Interf. Anal. 20 (1993) 621.
[118] H.J. Mathieu, Le Vide 52 (N°279) (1996) 81.
[119] S. Hofmann, Surf. Interf. Anal. 30 (2000) 228
[120] a : D.K. Skinner, Surf. Interf. Anal. 14 (1989) 567.
   b : D.K. Skinner, Microscopy, Microstructures, Microanalysis 6 (1995) 321.
[121] M.J.G. Wenham, I.R. Barkshire, M. Prutton, R.H. Roberts and D.K. Wilkinson, Surf. Interf. Anal. 23 (1995) 858.
[122] R. Watts, D.K. Wilkinson, I.R. Barkshire, M. Prutton, A. Chambers, Phys. Rev. B52 (1995) 451.
[123] I.R. Barkshire and M. Prutton, J. Appl. Phys. 77 (1995) 1082.
[124] P.C. Schamberger, G.L. Jones, J.A. Gardella Jr., P.J. Mc Keown, L.E. Davis, J. Vac. Sci. Technol. A14 (1996) 2289.
[125] K.Takanashi, H. Yu, Y. Kuramoto, Z.H. Cheng, T. Sakamoto, M. Owari, Y. Nihei, Surf. Interf. Anal. 30 (2000) 493.

SUBJECT INDEX

Alloys

An atomistic approach for stress relaxation in materials, Guy Tréglia	119
Ab initio study of the structural stability of thin films (only abstract and references in this book), Alain Pasturel	151

Atomic layers

An atomistic approach for stress relaxation in materials, Guy Tréglia	119
Stress measurements of atomic layers and at surfaces, Dirk Sander	221
Strain measurements in ultra-thin films using RHEED and X-ray techniques, Bruno Gilles	173

Atomic structure

Introduction to the atomic structure of surfaces: a theoretical point of view, Marie-Catherine Desjonqueres and Daniel Spanjaard	63
An atomistic approach for stress relaxation in materials, Guy Tréglia	119
STM spectroscopy on semiconductors, Didier Stievenard	243

Auger electron spectroscopy

Spatially resolved surface spectroscopy, Jacques Cazaux, and Jean Olivier	287

Band structure

Introduction to the atomic structure of surfaces: a theoretical point of view, Marie-Catherine Desjonqueres and Daniel Spanjaard	63

Catalysis

Stress, strain and chemical reactivity: a theoretical analysis, Philippe Sautet	155

Crystal growth, epitaxy

Some elastic effects in crystal growth, Pierre Müller and Raymond Kern	3
An atomistic approach for stress relaxation in materials, Guy Tréglia	119

Dislocation

Some elastic effects in crystal growth, Pierre Müller and Raymond Kern	3
Dislocations and stress relaxation in heteroepitaxial films, Ladislas Kubin	99
Measurements of displacement and strain by high-resolution transmission electron microscopy, Martin Hÿtch	201
An atomistic approach for stress relaxation in materials, Guy Tréglia	119

Elasticity

Some elastic effects in crystal growth, Pierre Müller and Raymond Kern — 3
Dislocations and stress relaxation in heteroepitaxial films, Ladislas Kubin — 99

Electron diffraction

Strain measurements in ultra-thin films using RHEED and
X-ray techniques, Bruno Gilles — 173
Measurements of displacement and strain by high-resolution
transmission electron microscopy, Martin Hÿtch — 201

Electron microscopy

Measurements of displacement and strain by high-resolution
transmission electron microscopy, Martin Hÿtch — 201

Electronic structure methods

Ab initio study of the structural stability of thin films
(only abstract and references in this book), Alain Pasturel — 151
An atomistic approach for stress relaxation in materials, Guy Tréglia — 119

Numeric simulation

An atomistic approach for stress relaxation in materials, Guy Tréglia — 119

Photoelectron spectroscopy

Spatially resolved surface spectroscopy, Jacques Cazaux, and Jean Olivier — 287

Scanning tunneling microscopy and spectroscopy

STM spectroscopy on semiconductors, Didier Stievenard — 243

Surface diffraction

Introduction to the atomic structure of surfaces: a theoretical
point of view, Marie-Catherine Desjonqueres and Daniel Spanjaard — 63
Strain measurements in ultra-thin films using RHEED and
X-ray techniques, Bruno Gilles — 173

Surface Relaxation

Introduction to the atomic structure of surfaces: a theoretical
point of view, Marie-Catherine Desjonqueres and Daniel Spanjaard — 63

Surface stress

Some elastic effects in crystal growth, Pierre Müller and Raymond Kern — 3

Strain

Some elastic effects in crystal growth, Pierre Müller and Raymond Kern	3
An atomistic approach for stress relaxation in materials, Guy Tréglia	119
Stress, strain and chemical reactivity: a theoretical analysis, Philippe Sautet	155
Strain measurements in ultra-thin films using RHEED and X-ray techniques, Bruno Gilles	173
Ab initio study of the structural stability of thin films (only abstract and references in this book), Alain Pasturel	151
Measurements of displacement and strain by high-resolution transmission electron microscopy, Martin Hÿtch	201

Stress

Some elastic effects in crystal growth, Pierre Müller and Raymond Kern	3
Dislocations and stress relaxation in heteroepitaxial films, Ladislas Kubin	99
An atomistic approach for stress relaxation in materials, Guy Tréglia	119
Stress, strain and chemical reactivity: a theoretical analysis, Philippe Sautet	155
Stress measurements of atomic layers and at surfaces, Dirk Sander	221

Stress relaxation

Some elastic effects in crystal growth, Pierre Müller and Raymond Kern	3
Dislocations and stress relaxation in heteroepitaxial films, Ladislas Kubin	99
An atomistic approach for stress relaxation in materials, Guy Tréglia	119
Ab initio study of the structural stability of thin films (only abstract and references in this book), Alain Pasturel	151

Surface

Some elastic effects in crystal growth, Pierre Müller and Raymond Kern	3
Introduction to the atomic structure of surfaces: a theoretical point of view, Marie-Catherine Desjonqueres and Daniel Spanjaard	63
An atomistic approach for stress relaxation in materials, Guy Tréglia	119
Strain measurements in ultra-thin films using RHEED and X-ray techniques, Bruno Gilles	173
Stress measurements of atomic layers and at surfaces, Dirk Sander	221

Surface segregation

An atomistic approach for stress relaxation in materials, Guy Tréglia	119

Surface spectroscopy

STM spectroscopy on semiconductors, Didier Stievenard	243

Spatially resolved surface spectroscopy, Jacques Cazaux, and Jean Olivier   287

Thin films

Some elastic effects in crystal growth, Pierre Müller and Raymond Kern   3
Dislocations and stress relaxation in heteroepitaxial films, Ladislas Kubin   99
An atomistic approach for stress relaxation in materials, Guy Tréglia   119
Ab initio study of the structural stability of thin films
(only abstract and references in this book), Alain Pasturel   151
Strain measurements in ultra-thin films using RHEED and
X-ray techniques, Bruno Gilles   173

Transmission electron microscopy

Measurements of displacement and strain by high-resolution
transmission electron microscopy, Martin Hÿtch   201

X-ray techniques

Strain measurements in ultra-thin films using RHEED and
X-ray techniques, Bruno Gilles   173